VOLUME THREE HUNDRED AND TWENTY NINE

INTERNATIONAL REVIEW OF CELL AND MOLECULAR BIOLOGY

Early Stage Protein Misfolding and Amyloid Aggregation

INTERNATIONAL REVIEW OF CELL AND MOLECULAR BIOLOGY

VOLUME THREE HUNDRED AND TWENTY NINE

INTERNATIONAL REVIEW OF CELL AND MOLECULAR BIOLOGY

Early Stage Protein Misfolding and Amyloid Aggregation

Edited by

MASSIMO SANDAL

ELSEVIER

AP

ACADEMIC PRESS

An imprint of Elsevier
elsevier.com

Academic Press is an imprint of Elsevier
50 Hampshire Street, 5th Floor, Cambridge, MA 02139, United States
525 B Street, Suite 1800, San Diego, CA 92101-4495, United States
The Boulevard, Langford Lane, Kidlington, Oxford OX5 1GB, United Kingdom
125 London Wall, London, EC2Y 5AS, United Kingdom

First edition 2017

ISBN: 978-0-12-812251-8
ISSN: 1937-6448

For information on all Academic Press publications visit our website at https://www.elsevier.com/

Publisher: Zoe Kruze
Acquisition Editor: Alex White
Editorial Project Manager: Fenton Coulthurst
Production Project Manager: Magesh Mahalingam
Cover Designer: Mark Rogers

CONTENTS

CONTRIBUTORS

F. Bemporad
Università degli Studi di Firenze, Firenze, Italy

L. Breydo
Byrd Alzheimer's Research Institute, Morsani College of Medicine, University of South Florida, Tampa, FL, United States

Alexander K. Buell
Institute of Physical Biology, University of Düsseldorf, Universitätsstr.1, 40225, Düsseldorf, Germany

S.W. Chen
University of Cambridge, Cambridge, United Kingdom

N. Cremades
Biocomputation and Complex Systems Physics Institute (BIFI)-Joint Unit BIFI-IQFR (CSIC), Universidad de Zaragoza, Zaragoza, Spain

C.M. Dobson
University of Cambridge, Cambridge, United Kingdom

A. Giorgetti
Applied Bioinformatics Group, University of Verona, Verona, Italy

K.S. MacLea
University of New Hampshire, Manchester, NH, United States

F. Musiani
Laboratory of Bioinorganic Chemistry, University of Bologna, Bologna, Italy

M. Ramazzotti
Università degli Studi di Firenze, Firenze, Italy

J.M. Redington
Byrd Alzheimer's Research Institute, Morsani College of Medicine, University of South Florida, Tampa, FL, United States

J.R. Requena
CIMUS Biomedical Research Institute, University of Santiago de Compostela-IDIS, Santiago de Compostela, Spain

V.N. Uversky
Byrd Alzheimer's Research Institute, Morsani College of Medicine, University of South Florida, Tampa, FL, United States; Laboratory of Structural Dynamics, Stability and Folding of Proteins, Institute of Cytology, Russian Academy of Sciences, St. Petersburg, Russia

E. Vázquez-Fernández
Centre for Prions and Protein Folding Diseases, University of Alberta, Edmonton, AB, Canada

H. Wille
Centre for Prions and Protein Folding Diseases, University of Alberta, Edmonton, AB, Canada

H.S. Young
University of Alberta, Edmonton, AB, Canada

CHAPTER ONE

From the Evolution of Protein Sequences Able to Resist Self-Assembly to the Prediction of Aggregation Propensity

F. Bemporad[1,2], M. Ramazzotti[1,2]
Università degli Studi di Firenze, Firenze, Italy
[2]Corresponding authors: e-mail address: francesco.bemporad@unifi.it; matteo.ramazzotti@unifi.it

Contents

[1] These authors contributed equally.

International Review of Cell and Molecular Biology, Volume 329
ISSN 1937-6448
http://dx.doi.org/10.1016/bs.ircmb.2016.08.008

Abstract

Folding of polypeptide chains into biologically active entities is an astonishingly complex process, determined by the nature and the sequence of residues emerging from ribosomes. While it has been long believed that evolution has pressed genomes so that specific sequences could adopt unique, functional three-dimensional folds, it is now clear that complex protein machineries act as quality control system and supervise folding. Notwithstanding that, events such as erroneous folding, partial folding, or misfolding are frequent during the life of a cell or a whole organism, and they can escape controls. One of the possible outcomes of this misbehavior is cross-β aggregation, a super secondary structure which represents the hallmark of self-assembled, well organized, and extremely ordered structures termed amyloid fibrils. What if evolution would have not taken into account such possibilities? Twenty years of research point toward the idea that, in fact, evolution has constantly supervised the risk of errors and minimized their impact. In this review we tried to survey the major findings in the amyloid field, trying to describe what the real pitfalls of protein folding are—from an evolutionary perspective—and how sequence and structural features have evolved to balance the need for perfect, dynamic, functionally efficient structures, and the detrimental effects implicit in the dangerous process of folding. We will discuss how the knowledge obtained from these studies has been employed to produce computational methods able to assess, predict, and discriminate the aggregation properties of protein sequences.

1. INTRODUCTION

In the crowded milieu of a living cell, the concentration of molecules can reach the astounding value of 400 mg/mL (Ellis, 2001; Zimmerman and Trach, 1991). Molecules/ions flow across cell membrane and depending on transient stimuli or on the conditions to which a cell is subjected, new proteins are continuously synthesized on ribosomes and processed in the appropriate cellular compartments, while old ones are degraded. The protein machinery must be perfectly functional under such constantly changing conditions. In order for this task to be accomplished, proteins need to populate an ensemble of biologically active conformations and to undergo productive molecular motions (Bahar et al., 2015); plus, they must be able to interact with metabolites and/or with their partners (Snider et al., 2015). However, they must be able to avoid the establishment of wrong interactions which may inactivate them or give rise to the formation of aberrant complexes endowed with undesired, possibly detrimental, capabilities. Among all the possible interactions, the formation of well-organized protein self-assemblies has emerged during the last decades as a constant risk that any living organisms must face, counteract, or cope with (Bemporad and Chiti, 2012; Chiti and Dobson, 2006). Indeed, it is now clear that the biologically active state of a protein does not necessarily correspond to the most thermodynamically stable conformation but, rather, that many, if not all, proteins possess the inherent ability to convert into a well-known type of β-rich self-assemblies (Chiti et al., 1999; Guijarro et al., 1998), usually referred to as amyloid aggregates, whose formation is irreversible (Jahn and Radford, 2005, 2008). The formation of such aggregates results into the loss of the physiological function of the protein and/or into the gain of a new toxic function (Bemporad and Chiti, 2012). Consequently, life is possible as long as proteins can accomplish their tasks and be removed before they eventually convert into amyloid aggregates (Bemporad and Chiti, 2013). Since the beginning of molecular evolution, primary sequences have evolved on the thin edge between the need of proteins to experience the motions they require to interact with their partners and the risk of misfolding, aggregation, and precipitation. The crucial role played by the maintenance of this equilibrium is shown by the evidence that protein misfolding and aggregation are related to the onset of more than 40 human pathological conditions, commonly referred to as protein misfolding diseases or protein conformational diseases (Chiti and Dobson, 2006). These diseases arise from the failure of a specific polypeptide chain to remain in its native state, an event which leads to loss of function (e.g., cystic fibrosis), improper trafficking (e.g., early onset emphysema), or formation of highly organized fibrillar aggregates (e.g., Parkinson's disease) (Chiti and Dobson, 2006). In this review we will try to survey the sequence features that promote amyloid and the strategies devised by evolution to keep aggregation phenomena under control. These features will be described at the primary sequence level and at the structural level. In the final section we will discuss how the parameters that will be described can be converted into algorithms able to rationalize and predict the ability of protein sequences to convert into amyloid aggregates.

2. EVOLUTION OF PROTEIN SEQUENCES: NEED TO BE FUNCTIONAL VS RISK OF AGGREGATION

2.1 Determinants of Order/Disorder as Strategies to Avoid Aggregation

It was shown that, in vitro, most proteins fold into a well-defined structure (F in Fig. 1) following a set of possible mechanisms which range from the so-called nucleation–condensation (Fersht, 1995), wherein a set of specific

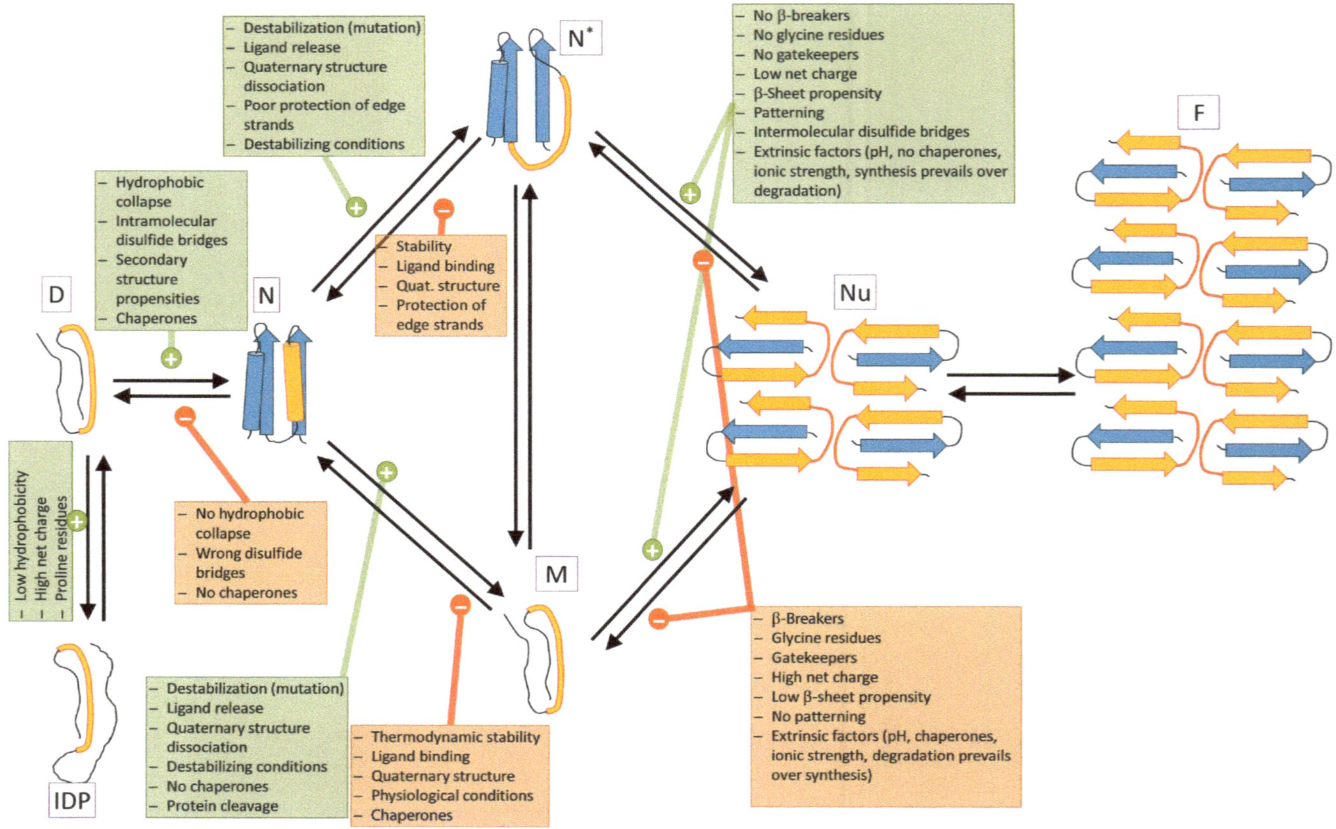

Fig. 1 See legend on opposite page.

native contacts formed early limits the number of conformations accessible to a polypeptide chain and triggers a hierarchical assembly, to the framework model, in which secondary structure elements fold locally and then assemble thanks to their diffusion and collision (Udgaonkar, 2013). In vivo, folding of globular proteins is a cotranslational process (O'Brien et al., 2014), i.e., nascent polypeptide chains begin the conformational motions that lead to the formation of their folded states during protein translation. In this process the ribosome acts as a pace maker of translation rates (see later) and as a docking site for the machinery that assists folding (Gloge et al., 2014). In both cases a set of properties inherent to the side-chains dominate the equilibrium between unfolded and folded state. The main force that underlies the folding process is the hydrophobic force (Fig. 1). Hydrophobic patches need to be masked from and inaccessible to the solvent as they are inherently prone to precipitate and/or self-assemble (Udgaonkar, 2013). Thus, the packing of hydrophobic sequences represents the turning point of protein folding/misfolding, i.e., an event that all living cells must necessarily cope with. The analysis of protein sequences reveals that two main strategies contribute to the avoidance of misfolding. The first strategy consists of sequences in which hydrophobic patches are carefully buried in a hydrophobic core or masked by an interacting partner. A second possible strategy is the evolution of sequences containing a fraction of hydrophobic residues below a certain threshold. Indeed, proteins with few hydrophobic/aromatic residues and a high number of proline and charged/hydrophilic residues do not need and are in fact unable to collapse into a folded state (Uversky and Dunker, 2010). This is the case of intrinsically disordered proteins (IDP in Fig. 1) (Berlow

Fig. 1 Conformational equilibria involved in the process of folding/misfolding with the parameters which induce (*green*) or inhibit (*red*) each passage. In brief, nascent polypeptide chains are initially disordered (D). Depending on the relative fraction of proline, hydrophobic, and charged residues, D can fold into a native (N) state or resist as an intrinsically disordered protein (IDP). In both cases, amyloidogenic stretches (depicted in orange) are sequestered from the solvent and/or folded. However, a set of parameters can lead N to undergo structural perturbations and convert into an aggregation-competent state. The latter can be globally folded and native-like (N*), or globally unfolded (misfolded, M). The aggregation-competent state possesses solvent-exposed amyloidogenic stretches and this leads to self-assembly, with the subsequent formation of an aggregation nucleus (Nu), i.e., the multimer of minimal size for which association rate exceeds dissociation rate. Once Nu is formed, aggregates grow in dimension by monomer addition and eventually convert into mature fibrils (F).

et al., 2015; Uversky and Dunker, 2010), polypeptide chains whose native biologically active conformation lacks persistent structure (Fig. 1). Although the description of the features of these proteins lies far beyond the scopes of this review, it is important to point out that the evolution of polypeptide chains devoid of regular structure presents many advantages (reviewed, for example, in Uversky and Dunker, 2010), including the possibility to couple high specificity with low affinity and to form fuzzy complexes (Sharma et al., 2015), the possibility of overlapping binding sites due to extended linear conformation, hence the possibility to bind several structured partners with one interacting site, and the possibility to fold into diverse structures upon binding (Gianni et al., 2015). These features make IDPs particularly useful systems for protein interaction networks. Consistently, it was proposed that IDPs are often the hubs of protein interaction networks (Rangarajan et al., 2015; Uversky and Dunker, 2010) and a series of bioinformatics studies proposed that IDPs represent 30% of eukaryotic proteins (Oldfield et al., 2005) and that 70% of signaling proteins have long disordered regions (Iakoucheva et al., 2002).

Hydrophobicity is not the only force that defines the equilibrium between folded, unfolded, and misfolded conformations. Although the identification of the so-called folding code is still elusive and this problem was included in the list of the most difficult scientific challenges (No authors, 2005), it is clear that any amino acid possesses a certain degree of structural preference and that certain types of residues promote the formation of specific secondary structure elements (Fig. 1). Thus, valine and isoleucine residues are more frequently found in β-strands, alanine and leucine residues in α-helices, and proline and glycine residues in β-turns (Creighton, 1992). However, while evolution exploited these preferences to promote structure formation, protein misfolding, and self-assembly lead to the formation of highly stables fibrils whose secondary structure is dominated by intermolecular β-sheets (Chiti and Dobson, 2006). It is therefore reasonable that sequences with a high propensity to form β-sheet structure are inherently prone to aggregate. Indeed, several studies showed that mutations that induce an increase in β-sheet propensity of a sequence enhance aggregation rates (Chiti et al., 2003; de Groot et al., 2006). Thus, as for hydrophobicity, although secondary structure propensities of amino acids underlie folding, sequences did not evolve to maximize the propensity of a sequence to form secondary structure, but rather to keep under control the equilibrium between folding and misfolding so that monomer folding dominated over intermolecular interactions that trigger the amyloid cascade.

A pivotal role in determining the conformation and the monomeric/multimeric state of a polypeptide chain is also played by cysteine residues and disulfide bridges (Fig. 1). While the major role played by these covalent bonds in determining the formation of folded or misfolded conformations was pointed out with the pioneer experiments by C. Anfinsen (Anfinsen et al., 1961; White, 1961), cases have been reported in which the formation of a disulfide bridge induces a strong increase in conformational stability of the folded state (Ramazzotti et al., 2006) due to an increase of the folding rate (Parrini et al., 2008) and a switch in the folding pathway (Parrini et al., 2008). However, intramolecular disulfide bridges can also have a complimentary role and contribute to inhibit aggregation (Parrini et al., 2008; Ramazzotti et al., 2006). Cyclization of the ABri peptide resulted in an inert monomer unable to aggregate (Cantlon et al., 2015). When compared to the wild-type, a cyclized variant of PI3-SH3 was found to fold faster and aggregate slower (Grana-Montes et al., 2012). Recently, a α-synuclein mutant has been designed carrying two cysteine residues; formation of an intramolecular disulfide bridge between these two residues leads to an aggregation-resistant monomer (Shaykhalishahi et al., 2015). Furthermore, intramolecular disulfide bonds have been proposed as stabilizers of amyloid fibrils, as they induce a decrease of the entropic penalty associated with the formation of this ordered type of protein structure (Mossuto et al., 2011). Since toxic aggregates are usually transiently populated and poorly compact self-assemblies, this stabilization can be concomitant with significant decreases in the toxicity of the resulting fibrils, which suggests that disulfide bonds might have coevolved with protein sequences to reduce toxic aggregation (Monsellier et al., 2008). Accordingly, in the case of PI3-SH3, presence of an intramolecular disulfide bridge determines differences in fibril morphology and induces a decrease of their hydrophobicity and toxicity (Grana-Montes et al., 2012). Remarkably, when formation of covalent bonds between two cysteine side-chains anchors two molecules, aggregation can become faster, as it was shown in some cases. The R17C mutant of the Ure2p protein exhibits faster aggregation under oxidizing conditions than the wild-type protein (Fei and Perrett, 2009). In the case of the ABri peptide, formation of oligomers stabilized by intermolecular disulfide bridges was found to be associated to an increase in aggregate toxicity (Cantlon et al., 2015).

In conclusion to this section, while the main forces that contribute to the collapse of polypeptide chains into folded states were identified some 50 years ago, the emergence of protein misfolding as a distinct field of investigation highlighted that the same features involved in folding can be the culprit of misfolding. Consequently, evolution may have tailored protein sequences on the thin edge that separates these two processes in order to minimize undesired chain collapse with respect to the formation of functional states of proteins (Monsellier and Chiti, 2007).

2.2 Rate-Limiting Steps of Aggregation

When a protein fails to take advantage of the aforementioned strategies, the risk becomes concrete to undergo misfolding and aggregation (Fig. 1). What are the mechanisms by which normally soluble proteins convert into insoluble aggregates? Decades of research illustrated a complicated scenario in which many, possibly concomitant, pathways lead to the conversion of proteins and polypeptide chains into well-ordered assemblies (Bemporad and Chiti, 2012, 2013; Chiti and Dobson, 2006).

2.2.1 Formation of Aggregation-Prone Conformations

The first rate-limiting step of self-assembly is the population of a conformation possessing aberrant features which enable it to aggregate. The features of a monomeric protein which enable it to aggregate have been the subject of intense studies. In general, folded states of globular proteins need to unfold, completely or partially, before aggregation can occur (Bemporad et al., 2006a; Chiti and Dobson, 2006). Indeed, protein unfolding leads segments, which are normally packed and buried within the hydrophobic core, to become exposed and highly dynamic (M in Fig. 1). Consequently, they can experience a larger amount of conformations and, being inherently sticky, they can give rise to intermolecular interactions. IDPs or disordered segments, such as the Aβ peptide (Bitan et al., 2003), the NM region of Sup35 (Scheibel and Lindquist, 2001), or α-synuclein under conditions in which the protein is intrinsically disordered (Alderson and Markley, 2013), also aggregate through this pathway, although they do not require the initial unfolding step. Importantly, fully unfolded conformations are not the only state able to aggregate and recent pieces of evidence lend support to the idea the folded states of globular proteins retain a small, albeit significant, tendency to aggregate in the absence of global and cooperative unfolding events (Chiti and Dobson, 2009). As folded states do not normally aggregate, the misfolded state of these proteins consists of a globally folded state possessing some aberrant features which enable it to aggregate. One such conformation is usually referred to as an aggregation-competent native-like state (N* in Fig. 1). The main features, identified so far, that enable N* to aggregate are an increase in conformational fluctuations and the local exposure of hydrophobic patches. These two features are strictly correlated and the increase in dynamics can be the consequence of destabilizing mutations (Plakoutsi et al., 2006) or solution conditions (Bemporad et al., 2012). Local exposure of hydrophobic stretches can be the consequence of (i) local unfolding events, which partially unmask hydrophobic side-chains (Canet et al., 2002; De Simone et al., 2011); (ii) the release of natural ligands, which exposes the binding site (Ferrolino et al., 2013; Masino et al., 2011) or destabilizes/distorts the folded state (Pertinhez et al., 2001; Soldi et al., 2006); or (iii) the release of monomers from quaternary structure, which possibly exposes a hydrophobic interacting surface, as it was observed in transthyretin (Johnson et al., 2012) and β2-microglobulin (Esposito et al., 2008; Halabelian et al., 2014; Platt et al., 2008).

2.2.2 Pathways to Amyloid Formation

Conversion of proteins into the aggregation-competent conformational ensemble triggers self-assembly. The initial aggregates usually resemble, in terms of monomer structure, the misfolded monomer. However, as aggregation proceeds, the early oligomers sequester more monomers and clump together forming multimers which exhibit increasingly larger size, compactness, β-sheet content, and hydrophobic burial (Bemporad and Chiti, 2012). Concomitantly, structural rearrangements occur within the oligomer, leading to a decrease in hydrophobic exposure, exposed surface per number of monomers, and structural flexibility. A set of mechanisms has been proposed for these mechanisms to occur. The first proposed mechanism was the so-called nucleation growth (Jarrett and Lansbury, 1993; Lomakin et al., 1996). According to this mechanism, the rate-limiting step of aggregation is the conversion of monomers into the multimer characterized by the minimal size for which the rate of monomer addition exceeds the rate of monomer association. This latter species is usually referred to as the aggregation nucleus (Nu in Fig. 1) and, once the nucleus is formed, aggregation is believed to be an irreversible process which eventually leads to the formation of large fibrils (F in Fig. 1). Although formation of the nucleus is energetically disfavored, after such species is formed aggregation becomes an irreversible process. Depending on the aggregating system, the nucleus can be a large multimer (Collins et al., 2004; Morris et al., 2008) or a monomer (Hurshman et al., 2004; Sandal et al., 2008). In the latter case aggregation kinetics may not exhibit the typical lag-phase which corresponds to the assembly of the nucleus and the aggregation has been proposed to occur through a down-hill polymerization (Hurshman et al., 2004). After the nucleation-growth model was proposed, other authors posited a different model to describe aggregation, termed "nucleated conformational conversion" (Serio et al., 2000). In this mechanism, monomers initially clump together into a molten oligomer lacking any regular and persistent structure (Fu et al., 2015; Gerber et al., 2007). This process is relatively fast, whereas the step limiting the rate of the overall aggregation reaction is the reorganization of this unstructured oligomer into an amyloid-competent oligomer (Cerda-Costa et al., 2007; Lee et al., 2011; Petty and Decatur, 2005; Thakur et al., 2009; Wei et al., 2011), a species which features template capability. This latter species corresponds to the nucleus and, following its formation, aggregation becomes fast and irreversible due to the template effect (Serio et al., 2000). Although these two models are in apparent contrast, it is now clear that aggregation can occur through parallel pathways and cases have been reported in which the aggregation mechanism can be switched, depending on specific amino acid substitutions (Bitan et al., 2003; de Rosa et al., 2014; Kumar and Udgaonkar, 2009) or on solution parameters such as protein concentration and pH (Bader et al., 2006; Gosal et al., 2005). Consequently, we recently proposed a possible unifying view in which the second rate-limiting step of aggregation is the formation of the nucleus, the amyloid-competent multimer able to sequester further monomers/oligomers into increasingly larger oligomers. When the collapse of misfolded monomers is faster than their conversion into an amyloid-competent fold, the nucleated conformational conversion prevails; if, instead, conversion of the monomer into the nucleus is faster than monomer build-up, then the nucleation growth prevails (Bemporad and Chiti, 2013).

The fact that most proteins need to undergo fluctuations to be functional and that an excess of fluctuations may trigger misfolding makes it reasonable the idea that evolution devised strategies to keep under control the two rate-limiting steps described above. On the one hand, sequences were tailored to the need to inhibit or slow down the population of aggregation-competent monomeric conformations; on the other hand, when formation of such conformers cannot be kept under control, strategies were evolved to counteract the formation of aggregates possessing features that make irreversible the cascade of reactions which eventually lead to toxic species. In the next section, we will survey the sequence and structural parameters that favor or disfavor self-assembly.

3. EVOLUTIONARY STRATEGIES TO INHIBIT AGGREGATION AT THE PRIMARY SEQUENCE LEVEL

Amyloid structure is now widely accepted as an inherent, thermodynamically favored property of polypeptide chains (Dobson, 1999). As mentioned above, structural and sequential determinants of protein folding are strictly related to those promoting intermolecular aggregation. Although most proteins can, under appropriate conditions, convert into amyloid-like structure and cases exist of native functional amyloid structures (Maury, 2009; Romero and Kolter, 2014), the vast majority of peptide sequences, under physiological conditions, do not form amyloid. The solution of this contradictory observation must be searched in evolution. In fact, genomes have obeyed to two main driving forces, one pointing to the formation of functional 3D structures, the other aimed at escaping from undesired aggregation, an event which leads to loss of the physiological function and/or to the gain of a possible toxic function. A wealth of literature during the last 20 years has illustrated many features of polypeptide sequences that lead to or favor amyloid. Consequently, strategies to disfavor amyloid have also been pointed out. Curiously enough, but not surprisingly from a biological perspective, studies concerning the comparative analysis of proteomes point toward the idea that evolution have optimized sequences with a set of inhibitory strategies to counteract amyloid as those identified by experimental approaches (Monsellier et al., 2008). It is important to point out here that the formation of fibrils is triggered by the self-assembly of misfolded proteins and that, consequently, any strategy aimed at keeping proteins in their physiological conformation must involve mechanisms that disfavor one of the aforementioned rate-limiting steps, misfolding, and self-assembly. Still, given that aggregation seems to be an inherent tendency of polypeptide chains (Rousseau et al., 2006a), important questions remain open as to what is required in terms of polypeptide sequence to form cross-β amyloid fibrils typical of amyloid, or amorphous aggregates with a high content of cross-β secondary structure.

While the major roles played by hydrophobicity, secondary structure propensities, and disulfide bridges in inducing folding/misfolding have been discussed above, several strategies that evolution may have exploited to counteract amyloid tendency at the primary sequence level have been reported, that can be ideally divided into four categories: (i) patterns, (ii) charges, (iii) β-breakers, (iv) gatekeepers. With reference to Fig. 1, here it will follow an in-depth description of all those four points.

3.1 Patterns

Amyloid formation is an ordering process where peptides and proteins juxtapose in cross-β secondary structures to form long fibers. Sequences that naturally form functional zippers (e.g., leucine zippers in transcription factors) are arranged so that paired monomers regularly interact, and this is a well-established example of how regular residue patterns can be used to create adhesive structures (Ryadnov et al., 2008). Remarkably, stretches of alternating polar and nonpolar residues were found to promote amyloid-like aggregation (Broome and Hecht, 2000). An analysis at the proteome-level demonstrated in fact that such patterns occurred significantly less often than other patterns with similar amino acid compositions, providing a first link between amyloid and evolutionary pressure toward amyloid. A retrospective analysis allowed such finding to be motivated and reinforced. The patterning of hydrophobic and hydrophilic residues is a well-established cause of secondary structure formation (Xiong et al., 1995) and it is generally associated to the amphipathic nature of secondary structures involved in the formation of globular proteins (West and Hecht, 1995). In the case of a β-strand exposed to the solvent on the surface of proteins, the periodicity was found to be two, implying that alternating polar and nonpolar residues favor the formation of such structures. Nevertheless, in most β-containing proteins, the number of exposed β-strands is much lower than that of buried ones, which instead have patterns enriched in hydrophobic residues (see later). In amyloid structures, the hydrophobic pairing of strands, the so-called steric-zipper (Nelson et al., 2005; Sawaya et al., 2007), was found to be a key step in protofilament formation, and this can be greatly favored if the facing edges of strands are hydrophobic (Kim and Hecht, 2006).

3.2 Charges

Surface repulsion/attraction is one of the main forces leading to polypeptide contacts. Protein structures tend to have conserved charged surface residue in order to accomplish specific functional tasks. Although charges have been demonstrated to be, in some circumstances, required for amyloid formation (Tjernberg et al., 2002), the presence of a high net charge, either global or local, has been generally found to increase solubility, as in the case of myoglobin from diving mammals (Mirceta et al., 2013), and inhibit aggregation/precipitation (Chiti et al., 2003). Conversely, several mutations that decrease the global net charge of proteins have been shown to be amyloidogenic (Chiti et al., 2003; Valleix et al., 2012). The inverse correlation between charge and aggregation tendency has been repeatedly demonstrated in studies that exploited mutagenesis (Chiti et al., 2002), neutralizing pH gradients (Schmittschmitt and Scholtz, 2003; Smirnova et al., 2015), and charge shielding compounds (Abdolvahabi et al., 2015; Fernandez-Escamilla et al., 2004a; Goers et al., 2003) to prevent/inhibit aggregation. This behavior is readily understandable if one considers the repulsion effects occurring when homopolymeric peptide chains have to interact during aggregation. Thus, evolution may have evolved sequences possessing a global charge sufficient to inhibit self-assembly. However, as it was observed for other parameters, sequences did not simply evolve to the highest fraction of charged residues. Indeed, as the case of hyperthermophilic enzymes recapitulates, electrostatic interactions lead to highly stable, yet poorly active, proteins (Corazza et al., 2006; Sterner and Liebl, 2001; Unsworth et al., 2007). Consequently, protein sequences evolved to bear a net charge sufficiently high to inhibit aggregation, yet they contain a number of electrostatic interactions sufficiently low to allow for biological activity. Recently, results on the systematic removal of negative charges in β2-microglobulin have contradicted the notion of net charge antiamyloid effect (de Rosa et al., 2015). These results raise questions on the universality of the charge theory and support the idea that different proteins may have evolved using different antiamyloid strategies, each of which may dominate or cooperate, also depending on the functional role of specific charges in the overall native structure.

3.3 β-Breakers

The notion that Pro and Gly residues are very rarely present in secondary structures has a long tradition in protein structure literature (Chou and Fasman, 1978a,b). The role of Pro as β-breaker (i.e., destabilizer of β-sheets) is also well described by most protein structure textbooks (Creighton, 1992). It is well known that the Pro residue, thanks to the rigidity inherent in its imino acid structure (backbone dihedral angle Φ blocked around -60 degree) and to its ability to interconvert from *cis* to *trans* conformations and vice versa, is very seldom found in β-sheets or α-helices. However, the same trend holds true, for complimentary reasons, for the Gly residue. The Gly residue lacks any side-chain and therefore is capable of accessing, without steric clashes, a number of combinations of the dihedral angles Φ and Ψ that would be virtually impossible to all other residues. From this perspective, it would seem reasonable that Gly can be found in any secondary structure without biases, yet this is false. Rather, it can be shown that Gly is almost never present in secondary structures. The extreme flexibility of Gly does not favor its involvement in a defined, rigid structure, and the inability of its side-chain to interact with other residues makes this residue unable to form tertiary interactions. Nevertheless, Pro and Gly play essential role in protein shapes and folding, primarily they participate in the formation of β-hairpins, which can be seen as a strategy to limit the extension of β-strands and protect their edges from aberrant pairings.

Although both Pro and Gly have the above-mentioned structurally relevant properties, they are never associated to amyloid core structures, reasonably because of the β-breaking ability of the former and the very low secondary structure propensity of the latter. In 2002, a study on fibronectin type III superfamily evidenced that Pro residues were fundamental antiaggregation systems. The alignment of protein sequences showed that three proline residues, namely at positions 5, 25, and 64, were highly conserved and that their replacement led to increased aggregation tendency (Steward et al., 2002). Similarly, in a 2005 study on acylphosphatase (AcP) proteins, Parrini and coworkers highlighted for the first time that Gly residues could have a protective effect against amyloid formation (Parrini et al., 2005). By comparing the multiple sequence alignment of the whole AcP family (encompassing proteins coming from all kingdoms of life), an extraordinary conservation was observed in several Gly residues (at positions 15, 19, 37, 45, 53, and 69). When systematically mutated to Ala, virtually none of the mutants showed issues in folding, conformational stability, or enzymatic activity, but their tendency to form amyloid-like fibrils was increased with respect to the wild-type protein (Parrini et al., 2005). Computational studies on proteomes further clarified that Gly and Pro were positively biased in the regions flanking the amyloid stretches identified by specifically designed engines (see Sections 3.4 and 6.1). Accordingly, Gly and Pro have been included in the list of gatekeeper residues, i.e., residues placed near to the boundaries of aggregation-prone stretches to prevent their elongation and therefore to decrease the risk of aggregation.

Glycine residues were shown in the past to be among the least variable amino acid residues (Uversky, 2005). In order to update and consolidate this observation, we analyzed the current release of the PFAM database (a function oriented, well-curated repository of multiple sequence alignments) searching for the residues that were most commonly found as extremely conserved. Fig. 2 summarizes the results of this survey. Protein alignments were scanned column-wise collecting the occurrences of residues with zero-entropy positions (i.e., a conservation of 100%). This analysis reveals

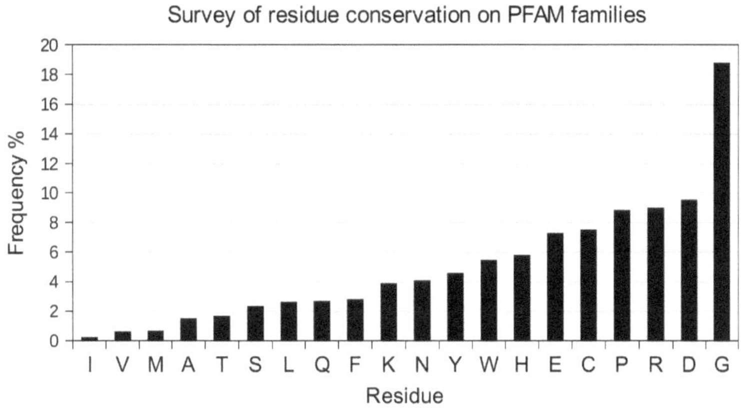

Fig. 2 Outcome of the survey aimed at identifying for the most conserved residues in PFAM. PFAM protein family alignments (with >50 members, a total of 9767 families) were scanned column-wise recording the number of occurrences for which a given residue is found as absolutely required at any position (sequence entropy equal to 0).

that Gly residues, although functionally irrelevant in terms of catalysis or ability to provide side-chain interactions, are the top-scoring residues in this peculiar contest. This also means that, when present and conserved, it plays fundamental roles in protein structures.

3.4 Gatekeepers

The notion of gatekeepers is related to the ability of specific residues to decrease or abolish the natural aggregation tendency of polypeptides if flanking highly aggregation-prone regions. Among such residues we may count the well-known β-breakers Pro and Gly (see above) and, more surprisingly, all charged residues, i.e., Lys, Arg, Glu, and Asp. The definition of gatekeepers was supported by several experimental and then theoretical/computational studies. The role of gatekeepers was first demonstrated some 20 years ago in the ribosomal protein S6 as a way to destabilize collapsed, off-folding pathway structures (Matysiak and Clementi, 2006; Otzen and Oliveberg, 1999; Stoycheva et al., 2004). Being sufficient to disfavor the accumulation of misfolded monomers, their removal was demonstrated to greatly increase S6 aggregation rate (Otzen and Oliveberg, 1999; Pedersen et al., 2004). More recently, mutations removing gatekeeper residues in PI3K-SH3 were found to lead to dramatic changes in its aggregation rates (Buell et al., 2009). A recent study further identified Lys35 as a fundamental gatekeeper which limits the aggregation potential of transthyretin (Sant'Anna et al., 2014). Gatekeepers were also shown to be of great relevance in functional amyloid. In *Escherichia coli*, the assembly of Curli by functional amyloid fibrils of the protein curlin is modulated by Asp residues flanking aggregation nuclei (Wang et al., 2010). Gatekeepers were also analyzed in terms of compositional bias in large-scale computational projects. Such studies are based on the application of algorithms able to identify amyloid stretches in protein sequences, such as TANGO (Fernandez-Escamilla et al., 2004b), Zyggregator (Pawar et al., 2005; Tartaglia and Vendruscolo, 2008), or the Sagg score (Monsellier et al., 2007). Rousseau and coworkers analyzed 28 complete proteomes and found that natural sequences contain many amyloidogenic protein segments, spanning 10–20% of the total residues of the proteomes. Although higher in prokaryotes than in eukaryotes, this size is considered a lower limit for the aggregation tendency of globular proteins, since proteomes are assumed to have evolved to minimize aggregation of the protein components. One of the most important results of the work is the observation that amyloid stretches are usually capped by small regions enriched in Arg, Lys, and Pro (frequently at the first position after the stretch) (Rousseau et al., 2006a). In a second survey, Monsellier and coworkers focused on the human proteome, identifying several features of polypeptide sequences that correlate with the estimated aggregation propensity (Monsellier et al., 2008). In particular, when the compositional biases were compared among aggregation peaks, regions flanking the peaks, and peak-free regions, the authors found strong biases in residues Pro, Arg, Lys, Asp, and Glu, i.e., in β-breaker and charged residues. Such biases exhibited a correlation with peak length: while Pro, Arg, and Lys (gatekeepers) frequencies at the flanks was found to increase significantly with peak length (increased aggregation risk), Asp and Glu frequencies decrease. This corroborated the role of basic residues Arg and Lys as gatekeepers for amyloid stretches, besides shedding a new, amyloid-oriented role in the β-breaking property of Pro.

The algorithms used in the proteome-scale studies described in the last paragraph were developed and trained with different strategies but were substantially aimed at identifying amyloid stretches following the "the Alzheimer's way," i.e., exploiting hydrophobicity as a major force (see more in Section 6). Nevertheless, when different algorithms were used to identify polyglutamine (polyQ) expansions in genomes, different results were obtained in terms of compositional biases (Ramazzotti et al., 2012). PolyQ proteins are proteins containing a long trait of glutamine residues. When the number of glutamines exceeds a certain threshold, the protein is prone to form amyloid-like aggregates and deposit into insoluble inclusions (Fan et al., 2014). Remarkably, despite their completely different origin, fibrils of polyQ-associated aggregates have the same compact β-sheet structure as proteins involved in well-characterized classic amyloid systems (e.g., Aβ peptide, α-synuclein) (Lyubchenko et al., 2012; Poirier et al., 2005). PolyQ stretches (termed "imperfect polyQs," i.e., polyQs with insertions) were recovered and analyzed in 30 eukaryotic, bacterial, and archaeal proteomes. Robust biases were found within and in the flanks of polyQ stretches: Pro, Leu, and His were found to be overrepresented while Asp, Cys, and Gly were underrepresented. The presence of Pro in the list of overrepresented residues seemed to indicate that a selection exists against amyloid aggregation even in the absence of hydrophobic regions, i.e., in the absence of protein folding determinants. Interestingly, this analysis on polyQ aggregation did not highlight enrichments in positively charged residues as in hydrophobic-based aggregation. The reason of one such finding could be rationalized if we consider assisted protein folding. In fact, many authors consider that the presence and evolutionary motivation of gatekeepers as antiaggregation tools have to be searched in the activity of chaperones (Hartl et al., 2011). Such proteins act by binding short, unstructured hydrophobic stretches in order to minimize aberrant contacts and, in turn, to prevent misfolding. Their recognition site has been identified to correspond to positively charged residues flanking hydrophobic stretches (Mayer and Bukau, 2005). During the last 10 years several proteins with a chaperon or chaperon-like activity have been associated to hydrophobic amyloid (see Landreh et al., 2015, for a review and a comprehensive list). Since polyQ stretches are not hydrophobic, chaperones are not expected to be recruited or involved in the folding of polyQ-containing proteins. In contrast to this prediction, chaperones have been demonstrated to slow down the aggregation of polyQ-containing polypeptides (Hay et al., 2004; Rujano et al., 2007). Recently, a detailed experimental analysis on the exon 1 of the huntingtin gene showed that the Hsp70 chaperone does not bind the polyQ stretch, neither is it affected by its length. Rather, Hsp70 binds the N-terminal flanks of the protein, which contain a hydrophobic stretch and positively charged residues that are required for an effective coupling (Monsellier et al., 2015). This clearly indicates that the rules previously established for chaperone binding hold true for polyQ-containing proteins, but the binding site for chaperones on them should not be searched within or in close proximity to the polyQ stretch, and that Pro and His only act and have been evolutionary selected as gatekeepers of polyQ aggregation. This body of experimental evidence was basically confirmed and corroborated by computational studies on proteomes illustrating that gatekeepers are required for an efficient binding of chaperones to aggregation-prone regions in order to decrease, mitigate, or abolish their intrinsic tendency to form amyloid fibrils.

4. EVOLUTIONARY STRATEGIES TO INHIBIT AGGREGATION AT THE STRUCTURAL LEVEL

Beside evolutionary strategies that can be directly observed at the primary sequence level, protein sequences possess also structural, thermodynamic, and dynamic features that allow for their biological tasks to be accomplished in the absence of any aggregation risk. Within this framework, one major problem is the maintenance of the delicate balance between the conformational fluctuations required by the biological function for which a given protein is evolved, and the risk of misfolding and aggregation. As mentioned in the previous sections, in order to be able to carry out their function, proteins need to undergo a certain amount of productive molecular fluctuations (Bahar et al., 2015; Best and Vendruscolo, 2006; Vendruscolo and Dobson, 2006). The intensity of these fluctuations usually exhibits an inverse correlation with thermodynamic stability. For example, hyperthermophilic enzymes are very stable at room temperature but their activity is very low; as temperature increases, such enzymes get increasingly destabilized but can undergo the fluctuations required for catalysis (Corazza et al., 2006; Karshikoff et al., 2015; Unsworth et al., 2007). However, protein fluctuations must not disrupt the folded state of globular proteins. Indeed, while only a few cases have been reported to date of enzymatically active disordered states (Bemporad et al., 2008a; D'Urzo et al., 2014; Pervushin et al., 2007; Vendruscolo, 2010), the disruption of the folded state can impair the ability of the protein to carry out its function and/or lead to the formation of aggregation-prone conformations. In this case, aggregation propensity increases as destabilization increases. For example, stabilization of folded states by ligand binding has been shown to inhibit aggregation of folded proteins (Bemporad et al., 2008b; Soldi et al., 2006). An inverse correlation was shown among the thermodynamic stability of a set of protein variants of Sso AcP and the rate of their aggregation (Plakoutsi et al., 2006). Consequently, a similar inverse correlation there exists between thermodynamic stability and the fluctuations required by biological functions, and between thermodynamic stability and aggregation propensity. The vast majority of globular proteins are able to experience the molecular fluctuations that allow for biological function in the absence of any risk of misfolding.

As mentioned in Section 1, protein sequences evolved so that functional interactions were favored, while deleterious interactions were disfavored. Within this framework, an important role is played by the possible establishment of intermolecular β-sheet interactions. The problem related to intermolecular β-sheets can be readily illustrated by the observation that, on the one hand, the edge strands of β-sheets populate a local conformation which is inherently prone to give rise to intermolecular interactions. On the other hand, it should be emphasized that the establishment of interactions between proteins through the formation of intermolecular β-sheet structure is not necessarily detrimental. Many biologically relevant complexes are formed through the donation of a β-strand from one of the partners and its interaction with a β-sheet or strand of the second partner, as reviewed in Remaut and Waksman (2006). The formation of these complexes can occur with at least three mechanisms: (i) β-sheet augmentation, in which one protein donates a strand and this segment interacts with an edge strand of the target protein, consequently augmenting a preformed β-sheet; (ii) β-strand insertion and fold complementation, in which a segment is donated to the target protein, with the consequence of a rearrangement of the latter and the formation of a new β-sheet; (iii) β-strand zippering, which consists in the pairing of peptide-like stretches in both ligand and receptor (Remaut and Waksman, 2006 and references therein). Consequently, evolution generated certain sequences bestowing on them the ability to form intermolecular β-sheets with specific partners, whereas the vast majority of the sequences were protected against unwanted intermolecular interactions of their β-sheets. A pioneering work illustrated the main structural features devised by evolution to make native states of proteins resistant to intermolecular β-sheet interactions (Richardson and Richardson, 2002). In brief, a survey of naturally occurring β-sheets led the authors to the identifications of common features that keep the propensity of edge strands under control. In certain cases edge strands are simply covered by segments adopting different types of structures. When instead edge strands are exposed to the solvent, they exhibit a set of structural features which render them unable to form intermolecular β-sheets (Richardson and Richardson, 2002). Among the most important ones, edge strands are usually significantly shorter. Second, they are often twisted or distorted by the presence of proline residues or β-bulges, i.e., local disruptions of the regular bonding between two strands, consequence of the insertion of an amino acid into one strand. Third, some edge strands possess inward-pointing charges which anchor them to inner strands. It is reasonable to speculate that evolution adopted these strategies in order to prevent aggregation of folded states and that proteins possessing poorly protected edge strands are more prone to aggregate from their N* state. Indeed, as mentioned above, it was shown that while the majority of the proteins aggregate from largely unfolded conformations, the folded state of globular proteins retain a small, yet significant, tendency to form amyloid with no need of transitions across the major energy barrier for folding/unfolding, but from a conformation which is globally folded but possesses some aberrant features which render it able to aggregate (Chiti and Dobson, 2009).

4.1 Case Study: Native-Like Aggregation in the Acylphosphatase Structural Family

The applicability of the aforementioned strategies was promptly shown by a case in which the use of the inward-pointing charge strategy turned a highly aggregated β-sandwich design into soluble monomers (Wang and Hecht, 2002). However, perhaps the most paradigmatic case to illustrate the importance of protecting edge strands is represented by two proteins belonging to the acylphosphatase (AcP) structural family. AcPs are small α/β enzymes expressed by organisms from all three cellular domains (Stefani et al., 1997). Their structure is characterized by a ferredoxin-like fold, consisting of a 5-stranded β-sheet facing against two α-helices, with a typical βαββαβ topology (Cheung et al., 2005; Corazza et al., 2006; Fusco et al., 2012; Pagano et al., 2006; Pastore et al., 1992; Thunnissen et al., 1997; Zuccotti et al., 2004, 2005) (Fig. 3A and B). In addition to this conserved globular unit, the AcP from the archebacterium *Sulfolobus solfataricus* (Sso AcP) possesses an 11 residue long-unstructured segment, located at the N-terminus of the protein (Corazza et al., 2006). Although AcP misfolding is not linked to any known pathological condition, the protein has been largely employed as a model for folding and misfolding studies. Importantly, AcP structure bears two edge strands. A structural investigation of several AcPs revealed some important differences in edge strand protection among the members of the family (Soldi et al., 2008). Members of the family which had been shown to aggregate only when incubated under conditions that promoted their conversion into largely unfolded conformations, exhibited a high degree of structural protection on their edge strands. In stark contrast, Sso AcP and the second AcP from *Drosophila melanogaster* (AcP Dro2) exhibited a significantly lower protection of such structural elements. In particular, two members of the family which are not able to aggregate from a native-like state were shown to possess roughly three protecting structural features on their edge strands, including the fact that strands are very short,

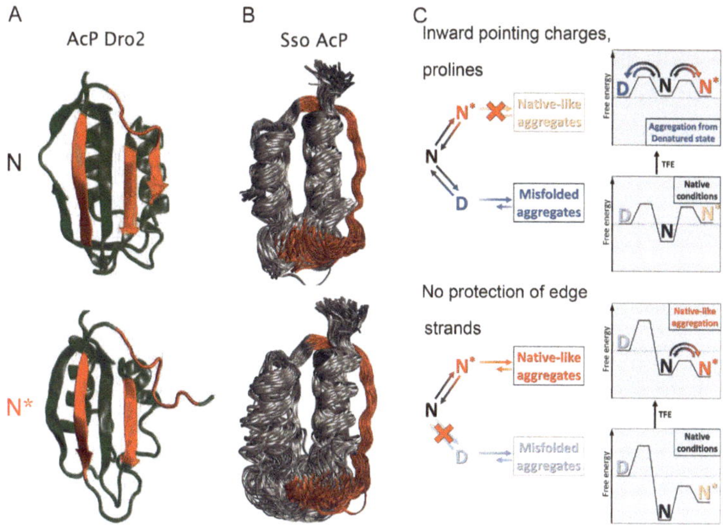

Fig. 3 (A) Native (N, *top*) and native-like (N*, *bottom*) conformational ensembles populated by AcP Dro2 (De Simone et al., 2011). N* exhibits a displaced edge strand. Regions colored in *red* correspond to the most affected regions upon dilution of the protein into aggregation-promoting conditions. (B) Globular domain of Sso AcP populating native (N, *top*) and native-like (N*, *bottom*) conformational ensembles (Bemporad et al., 2012). N* exhibits increased dynamics, larger hydrodynamic residue, lower number of residues involved in secondary structure. Region colored in *red* corresponds to the fourth strand (S4), which was shown to underlie the initial steps of aggregation. (C) Conformational equilibria of Sso AcP in the absence (*bottom*) and in the presence (*top*) of mutations designed to add protection on edge strand (de Rosa et al., 2014). In the absence of these protecting mutations, the protein can convert from N to N* and the latter conformation is able to self-assemble. Native-like aggregation is therefore faster than aggregation from the unfolded state. In the presence of such mutations N* can still be populated, but it is protected against native-like aggregation. Thus, the only possibility for the protein to aggregate is following a completely different pathway which probably involves full unfolding. *Modified from Bemporad, F., De Simone, A., Chiti, F., Dobson, C.M., 2012. Characterizing intermolecular interactions that initiate native-like protein aggregation. Biophys. J. 102, 2595–2604; de Rosa, M., Bemporad, F., Pellegrino, S., Chiti, F., Bolognesi, M., Ricagno, S., 2014. Edge strand engineering prevents native-like aggregation in Sulfolobus solfataricus acylphosphatase. FEBS J. 281, 4072–4084; De Simone, A., Dhulesia, A., Soldi, G., Vendruscolo, M., Hsu, S.T., Chiti, F., et al., 2011. Experimental free energy surfaces reveal the mechanisms of maintenance of protein solubility. Proc. Natl. Acad. Sci. U.S.A. 108, 21057–21062.*

possess β-bulges, proline residues, and inward-pointing charges. As opposed to that, Sso AcP and AcP Dro2 have evolved only one protecting feature per edge strand, a β-bulge on one strand and a very short strand-5. This raises the concrete possibility that Sso AcP and AcP Dro2 are not well protected against native-like aggregation.

Aggregation of AcP Dro2 (Degl'Innocenti et al., 2003) was studied in the presence of small amounts of 2,2,2-trifluoroethanol (TFE) (Soldi et al., 2005). Under such conditions, the protein converts into amyloid-like fibrillar aggregates. Several pieces of evidence support the idea that AcP Dro2 aggregates in the absence of global unfolding. First, the addition of TFE leaves hydrodynamic diameter, secondary structure content, and packing around aromatic residues substantially unaffected. Second, the protein is enzymatically active before aggregation occurs (Soldi et al., 2005). Furthermore, NMR revealed only small changes in the chemical shifts when the protein is moved from native to aggregation-promoting conditions (De Simone et al., 2011).

In an effort to get insight into the structural features that enable native-like AcP Dro2 to self-assemble, an investigation was carried out exploiting amide exchange NMR protection factors (De Simone et al., 2011). The results of such analysis revealed that, in the presence of TFE, the interaction between β-strand 5 and its inner partner is impaired and destabilized (Fig. 3A). This effect is inhibited following addition of phosphate, which binds to the catalytic site of AcP Dro2 (De Simone et al., 2011) and slows down aggregation (Soldi et al., 2006). Molecular dynamics (MD) simulations restrained by experimental protection factors revealed that, under conditions that promote aggregation, AcP Dro2 can access a N* conformational ensemble which is not significantly accessible under native conditions (De Simone et al., 2011). When compared with the fully folded state, N* exhibits a lower fraction of native contacts, a larger radius of gyration, increased exposure of regions of hydrophobic surface and main-chain atoms, and a lower content of secondary structure (De Simone et al., 2011). Furthermore, the β-bulge observed in the strand-4 is partially disrupted in N*. The interface between strands S2 and S5 is also disrupted, giving rise to an enhancement in solvent accessible surface. The displacement of the latter segment induces an increase in hydrophobic exposure.

Sso AcP aggregation was initially investigated in the presence of 15–25% (v/v) TFE at pH 5.5 (Bemporad and Chiti, 2009). Under such conditions, the protein converts, within a few hours, into spheroidal assemblies or thin short amyloid-like fibrils, exhibiting a diameter of 3–5 nm, as assessed by tinctorial assays and from transmission electron microscopy images (Plakoutsi et al., 2004, 2005). Before aggregation occurs, Sso AcP adopts a prevalently folded monomeric conformation. This conclusion was supported by a set of experimental observations. First, under aggregation-promoting conditions, Sso AcP folding is manifold faster than unfolding

(Plakoutsi et al., 2004). Second, the [^{1}H, ^{1}5N] HSQC (Bemporad et al., 2012; Pagano et al., 2010) and the circular dichroism (Plakoutsi et al., 2006) spectra of Sso AcP do not change substantially following addition of TFE at pH 5.5, with only minor changes being evident. Third, Sso AcP is enzymatically active under conditions promoting aggregation, before aggregation occurs (Plakoutsi et al., 2004). Fourth, the pattern of sites cleaved by proteases under conditions of limited proteolysis is similar when the protein is investigated in the absence or in the presence of TFE (Plakoutsi et al., 2006).

The regions of the protein involved in the establishment of the intermolecular interactions which lead to the aggregation of the protein were investigated. It was found that a major role is played by the unfolded N-terminal segment (Bemporad et al., 2008b; Plakoutsi et al., 2006). Indeed, a mutant of the protein lacking this segment is unable to aggregate (Bemporad and Chiti, 2012; Plakoutsi et al., 2006). A second region identified as a fundamental element for Sso AcP aggregation was the fourth edge β-strand (hereinafter β4). The important role played by this segment was supported by several pieces of experimental evidence. First, limited proteolysis experiments revealed that, both in the absence and presence of TFE, the regions of Sso AcP exposed to the solvent and/or flexible span the N-terminal segment, the β-hairpin between β-strand 2 and 3 and β4 (Plakoutsi et al., 2006). Second, a mutagenesis study revealed that a significant correlation there exist between the decrease in thermodynamic stability of the native state induced by a given mutation and the rate constant for aggregation of the corresponding protein variant: the more destabilized is the variant, the faster is the aggregation. This confirms the importance of flexibility within the native-like state of Sso AcP in promoting aggregation (Plakoutsi et al., 2006). Remarkably, a few residues significantly deviate from such correlation. These correspond to the mutants of residues located in β4 and in the unstructured N-terminal segment, suggesting that these regions play an additional, different role (Plakoutsi et al., 2006).

In order to pinpoint the role played by these two segments, a kinetic study was undertaken which combined aggregation kinetic experiments with a mutagenesis strategy (Bemporad et al., 2008b). This study led to a model in which the main event that triggers aggregation of Sso AcP is the establishment of an intermolecular interaction between the N-terminal segment of one Sso AcP molecule and β4 of another Sso AcP molecule (Bemporad et al., 2008b). The occurrence of this interaction was confirmed by means of intermolecular paramagnetic relaxation enhancements, which also illustrated that the interaction is formed with an antiparallel arrangement (Bemporad et al., 2012). The interaction does not alter the global fold of the Sso AcP globular unit but forces molecules to collapse and clump together into a multimer of native-like monomers (Bemporad and Chiti, 2009).

More recently, a strategy which exploits experimental information obtained by means of hydrogen/deuterium experiments as restraints for computer simulations was employed to study aggregation of the protein (Bemporad et al., 2012). Under conditions that promote aggregation, Sso AcP does not undergo any major transition across the energy barrier for folding/unfolding but experiences the conversion into an N* conformation exhibiting an increase in hydrodynamic radius and dynamics, in the absence of local unfolding (Fig. 3B). This transition leads β4 to experience an increased affinity for the N-terminus as, probably, conformations in which the interaction is possible can be transiently populated in the aggregation-prone native-like state (Bemporad et al., 2012). Taken together, the experimental evidence described so far indicates that Sso AcP possesses an edge strand poorly protected against aggregation and that, under mildly destabilizing conditions, this segment experiences an increase in molecular fluctuations which make possible the establishment of new aberrant intermolecular interactions. Consistently, it was shown that the addition of inorganic phosphate, which is able to bind the catalytic site of AcP, results in a strong decrease of dynamics of Sso AcP (Bemporad et al., 2012), with the consequence of slowing both the formation of the early native-like aggregates and their conversion into amyloid-like assemblies (Soldi et al., 2006).

The question arises as to whether it is possible to take advantage of this knowledge to rationally design mutants of Sso AcP engineered to be more protected against native-like aggregation. This strategy was exploited (Fig. 3C) and three mutants of Sso AcP were designed carrying one single point mutation each in β4 (de Rosa et al., 2014; Soldi et al., 2008): in particular, Y86E and V84D carry an additional negative charge in the edge strand. V84P possesses an additional proline residue. The crystal structure of these mutants was recently solved (de Rosa et al., 2014). The β4 of Y86E harbors a β-bulge, also present in wild-type Sso AcP, and the newly inserted inward-pointing charge. A strong twist is inserted in β4 of V84P, resulting in the formation of a strand less prone to form intermolecular sheets. The D84-charged side-chain of V84D points toward a potential edge-to-edge interface inhibiting such interaction. As expected, the folded states of these mutants are destabilized, as the mutations were not designed to increase conformational stability (Soldi et al., 2008). However, none of the three mutants was able to aggregate starting from the folded state (Fig. 3C). Indeed, although these variants can still form β-structured oligomers able to bind thioflavin-T, formation of these latter species proceeds via an alternative mechanism that is independent of the interaction between β4 and the unstructured N-terminal segment and may involve protein unfolding (de Rosa et al., 2014; Soldi et al., 2008).

Consequently, the experimental evidence described here illustrates a scenario in which evolution had to handle the possibility that the folded state of globular proteins give rise to unwanted intermolecular interactions. To face such risk, some specific interactions have been evolved, whereas the majority of naturally occurring β-proteins evolved so that aberrant and undesired interactions were inhibited by a set of specific strategies. The failure of such mechanisms generates proteins that are folded but can access ensembles of native-like misfolded states able to self-assemble with no need of unfolding.

5. EVOLUTIONARY STRATEGIES TO INHIBIT AGGREGATION AT THE CELLULAR LEVEL

Since the natural tendency of polypeptide sequences to form amyloid structures represents a risk for a living cell, the equilibrium among monomers, oligomers, and aggregates is subjected to an evolutionary functional control in the cellular environment that has been widely investigated in the recent past. Tartaglia and coworkers proposed that proteins have coevolved with the quality-control mechanisms present in the cellular environment so that they possess the highest aggregation propensity that enables an organism to function optimally (Tartaglia et al., 2007b). Proteins in the cell are subject to a proteostasis machinery so that their concentration is tightly correlated with their tendency to aggregate. A transcriptional study published in 2009 found that the expression level of genes in *E. coli* inversely correlate with the aggregation propensity of their protein product (Tartaglia et al., 2009). The same result was obtained in humans (Tartaglia et al., 2007b). Furthermore, an important link was found between the evolution of coding sequences in the genome, particularly the codon usage and protein aggregation. The connection was settled at the protein folding level, a process in which the functional needs are translated in terms of proteins that need to fold into a well-defined structure to accomplish specific tasks. In 1996 it was observed that protein folding was assisted by a sequence-encoded timing at the ribosome level: less abundant codons slow down the translation machinery (Thanaraj and Argos, 1996) and allow protein domains whose folding requires longer time to completely and accurately fold, with no risk of aggregation pitfalls. Perhaps more important, the same authors investigated the relationships between translational rate and secondary structure formation of the nascent polypeptide chain, revealing that α-helices of proteins tend to be preferentially encoded by translationally fast mRNA regions while the slow segments often encode β-strands and coil regions (Thanaraj and Argos, 1996).

Beside this optimization at the translational level, other mechanisms concur to the correct folding of proteins, the most important of which are molecular chaperones. As reviewed by Nerelius et al. (2010), much proof has been collected in the past to support the idea that chaperones may have antiamyloid properties (Landreh et al., 2015; Rousseau et al., 2006b). Chaperones bind to nascent protein sequences and delay their hydrophobic collapse until other stretches required for the completion of the hydrophobic core are released. Chaperones also play a pivotal role in the prevention of damage due to misfolding induced by stress. Most chaperones belong to the heat shock protein family, i.e., proteins whose expression is strongly enhanced when physical or chemical alterations perturb cell functionality, possibly triggering protein unfolding and the exposure of hydrophobic stretches, which are the main culprit of amyloid aggregation (Kim et al., 2013). For example, Hsp70 and Hsp90 have been shown to affect fibril formation of the Aβ peptide in vitro (Evans et al., 2006); αB-crystallin is able to sequester Aβ oligomeric forms (Narayan et al., 2012); clusterin is able to sequester and scavenge Aβ and lysozyme aggregates (Kumita et al., 2007; Narayan et al., 2012); Hsp70 affects the early stages of aggregation of α-synuclein (Chaari et al., 2016), haptoglobin inhibits fibril formation by β2-microglobulin (Sultan et al., 2013). Recent advances in chaperone research have led to propose chaperones as promising treatment strategies for misfolding diseases (Huang et al., 2014; Sablon-Carrazana et al., 2015). The fact that chaperones are also directly involved in mitigating the aggregation propensity of polypeptide chains is further demonstrated by large-scale proteomic studies showing that amyloid stretches are flanked by positively charged residues, termed gatekeepers (see later for more on gatekeepers), that represent, together with hydrophobic stretches, the substrate preference of many chaperone classes (Rodriguez et al., 2008; Rudiger et al., 2001).

Another important correlation has been highlighted between protein turnover and aggregation. In 2008 experimental protein lifetime was measured for 611 proteins (Yen et al., 2008); this was correlated in 2011 with gene expression data, interaction networks, and propensity to aggregate

(De Baets et al., 2011). The authors observed that, when compared to long-living proteins, short-living proteins have longer and stronger aggregating regions, lower conservation of gatekeeper residues, and lower chaperone-assisted folding. The consequences emerging from such observations are that short-living proteins are more dangerous in terms of aggregation than long-living proteins, especially when aging increases its detrimental effects on proteasomal function and the quality control machinery of the cell is less accurate (Li et al., 2008; Low, 2011; Ross et al., 2015). Intriguingly, almost all proteins reported as associated to deposition diseases can be classified as short living.

Last, when accounting all the cellular strategies devised to cope with the amyloid formation process, one may include functional amyloid (Chiti and Dobson, 2006; Maury, 2009; Romero and Kolter, 2014). Indeed, while the amyloid phenomenon was initially discovered within its pathological context, it is now clear that amyloid structure possesses some potentially useful characteristics, such as high resistance to proteolysis, self-replication, and the ability to transmit information. These features have been exploited by several organisms, from bacteria to mammals, to perform physiological functions such as formation of biofilms, development of aerial structures, scaffolding, regulation of melanin synthesis, epigenetic control of polyamines, and information transfer (Maury, 2009). In these cases, the inherent toxicity of amyloid structure is moderated via two main strategies: (i) compartmentalization, that is the formation of amyloid only in districts where it is not harmful (e.g., the extracellular space) and (ii) modulation of aggregation propensity by means of chaperone molecules. Thus, bacterial *curli* are assembled by enteric bacteria starting from the amyloid precursor protein CsgA. This protein is maintained in its soluble form in the cytoplasm (Evans et al., 2015), secreted to the cell surface and assembled into amyloid fibrils in the presence of the membrane-bound protein CsgB (Barnhart and Chapman, 2006). The astonishing resistance of spider silk is conferred by a peculiar form of amyloid that develops once fibroin components (spidroins) are released (Kenney et al., 2002; Romer and Scheibel, 2008). In humans, amyloid formed by the protein Pmel 17 templates the covalent polymerisation of precursors to melanin (Harper et al., 2008). Its toxicity is moderated by a dual mechanism involving both compartmentalization and control of aggregation propensity. Full-length Pmel17 is not amyloidogenic per se. However, when transported to the melanosome, Pmel17 aggregates following a single proteolytic cleavage that releases a lumenal fragment. In this case, risks of toxicity are limited by the fact that the latter is so highly aggregation prone that its aggregation into nontoxic fibrils is extremely fast (Harper et al., 2008). The [PSI$^+$] and [Het-s] fungal prions are able to populate amyloid conformations involved in polyamine regulation and heterokaryon formation, respectively (Wickner et al., 2015). In the yeast prion case, the full-length form is not amyloidogenic. Rather, the self-propagating capabilities (and toxicity) are controlled by chaperones (Hsp70 and Hsp104) that mediate prion fragmentation, therefore, controlling their infectivity and their tendency to from amyloid, a sort of modulation of functional amyloid formation (Allen et al., 2005; Shorter and Lindquist, 2006). Although a complete description of functional amyloid is beyond the scopes of this manuscript, from a biological perspective functional amyloid may be interpreted as the last strategy evolved to keep under control protein aggregation: a strategy which does not consist in avoiding aggregation, but rather in exploiting it.

6. DEVELOPMENT OF THE ALGORITHMS ABLE TO IDENTIFY AMYLOID HOT SPOTS AND AMYLOID PROPENSITY

If the information emerged from the body of literature described in the previous section is sufficient to rationalize the aggregation behavior of a given protein, one may speculate that the parameters identified thanks to this effort should be able to predict the propensity of proteins to form amyloid structure. Indeed, despite the unique and peculiar structure of amyloids, proteins, and peptides found in vivo or in vitro to be able to generate such structures (under appropriate conditions Chiti et al., 1999) are widely different both in terms of primary sequence and native three-dimensional structures, involving, with about the same frequency, proteins which belong to all the classes defined in SCOP (Lo Conte et al., 2000), namely α, β, $\alpha+\beta$, and α/β (Linding et al., 2004).

Since the pioneer work by the group of Chiti in 2003, which first tried to derive a function able to rationalize the contributions of different residues to the overall aggregation propensity of a polypeptide chain (Chiti et al., 2003), computational algorithms have flourished trying to predict solubility (Sormanni et al., 2015), the amyloidogenic potential of proteins (Belli et al., 2011), and identify the regions in the sequence most contributing to the amyloid cross-β interactions.

As detailed above, the propensity to form amyloid structures is primarily related to sequence parameters. Algorithms that aim at predicting amyloidogenic features receive protein/peptide sequences as input and emit coefficients, values, or profiles as output. Physio-chemical properties derived from the side-chains such as charge, hydrophobicity, secondary structure propensity, and the distribution of different residues along the sequence have been clearly taken into account in the initial developments. Such basic considerations have been refined, integrated with structural details, and supplemented with novel statistical approaches.

Surveys of computational methods have been recently published (Dovidchenko and Galzitskaya, 2015; Redler et al., 2014) and the ability of the different methods to predict aggregation in vivo has also been evaluated and reviewed (Belli et al., 2011). Although methods have been developed for predicting the presence of prions or prion-like sequences in proteins (Alberti et al., 2009; Toombs et al., 2010, 2012), going into details of this kind of aggregates is beyond the scopes of this review.

Hereinafter, we will briefly describe how computational approaches have been developed during time to exploit the sequence parameters and/or the strategies described in the previous sections, trying to recapitulate the progress brought by the different algorithms. For a systematic classification, methods can be roughly divided into four main classes, that we will here name as "empirical," "structure based," "structure predicting," and "statistical." Empirical tools try to explain experimental results and make predictions by combining, both locally and globally, the physico-chemical properties of amino acids constituting the protein sequence. Structure-based tools are based on statistical observations of 3D structures and thermodynamic optimizations to create functions that predict cross-β aggregation as a possible fold of the sequence under investigation. Structure-predicting methods are based on threading or MD to construct favorable aggregates from small stretches of residues taken from candidate protein sequences. Last, statistical methods try to capture the relationships among residues in experimentally validated aggregating stretches, learning lessons from them and applying what they have learned to unknown test sequences.

6.1 Empirical Methods

In 2003 Chiti and coworkers analyzed single point mutants in the acylphosphatase (AcP) protein and, by analyzing separately hydrophobicity, charge, and β-propensity, derived the first (empirical) formula able to predict whether and to which extent a mutation could increase the amyloid propensity of sequence (Chiti et al., 2003). This algorithm represented the first demonstration that the same forces involved in protein folding also dominate protein misfolding. In 2004 Tartaglia and coworkers presented a second empirical method that gathered changes in polarity, aromaticity, dipole moment, ratio of accessible surface area, and β-sheet propensity following a mutation in a scale-free equation able to predict the changes in aggregation rates (Tartaglia et al., 2004). Although the importance of aromatic residues in inducing aggregation has been later underscored by the observation that the ability of aromatic residues to induce aggregation can be rationalized in terms of their hydrophobicity (Bemporad et al., 2006b) and β-sheet propensity (Street and Mayo, 1999), updates to both methods were then presented, able to predict absolute aggregation rates rather than relative variation upon mutation, thanks to extrinsic factors such as ionic strength and pH (DuBay et al., 2004) or temperature and concentration (Tartaglia et al., 2005). Latest development of algorithms which exploit simple physico-chemical properties of amino acids focused on the prediction and improvement of protein solubility (Agostini et al., 2012; Sormanni et al., 2015).

In the meanwhile, the idea that short, highly amyloid prone, stretches inside proteins could be the real responsible of amyloid aggregation was demonstrated by transplanting a 6-residue stretch from Src homology 3 (SH3) domain of the p58α subunit of bovine phosphatidyl-inositol-3′-kinase (PI3-SH3), which could be induced to form amyloid-like fibrils, into the homologous protein α-spectrin-SH3, resistant to fibril formation, giving to the latter amyloid properties (Ventura et al., 2004). The quest for the features of such stretch, both in terms of sequence and structure determinants led to intense research and the amyloid transplantation experiment was successfully validated in several systems, both in vitro and in vivo. An extreme rationalization of the so-called amyloid stretch hypothesis led the group of Serrano in 2007 to perform a saturation mutagenesis on the de novo-designed amyloid peptide STVIIE and associating the position and the nature of the residues with an amyloidogenic potential (Lopez De La Paz et al., 2002). This is suggested a residue pattern that could be defined an amyloid stretch with the following six residue long signature: $\langle \mathrm{P}\rangle_{1}$-$\langle \mathrm{PKRHW}\rangle_{2}$-$[\mathrm{VLS(C)WFNQ}]_{3}$-$[\mathrm{ILTYWFN}]_{4}$-$[\mathrm{FIY}]_{5}$-$\langle \mathrm{PKRH}\rangle_{6}$ (according to PROSITE rules, with squared brackets delimiting allowed residues and chevron brackets delimiting forbidden residues) (Lopez de la Paz and Serrano, 2004; Pastor et al., 2007). Such definition of the pattern was somehow vague, reflecting the high variability of amyloid stretches identified in aggregation-prone proteins, yet in fact it allowed amyloid-related proteins to be captured in data banks. Although the amyloid pattern cannot be strictly defined as a computational method, having such a definition provided a first tool to easily access large proteomics studies. In agreement with the evolutionary strategies described above, it is important to point out that in this stretch definition the β-breaker Pro is never allowed, Gly is absent form core positions (3–5) and the gatekeeper residues Arg and Lys must be absent in the flanks of the core segment.

In 2006 a mutational study on the Aβ42 Alzheimer's peptide allowed the amyloid aggregation propensity of individual side-chains to be estimated (de Groot et al., 2006). Given amino acid-level propensities, it was natural for bioinformatics to implement, for most of the above-mentioned sources, sliding window methods to predict, besides relative or absolute aggregation scores, the regions inside the sequence that most contribute to the aggregation tendency, generating the so-called amyloid profiles able to identify "amyloid hot spots" (Conchillo-Sole et al., 2007; Pawar et al., 2005; Sanchez de Groot et al., 2005). The above-mentioned methods combined weighted contributions derived from the entire sequence under consideration, so they implicitly assume that aggregation starts from a fully or partially denatured state. To overcome this limitation, in 2008 the Zyggregator algorithm was created (Tartaglia and Vendruscolo, 2008). This algorithm combined intrinsic aggregation propensity as in Pawar et al. and solvent accessibility as predicted by the CamP method (Tartaglia et al., 2007a) to locate aggregation hot spots. This optimization was proposed as a way to introduce structural information into the calculations, i.e., to moderate the propensity scores with the structural protection toward aggregation offered by the burial in the core of the structure of highly amyloidogenic segments. The computational simplicity of empirical approaches allowed the research on amyloid determinants to be moved to a much broader level and the aggregation tendency and the related features of whole proteomes to be investigated (Monsellier et al., 2008; Tartaglia and Vendruscolo, 2010).

6.2 Structure-Based Methods

Methods such as Zyggregator represent a sort of transition toward a completely structure-based algorithms, the first of which, still widely used, is TANGO. Originally published in 2004 (Fernandez-Escamilla et al., 2004b), TANGO was based on the structural energy prediction ability of FoldX (Guerois et al., 2002) to determine the propensity of a particular residue (given its neighborhood) to exist in a given secondary structure, namely, the cross-β structure. The terms considered here are secondary structure propensity, solvation penalty, and charge interactions. TANGO proved to be effective in determining the exact regions of a fully folded protein that would likely be involved in the cross-β structure, but suffered from biases due to the limited train set. The TANGO algorithm was also successfully used to investigate chaperone specificity in genomes (Rousseau et al., 2006b). In 2006 the PASTA method was proposed, gathering structural information by accurately analyzing the pairing of residues in well-annotated β-strands, obtaining energy values for sequence scanning (Trovato et al., 2006, 2007). PASTA used these β-energies to identify consecutive stretches of residues with a minimal energy (maximal stability) and is capable of identifying amyloid stretches and of indicating the parallel/antiparallel arrangement of the predicted core aggregate. PASTA has been recently updated and it has been proposed as a prediction server that incorporates additional factors to improve the prediction (Walsh et al., 2014). In the same year the FoldAmyloid method was also introduced; this algorithm takes into account as determinants for aggregation the packing density and the propensity to form of hydrogen bonds of a given residue (both from backbone and side-chains) (Galzitskaya et al., 2006; Garbuzynskiy et al., 2010). These two parameters were recovered from a collection of nonredundant protein structures spanning the four main categories of SCOP and used as factors in a classic sliding-window averaging technique to identify aggregation hot spots in candidate primary sequences. To the best of our knowledge, 11 tools for amyloid propensity prediction from the two classes "empirical" and "structure based" are currently used, updated, and actively maintained. Since all of them have pitfalls and benefits, a consensus approach termed AMYLPRED is highly valuable and under development since 2009 (Frousios et al., 2009). The most updated version is currently accessible at the AMYLPRED2 server (Tsolis et al., 2013).

6.3 Structure-Predicting Methods

The third class of methods is termed structure-predicting because they are entirely based on 3D structure prediction. In 2006 equilibrium implicit solvent MD was applied to short stretches of residues derived from a window of residues scanning a test sequence (Cecchini et al., 2006). The MD served as a proxy to obtain the structural stability of a hypothetical in-register parallel amyloid aggregate formed by several identical peptides. Such stability was used as shifting β-propensity values so that aggregation hot spots could be identified in several model proteins, obtaining in most cases a significant agreement with experimental results. In the same year, a threading approach was also implemented to predict how well a given stretch or residues could fit into a steric zipper promoting the β-spline structure of an amyloid fibril (Thompson et al., 2006). The method was a derivation of the well-known 3D profiles threading method developed by Bowie et al. (1991) where sequence/structure threads were evaluated by the ROSETTADESIGN program (Kuhlman and Baker, 2000) to find a minimum of energy. A faster and refined implementation of this method was then developed and applied to proteomes (Goldschmidt et al., 2010). These methods, which take into account the structures of amyloid fibrils obtained from experimental approaches and consider fundamental features such as twisting of β-strands, 3D contacts between amino acids (including electrostatic and steric effects), come with high expectations in terms of accuracy or prediction. Indeed, the detachment of predictions from biophysical properties of residues allows to be taken into account as structurally feasible also aggregates from peptide sequences which would not be scored as amyloidogenic by empirical methods (e.g., amyloid structures form hydrophilic peptides). Nevertheless, despite the promises and the high computational burden associated to such approaches, their performance and accuracy does not outperform less complex methods: this is reasonably due to the small number of currently available detailed 3D structures of fibrils solved at high resolution.

6.4 Statistical Methods

Another completely different strategy to locate amyloid hot spots in polypeptide sequences is offered by statistical methods. In 2010 the diversity of more than 200 experimentally validated amyloid hexapeptides was used to calibrate a position-specific scoring matrix that was used as an engine to find amyloid stretches inside sequences. The method was named Waltz (Maurer-Stroh et al., 2010) and proved to be highly effective in identifying aggregation hot spots and, importantly, allowed to better distinguish between amyloid sequences and amorphous aggregates, whose formation is frequently associated to high concentration of proteins in solution. In 2010 a naïve Bayesian classifier combined to a weighted decision tree was proposed as a method to predict amyloidogenesis in antibodies (David et al., 2010). Further, in 2013 machine-learning methods were systematically evaluated based on position-specific frequencies of an energy-based dataset of hexapeptides classified as amyloid or nonamyloid (Stanislawski et al., 2013). Among the many methods available, alternating decision tree showed the best accuracy of prediction (as well as the easiest biological interpretation). Latest, in 2014 the FISH Amyloid method was proposed based on site specific cooccurrence of amino acids (Gasior and Kotulska, 2014) and relying on the assumption that amino acid sequences (such as amyloidogenic fragments) exhibit patterns of residue distribution which is position specific. Trained using a six residue long-sliding window on experimentally verified amyloid sequences, FISH Amyloid showed a performance similar to that achieved with Waltz or FoldAmyloid, lending further support to the notion that amyloid can be predicted from properly trained residue patterns.

6.5 Amyloid Datasets

Whatever prediction method has been or will be developed, it is important here to enumerate the major efforts put so far to create the so-called amyloid datasets, that constitute fundamental testing grounds or learning set for prediction, to be opposed to whole proteomes or scrambled sequences. The first dataset was compiled in 2004 for testing the TANGO algorithm and was entirely derived from literature, consisting in 179 segments from 21 proteins (segment length ranging from 4 to 43), 67 of which experimentally proved to undergo amyloid aggregation in vitro (Fernandez-Escamilla et al., 2004b). Then, the AmylEx dataset was compiled, containing a set of 158 hexapeptides of which 67 have been shown to form fibrils and 91 have yielded negative results in fibril-forming assays (Thompson et al., 2006). In the same work authors also compiled a set of 45 experimentally validated amyloidogenic fragments of proteins, constituting the AmylFrag dataset. The latest resource of experimentally validated amyloid stretches is represented by the WaltzDB (Beerten et al., 2015). After the advent of amyloid spline predictors (Thompson et al., 2006), large computationally derived sets were also released, collected in datasets such as the ZipperDB (Goldschmidt et al., 2010), and a recent dataset compiled by Stanislawski and coworkers featuring a faster 3D profile method (Stanislawski et al., 2013) and composed by 825 positively and 3656 negatively classified segments (4481 hexapeptides in total).

7. CONCLUSIONS AND FUTURE PERSPECTIVES

Decades of research and a gargantuan effort by the protein community led to the identification of the main features that enable normally soluble proteins to convert into misfolded states. The importance of these features has been validated by the development of algorithms which, by taking into account combinations of such parameters, proved to be able to predict and rationalize the amyloid phenomenon. While features that induce or reduce aggregation can be found at the sequence, structural, and even cellular level, it is now clear that the forces that regulate the equilibrium among folding, biological function, misfolding, and pathological function are interconnected with each other. For example, we illustrated that folding and misfolding are dominated by similar forces. On a different line, we surveyed the strategies or sequence features that evolution might have exploited to inhibit misfolding and aggregation, and we pointed out that the same features also tend to impair the conformational motions that are required for biological function. Thus, the current consensus can be summarized as follows: the consequence of protein evolution is that proteins possess the maximum plasticity that allows for their function to be carried out in the absence of misfolding risks, or, from the complimentary point of view, proteins possess the minimal protection against misfolding that does not jeopardize the plasticity required by physiological function.

While this general trend has been clarified, important questions still seek an answer. First, it is not yet clear the reason why, although it was shown that many, possibly all polypeptide chains can convert into amyloid aggregates, only a few of them have been observed to actually deposit in vivo. If the parameters that underlie misfolding are now clear, none of the algorithms so far proposed seems to be able to highlight the proteins whose aggregation propensity cannot be kept under control. This is probably related to the difficulty of including in the algorithms those strategies that cells evolved to cope with aggregation (see Section 5). Another possible reason is related to kinetic aspects of ordered deposition. Amyloid fibrils can be ideally approximated to crystals: under delicate and fragile conditions, fibril formation is thermodynamically favored as the large entropic reduction of monomer assembly is surpassed by the enthalpy contribution. However, kinetics dominate the process and ordered assembly is in competition with disordered assembly, the latter being the real tendency of misfolded proteins, as demonstrated by the high content of cross-β secondary structure found in highly concentrated protein solutions. Since taking into account kinetics into algorithms is inherently difficult and currently unexplored, this limit may lead to false conclusions, both in algorithmic and experimental set up. Accordingly, finding (or computationally predicting) conditions for fibrils formation is about the intricate search for favorable kinetic conditions, and the identification of the solution parameters that determine the aggregation pathway is still not complete (Yoshimura et al., 2012). Another problem is the need to identify with the highest accuracy and precision the regions involved in aggregation of disease-involved protein systems. The identification of such segments corresponds to the identification of the molecular targets for the search of molecules and drugs able to inhibit the aggregation phenomenon and to tackle the pathological conditions related to this phenomenon. To date, only a few cases have seen the successful design of specific drugs able to inhibit aggregation of specific proteins (Johnson et al., 2012). In conclusion, although the general parameters able to promote amyloid formation have been clarified, much must still be done before a full knowledge of the aggregation phenomenon can be accomplished and this information can contribute to a complete understanding of life at the cellular level.

REFERENCES

Abdolvahabi, A., Shi, Y., Rhodes, N.R., Cook, N.P., Marti, A.A., Shaw, B.F., 2015. Arresting amyloid with coulomb's law: acetylation of ALS-linked SOD1 by aspirin impedes aggregation. Biophys. J. 108, 1199–1212.

Agostini, F., Vendruscolo, M., Tartaglia, G.G., 2012. Sequence-based prediction of protein solubility. J. Mol. Biol. 421, 237–241.

Alberti, S., Halfmann, R., King, O., Kapila, A., Lindquist, S., 2009. A systematic survey identifies prions and illuminates sequence features of prionogenic proteins. Cell 137, 146–158.

Alderson, T.R., Markley, J.L., 2013. Biophysical characterization of alpha-synuclein and its controversial structure. Intrinsically Disord. Proteins 1, 18–39.

Allen, K.D., Wegrzyn, R.D., Chernova, T.A., Muller, S., Newnam, G.P., Winslett, P.A., et al., 2005. Hsp70 chaperones as modulators of prion life cycle: novel effects of Ssa and Ssb on the Saccharomyces cerevisiae prion [PSI+]. Genetics 169, 1227–1242.

Anfinsen, C.B., Haber, E., Sela, M., White Jr., F.H., 1961. The kinetics of formation of native ribonuclease during oxidation of the reduced polypeptide chain. Proc. Natl. Acad. Sci. U.S.A. 47, 1309–1314.

Bader, R., Bamford, R., Zurdo, J., Luisi, B.F., Dobson, C.M., 2006. Probing the mechanism of amyloidogenesis through a tandem repeat of the PI3-SH3 domain suggests a generic model for protein aggregation and fibril formation. J. Mol. Biol. 356, 189–208.

Bahar, I., Cheng, M.H., Lee, J.Y., Kaya, C., Zhang, S., 2015. Structure-encoded global motions and their role in mediating protein-substrate interactions. Biophys. J. 109, 1101–1109.

Barnhart, M.M., Chapman, M.R., 2006. Curli biogenesis and function. Annu. Rev. Microbiol. 60, 131–147.

Beerten, J., Van Durme, J., Gallardo, R., Capriotti, E., Serpell, L., Rousseau, F., et al., 2015. WALTZ-DB: a benchmark database of amyloidogenic hexapeptides. Bioinformatics 31, 1698–1700.

Belli, M., Ramazzotti, M., Chiti, F., 2011. Prediction of amyloid aggregation in vivo. EMBO Rep. 12, 657–663.

Bemporad, F., Chiti, F., 2009. "Native-like aggregation" of the acylphosphatase from Sulfolobus solfataricus and its biological implications. FEBS Lett. 583, 2630–2638.

Bemporad, F., Chiti, F., 2012. Protein misfolded oligomers: experimental approaches, mechanism of formation, and structure-toxicity relationships. Chem. Biol. 19, 315–327.

Bemporad, F., Chiti, F., 2013. Pathways of amyloid formation. In: Otzen, D.E. (Ed.), In: Amyloid Fibrils and Prefibrillar Aggregates: Molecular and Biological Properties, vol. 8. Wiley VCH, Berlin, p. 464.

Bemporad, F., Calloni, G., Campioni, S., Plakoutsi, G., Taddei, N., Chiti, F., 2006a. Sequence and structural determinants of amyloid fibril formation. Acc. Chem. Res. 39, 620–627.

Bemporad, F., Taddei, N., Stefani, M., Chiti, F., 2006b. Assessing the role of aromatic residues in the amyloid aggregation of human muscle acylphosphatase. Protein Sci. 15, 862–870.

Bemporad, F., Gsponer, J., Hopearuoho, H.I., Plakoutsi, G., Stati, G., Stefani, M., et al., 2008a. Biological function in a non-native partially folded state of a protein. EMBO J. 27, 1525–1535.

Bemporad, F., Vannocci, T., Varela, L., Azuaga, A.I., Chiti, F., 2008b. A model for the aggregation of the acylphosphatase from Sulfolobus solfataricus in its native-like state. Biochim. Biophys. Acta 1784, 1986–1996.

Bemporad, F., De Simone, A., Chiti, F., Dobson, C.M., 2012. Characterizing intermolecular interactions that initiate native-like protein aggregation. Biophys. J. 102, 2595–2604.

Berlow, R.B., Dyson, H.J., Wright, P.E., 2015. Functional advantages of dynamic protein disorder. FEBS Lett. 589, 2433–2440.

Best, R.B., Vendruscolo, M., 2006. Structural interpretation of hydrogen exchange protection factors in proteins: characterization of the native state fluctuations of CI2. Structure 14, 97–106.

Bitan, G., Kirkitadze, M.D., Lomakin, A., Vollers, S.S., Benedek, G.B., Teplow, D.B., 2003. Amyloid beta-protein (Abeta) assembly: Abeta 40 and Abeta 42 oligomerize through distinct pathways. Proc. Natl. Acad. Sci. U.S.A. 100, 330–335.

Bowie, J.U., Luthy, R., Eisenberg, D., 1991. A method to identify protein sequences that fold into a known three-dimensional structure. Science 253, 164–170.

Broome, B.M., Hecht, M.H., 2000. Nature disfavors sequences of alternating polar and non-polar amino acids: implications for amyloidogenesis. J. Mol. Biol. 296, 961–968.

Buell, A.K., Tartaglia, G.G., Birkett, N.R., Waudby, C.A., Vendruscolo, M., Salvatella, X., et al., 2009. Position-dependent electrostatic protection against protein aggregation. Chembiochem 10, 1309–1312.

Canet, D., Last, A.M., Tito, P., Sunde, M., Spencer, A., Archer, D.B., et al., 2002. Local cooperativity in the unfolding of an amyloidogenic variant of human lysozyme. Nat. Struct. Biol. 9, 308–315.

Cantlon, A., Frigerio, C.S., Freir, D.B., Boland, B., Jin, M., Walsh, D.M., 2015. The familial British dementia mutation promotes formation of neurotoxic cystine cross-linked amyloid Bri (ABri) oligomers. J. Biol. Chem. 290, 16502–16516.

Cecchini, M., Curcio, R., Pappalardo, M., Melki, R., Caflisch, A., 2006. A molecular dynamics approach to the structural characterization of amyloid aggregation. J. Mol. Biol. 357, 1306–1321.

Cerda-Costa, N., Esteras-Chopo, A., Aviles, F.X., Serrano, L., Villegas, V., 2007. Early kinetics of amyloid fibril formation reveals conformational reorganisation of initial aggregates. J. Mol. Biol. 366, 1351–1363.

Chaari, A., Eliezer, D., Ladjimi, M., 2016. The C-terminal alpha-helices of mammalian Hsc70 play a critical role in the stabilization of alpha-synuclein binding and inhibition of aggregation. Int. J. Biol. Macromol. 83, 433–441.

Cheung, Y.Y., Lam, S.Y., Chu, W.K., Allen, M.D., Bycroft, M., Wong, K.B., 2005. Crystal structure of a hyperthermophilic archaeal acylphosphatase from Pyrococcus horikoshii—structural insights into enzymatic catalysis, thermostability, and dimerization. Biochemistry 44, 4601–4611.

Chiti, F., Dobson, C.M., 2006. Protein misfolding, functional amyloid, and human disease. Annu. Rev. Biochem. 75, 333–366.

Chiti, F., Dobson, C.M., 2009. Amyloid formation by globular proteins under native conditions. Nat. Chem. Biol. 5, 15–22.

Chiti, F., Webster, P., Taddei, N., Clark, A., Stefani, M., Ramponi, G., et al., 1999. Designing conditions for in vitro formation of amyloid protofilaments and fibrils. Proc. Natl. Acad. Sci. U.S.A. 96, 3590–3594.

Chiti, F., Calamai, M., Taddei, N., Stefani, M., Ramponi, G., Dobson, C.M., 2002. Studies of the aggregation of mutant proteins in vitro provide insights into the genetics of amyloid diseases. Proc. Natl. Acad. Sci. U.S.A. 99 (Suppl. 4), 16419–16426.

Chiti, F., Stefani, M., Taddei, N., Ramponi, G., Dobson, C.M., 2003. Rationalization of the effects of mutations on peptide and protein aggregation rates. Nature 424, 805–808.

Chou, P.Y., Fasman, G.D., 1978a. Empirical predictions of protein conformation. Annu. Rev. Biochem. 47, 251–276.

Chou, P.Y., Fasman, G.D., 1978b. Prediction of the secondary structure of proteins from their amino acid sequence. Adv. Enzymol. Relat. Areas Mol. Biol. 47, 45–148.

Collins, S.R., Douglass, A., Vale, R.D., Weissman, J.S., 2004. Mechanism of prion propagation: amyloid growth occurs by monomer addition. PLoS Biol. 2, e321.

Conchillo-Sole, O., de Groot, N.S., Aviles, F.X., Vendrell, J., Daura, X., Ventura, S., 2007. AGGRESCAN: a server for the prediction and evaluation of "hot spots" of aggregation in polypeptides. BMC Bioinformatics 8, 65.

Corazza, A., Rosano, C., Pagano, K., Alverdi, V., Esposito, G., Capanni, C., et al., 2006. Structure, conformational stability, and enzymatic properties of acylphosphatase from the hyperthermophile Sulfolobus solfataricus. Proteins 62, 64–79.

Creighton, T.E., 1992. Proteins: Structures and Molecular Properties. W H Freeman & Co (Sd), New York.

David, M.P., Concepcion, G.P., Padlan, E.A., 2010. Using simple artificial intelligence methods for predicting amyloidogenesis in antibodies. BMC Bioinformatics 11, 79.

De Baets, G., Reumers, J., Delgado, Blanco J., Dopazo, J., Schymkowitz, J., Rousseau, F., 2011. An evolutionary trade-off between protein turnover rate and protein aggregation favors a higher aggregation propensity in fast degrading proteins. PLoS Comput. Biol. 7, e1002090.

de Groot, N.S., Aviles, F.X., Vendrell, J., Ventura, S., 2006. Mutagenesis of the central hydrophobic cluster in Abeta42 Alzheimer's peptide. Side-chain properties correlate with aggregation propensities. FEBS J. 273, 658–668.

de Rosa, M., Bemporad, F., Pellegrino, S., Chiti, F., Bolognesi, M., Ricagno, S., 2014. Edge strand engineering prevents native-like aggregation in Sulfolobus solfataricus acylphosphatase. FEBS J. 281, 4072–4084.

de Rosa, M., Barbiroli, A., Giorgetti, S., Mangione, P.P., Bolognesi, M., Ricagno, S., 2015. Decoding the structural bases of D76N ss2-microglobulin high amyloidogenicity through crystallography and Asn-scan mutagenesis. PLoS One 10, e0144061.

De Simone, A., Dhulesia, A., Soldi, G., Vendruscolo, M., Hsu, S.T., Chiti, F., et al., 2011. Experimental free energy surfaces reveal the mechanisms of maintenance of protein solubility. Proc. Natl. Acad. Sci. U.S.A. 108, 21057–21062.

Degl'Innocenti, D., Ramazzotti, M., Marzocchini, R., Chiti, F., Raugei, G., Ramponi, G., 2003. Characterization of a novel *Drosophila melanogaster* acylphosphatase. FEBS Lett. 535, 171–174.

Dobson, C.M., 1999. Protein misfolding, evolution and disease. Trends Biochem. Sci. 24, 329–332.

Dovidchenko, N.V., Galzitskaya, O.V., 2015. Computational approaches to identification of aggregation sites and the mechanism of amyloid growth. Adv. Exp. Med. Biol. 855, 213–239.

DuBay, K.F., Pawar, A.P., Chiti, F., Zurdo, J., Dobson, C.M., Vendruscolo, M., 2004. Prediction of the absolute aggregation rates of amyloidogenic polypeptide chains. J. Mol. Biol. 341, 1317–1326.

D'Urzo, A., Santambrogio, C., Grandori, R., Ciurli, S., Zambelli, B., 2014. The conformational response to Zn(II) and Ni(II) binding of Sporosarcina pasteurii UreG, an intrinsically disordered GTPase. J. Biol. Inorg. Chem. 19, 1341–1354.

Ellis, R.J., 2001. Macromolecular crowding: obvious but underappreciated. Trends Biochem. Sci. 26, 597–604.

Esposito, G., Ricagno, S., Corazza, A., Rennella, E., Gumral, D., Mimmi, M.C., et al., 2008. The controlling roles of Trp60 and Trp95 in beta2-microglobulin function, folding and amyloid aggregation properties. J. Mol. Biol. 378, 887–897.

Evans, C.G., Wisen, S., Gestwicki, J.E., 2006. Heat shock proteins 70 and 90 inhibit early stages of amyloid beta-(1–42) aggregation in vitro. J. Biol. Chem. 281, 33182–33191.

Evans, M.L., Chorell, E., Taylor, J.D., Aden, J., Gotheson, A., Li, F., et al., 2015. The bacterial curli system possesses a potent and selective inhibitor of amyloid formation. Mol. Cell 57, 445–455.

Fan, H.C., Ho, L.I., Chi, C.S., Chen, S.J., Peng, G.S., Chan, T.M., et al., 2014. Polyglutamine (PolyQ) diseases: genetics to treatments. Cell Transplant. 23, 441–458.

Fei, L., Perrett, S., 2009. Disulfide bond formation significantly accelerates the assembly of Ure2p fibrils because of the proximity of a potential amyloid stretch. J. Biol. Chem. 284, 11134–11141.

Fernandez-Escamilla, A.M., Rousseau, F., Schymkowitz, J., Serrano, L., 2004a. Prediction of sequence-dependent and mutational effects on the aggregation of peptides and proteins. Nat. Biotechnol. 22, 1302–1306.

Fernandez-Escamilla, A.M., Cheung, M.S., Vega, M.C., Wilmanns, M., Onuchic, J.N., Serrano, L., 2004b. Solvation in protein folding analysis: combination of theoretical and experimental approaches. Proc. Natl. Acad. Sci. U.S.A. 101, 2834–2839.

Ferrolino, M.C., Zhuravleva, A., Budyak, I.L., Krishnan, B., Gierasch, L.M., 2013. Delicate balance between functionally required flexibility and aggregation risk in a beta-rich protein. Biochemistry 52, 8843–8854.

Fersht, A.R., 1995. Optimization of rates of protein folding: the nucleation-condensation mechanism and its implications. Proc. Natl. Acad. Sci. U.S.A. 92, 10869–10873.

Frousios, K.K., Iconomidou, V.A., Karletidi, C.M., Hamodrakas, S.J., 2009. Amyloidogenic determinants are usually not buried. BMC Struct. Biol. 9, 44.

Fu, Z., Aucoin, D., Davis, J., Van Nostrand, W.E., Smith, S.O., 2015. Mechanism of nucleated conformational conversion of Abeta42. Biochemistry 54, 4197–4207.

Fusco, G., De Simone, A., Hsu, S.T., Bemporad, F., Vendruscolo, M., Chiti, F., et al., 2012. (1)H, (1)(3)C and (1)(5)N resonance assignments of human muscle acylphosphatase. Biomol. NMR Assign. 6, 27–29.

Galzitskaya, O.V., Garbuzynskiy, S.O., Lobanov, M.Y., 2006. Prediction of amyloidogenic and disordered regions in protein chains. PLoS Comput. Biol. 2, e177.

Garbuzynskiy, S.O., Lobanov, M.Y., Galzitskaya, O.V., 2010. FoldAmyloid: a method of prediction of amyloidogenic regions from protein sequence. Bioinformatics 26, 326–332.

Gasior, P., Kotulska, M., 2014. FISH Amyloid—a new method for finding amyloidogenic segments in proteins based on site specific co-occurrence of aminoacids. BMC Bioinformatics 15, 54.

Gerber, R., Tahiri-Alaoui, A., Hore, P.J., James, W., 2007. Oligomerization of the human prion protein proceeds via a molten globule intermediate. J. Biol. Chem. 282, 6300–6307.

Gianni, S., Dogan, J., Jemth, P., 2015. Coupled binding and folding of intrinsically disordered proteins: what can we learn from kinetics? Curr. Opin. Struct. Biol. 36, 18–24.
Gloge, F., Becker, A.H., Kramer, G., Bukau, B., 2014. Co-translational mechanisms of protein maturation. Curr. Opin. Struct. Biol. 24, 24–33.
Goers, J., Uversky, V.N., Fink, A.L., 2003. Polycation-induced oligomerization and accelerated fibrillation of human alpha-synuclein in vitro. Protein Sci. 12, 702–707.
Goldschmidt, L., Teng, P.K., Riek, R., Eisenberg, D., 2010. Identifying the amylome, proteins capable of forming amyloid-like fibrils. Proc. Natl. Acad. Sci. U.S.A. 107, 3487–3492.
Gosal, W.S., Morten, I.J., Hewitt, E.W., Smith, D.A., Thomson, N.H., Radford, S.E., 2005. Competing pathways determine fibril morphology in the self-assembly of beta2-microglobulin into amyloid. J. Mol. Biol. 351, 850–864.
Grana-Montes, R., de Groot, N.S., Castillo, V., Sancho, J., Velazquez-Campoy, A., Ventura, S., 2012. Contribution of disulfide bonds to stability, folding, and amyloid fibril formation: the PI3-SH3 domain case. Antioxid. Redox Signal. 16, 1–15.
Guerois, R., Nielsen, J.E., Serrano, L., 2002. Predicting changes in the stability of proteins and protein complexes: a study of more than 1000 mutations. J. Mol. Biol. 320, 369–387.
Guijarro, J.I., Sunde, M., Jones, J.A., Campbell, I.D., Dobson, C.M., 1998. Amyloid fibril formation by an SH3 domain. Proc. Natl. Acad. Sci. U.S.A. 95, 4224–4228.
Halabelian, L., Ricagno, S., Giorgetti, S., Santambrogio, C., Barbiroli, A., Pellegrino, S., et al., 2014. Class I major histocompatibility complex, the trojan horse for secretion of amyloidogenic beta2-microglobulin. J. Biol. Chem. 289, 3318–3327.
Harper, D.C., Theos, A.C., Herman, K.E., Tenza, D., Raposo, G., Marks, M.S., 2008. Premelanosome amyloid-like fibrils are composed of only golgi-processed forms of Pmel17 that have been proteolytically processed in endosomes. J. Biol. Chem. 283, 2307–2322.
Hartl, F.U., Bracher, A., Hayer-Hartl, M., 2011. Molecular chaperones in protein folding and proteostasis. Nature 475, 324–332.
Hay, D.G., Sathasivam, K., Tobaben, S., Stahl, B., Marber, M., Mestril, R., et al., 2004. Progressive decrease in chaperone protein levels in a mouse model of Huntington's disease and induction of stress proteins as a therapeutic approach. Hum. Mol. Genet. 13, 1389–1405.
Huang, F., Wang, J., Qu, A., Shen, L., Liu, J., Zhang, Z., et al., 2014. Maintenance of amyloid beta peptide homeostasis by artificial chaperones based on mixed-shell polymeric micelles. Angew. Chem. Int. Ed. Engl. 53, 8985–8990.
Hurshman, A.R., White, J.T., Powers, E.T., Kelly, J.W., 2004. Transthyretin aggregation under partially denaturing conditions is a downhill polymerization. Biochemistry 43, 7365–7381.
Iakoucheva, L.M., Brown, C.J., Lawson, J.D., Obradovic, Z., Dunker, A.K., 2002. Intrinsic disorder in cell-signaling and cancer-associated proteins. J. Mol. Biol. 323, 573–584.
Jahn, T.R., Radford, S.E., 2005. The Yin and Yang of protein folding. FEBS J. 272, 5962–5970.
Jahn, T.R., Radford, S.E., 2008. Folding versus aggregation: polypeptide conformations on competing pathways. Arch. Biochem. Biophys. 469, 100–117.
Jarrett, J.T., Lansbury Jr., P.T., 1993. Seeding "one-dimensional crystallization" of amyloid: a pathogenic mechanism in Alzheimer's disease and scrapie? Cell 73, 1055–1058.
Johnson, S.M., Connelly, S., Fearns, C., Powers, E.T., Kelly, J.W., 2012. The transthyretin amyloidoses: from delineating the molecular mechanism of aggregation linked to pathology to a regulatory-agency-approved drug. J. Mol. Biol. 421, 185–203.
Karshikoff, A., Nilsson, L., Ladenstein, R., 2015. Rigidity versus flexibility: the dilemma of understanding protein thermal stability. FEBS J. 282, 3899–3917.

Kenney, J.M., Knight, D., Wise, M.J., Vollrath, F., 2002. Amyloidogenic nature of spider silk. Eur.J. Biochem. 269, 4159–4163.
Kim, W., Hecht, M.H., 2006. Generic hydrophobic residues are sufficient to promote aggregation of the Alzheimer's Abeta42 peptide. Proc. Natl. Acad. Sci. U.S.A. 103, 15824–15829.
Kim, Y.E., Hipp, M.S., Bracher, A., Hayer-Hartl, M., Hartl, F.U., 2013. Molecular chaperone functions in protein folding and proteostasis. Annu. Rev. Biochem. 82, 323–355.
Kuhlman, B., Baker, D., 2000. Native protein sequences are close to optimal for their structures. Proc. Natl. Acad. Sci. U.S.A. 97, 10383–10388.
Kumar, S., Udgaonkar, J.B., 2009. Conformational conversion may precede or follow aggregate elongation on alternative pathways of amyloid protofibril formation. J. Mol. Biol. 385, 1266–1276.
Kumita, J.R., Poon, S., Caddy, G.L., Hagan, C.L., Dumoulin, M., Yerbury, J.J., et al., 2007. The extracellular chaperone clusterin potently inhibits human lysozyme amyloid formation by interacting with prefibrillar species. J. Mol. Biol. 369, 157–167.
Landreh, M., Rising, A., Presto, J., Jornvall, H., Johansson, J., 2015. Specific chaperones and regulatory domains in control of amyloid formation. J. Biol. Chem. 290, 26430–26436.
Lee, J., Culyba, E.K., Powers, E.T., Kelly, J.W., 2011. Amyloid-beta forms fibrils by nucleated conformational conversion of oligomers. Nat. Chem. Biol. 7, 602–609.
Li, Y., Wang, Y.S., Shen, X.F., Hui, Y.N., Han, J., Zhao, W., et al., 2008. Alterations of activity and intracellular distribution of the 20S proteasome in ageing retinal pigment epithelial cells. Exp. Gerontol. 43, 1114–1122.
Linding, R., Schymkowitz, J., Rousseau, F., Diella, F., Serrano, L., 2004. A comparative study of the relationship between protein structure and beta-aggregation in globular and intrinsically disordered proteins. J. Mol. Biol. 342, 345–353.
Lo Conte, L., Ailey, B., Hubbard, T.J., Brenner, S.E., Murzin, A.G., Chothia, C., 2000. SCOP: a structural classification of proteins database. Nucleic Acids Res. 28, 257–259.
Lomakin, A., Chung, D.S., Benedek, G.B., Kirschner, D.A., Teplow, D.B., 1996. On the nucleation and growth of amyloid beta-protein fibrils: detection of nuclei and quantitation of rate constants. Proc. Natl. Acad. Sci. U.S.A. 93, 1125–1129.
Lopez de la Paz, M., Serrano, L., 2004. Sequence determinants of amyloid fibril formation. Proc. Natl. Acad. Sci. U.S.A. 101, 87–92.
Lopez De La Paz, M., Goldie, K., Zurdo, J., Lacroix, E., Dobson, C.M., Hoenger, A., et al., 2002. De novo designed peptide-based amyloid fibrils. Proc. Natl. Acad. Sci. U.S.A. 99, 16052–16057.
Low, P., 2011. The role of ubiquitin-proteasome system in ageing. Gen. Comp. Endocrinol. 172, 39–43.
Lyubchenko, Y.L., Krasnoslobodtsev, A.V., Luca, S., 2012. Fibrillogenesis of huntingtin and other glutamine containing proteins. Subcell. Biochem. 65, 225–251.
Masino, L., Nicastro, G., Calder, L., Vendruscolo, M., Pastore, A., 2011. Functional interactions as a survival strategy against abnormal aggregation. FASEB J. 25, 45–54.
Matysiak, S., Clementi, C., 2006. Minimalist protein model as a diagnostic tool for misfolding and aggregation. J. Mol. Biol. 363, 297–308.
Maurer-Stroh, S., Debulpaep, M., Kuemmerer, N., Lopez de la Paz, M., Martins, I.C., Reumers, J., et al., 2010. Exploring the sequence determinants of amyloid structure using position-specific scoring matrices. Nat. Methods 7, 237–242.
Maury, C.P., 2009. The emerging concept of functional amyloid. J. Intern. Med. 265, 329–334.
Mayer, M.P., Bukau, B., 2005. Hsp70 chaperones: cellular functions and molecular mechanism. Cell. Mol. Life Sci. 62, 670–684.
Mirceta, S., Signore, A.V., Burns, J.M., Cossins, A.R., Campbell, K.L., Berenbrink, M., 2013. Evolution of mammalian diving capacity traced by myoglobin net surface charge. Science 340, 1234192.

Monsellier, E., Chiti, F., 2007. Prevention of amyloid-like aggregation as a driving force of protein evolution. EMBO Rep. 8, 737–742.

Monsellier, E., Ramazzotti, M., de Laureto, P.P., Tartaglia, G.G., Taddei, N., Fontana, A., et al., 2007. The distribution of residues in a polypeptide sequence is a determinant of aggregation optimized by evolution. Biophys. J. 93, 4382–4391.

Monsellier, E., Ramazzotti, M., Taddei, N., Chiti, F., 2008. Aggregation propensity of the human proteome. PLoS Comput. Biol. 4, e1000199.

Monsellier, E., Redeker, V., Ruiz-Arlandis, G., Bousset, L., Melki, R., 2015. Molecular interaction between the chaperone Hsc70 and the N-terminal flank of huntingtin exon 1 modulates aggregation. J. Biol. Chem. 290, 2560–2576.

Morris, A.M., Watzky, M.A., Agar, J.N., Finke, R.G., 2008. Fitting neurological protein aggregation kinetic data via a 2-step, minimal/"Ockham's razor" model: the Finke-Watzky mechanism of nucleation followed by autocatalytic surface growth. Biochemistry 47, 2413–2427.

Mossuto, M.F., Bolognesi, B., Guixer, B., Dhulesia, A., Agostini, F., Kumita, J.R., et al., 2011. Disulfide bonds reduce the toxicity of the amyloid fibrils formed by an extracellular protein. Angew. Chem. Int. Ed. Engl. 50, 7048–7051.

Narayan, P., Meehan, S., Carver, J.A., Wilson, M.R., Dobson, C.M., Klenerman, D., 2012. Amyloid-beta oligomers are sequestered by both intracellular and extracellular chaperones. Biochemistry 51, 9270–9276.

Nelson, R., Sawaya, M.R., Balbirnie, M., Madsen, A.O., Riekel, C., Grothe, R., et al., 2005. Structure of the cross-beta spine of amyloid-like fibrils. Nature 435, 773–778.

Nerelius, C., Fitzen, M., Johansson, J., 2010. Amino acid sequence determinants and molecular chaperones in amyloid fibril formation. Biochem. Biophys. Res. Commun. 396, 2–6.

No authors, 2005. So much more to know. Science 309, 78–102.

O'Brien, E.P., Ciryam, P., Vendruscolo, M., Dobson, C.M., 2014. Understanding the influence of codon translation rates on cotranslational protein folding. Acc. Chem. Res. 47, 1536–1544.

Oldfield, C.J., Cheng, Y., Cortese, M.S., Brown, C.J., Uversky, V.N., Dunker, A.K., 2005. Comparing and combining predictors of mostly disordered proteins. Biochemistry 44, 1989–2000.

Otzen, D.E., Oliveberg, M., 1999. Salt-induced detour through compact regions of the protein folding landscape. Proc. Natl. Acad. Sci. U.S.A. 96, 11746–11751.

Pagano, K., Ramazzotti, M., Viglino, P., Esposito, G., Degl'Innocenti, D., Taddei, N., et al., 2006. NMR solution structure of the acylphosphatase from Escherichia coli. J. Biomol. NMR 36, 199–204.

Pagano, K., Bemporad, F., Fogolari, F., Esposito, G., Viglino, P., Chiti, F., et al., 2010. Structural and dynamics characteristics of acylphosphatase from Sulfolobus solfataricus in the monomeric state and in the initial native-like aggregates. J. Biol. Chem. 285, 14689–14700.

Parrini, C., Taddei, N., Ramazzotti, M., Degl'Innocenti, D., Ramponi, G., Dobson, C.M., et al., 2005. Glycine residues appear to be evolutionarily conserved for their ability to inhibit aggregation. Structure 13, 1143–1151.

Parrini, C., Bemporad, F., Baroncelli, A., Gianni, S., Travaglini-Allocatelli, C., Kohn, J.E., et al., 2008. The folding process of acylphosphatase from Escherichia coli is remarkably accelerated by the presence of a disulfide bond. J. Mol. Biol. 379, 1107–1118.

Pastor, M.T., Esteras-Chopo, A., Serrano, L., 2007. Hacking the code of amyloid formation: the amyloid stretch hypothesis. Prion 1, 9–14.

Pastore, A., Saudek, V., Ramponi, G., Williams, R.J., 1992. Three-dimensional structure of acylphosphatase. Refinement and structure analysis. J. Mol. Biol. 224, 427–440.

Pawar, A.P., Dubay, K.F., Zurdo, J., Chiti, F., Vendruscolo, M., Dobson, C.M., 2005. Prediction of "aggregation-prone" and "aggregation-susceptible" regions in proteins associated with neurodegenerative diseases. J. Mol. Biol. 350, 379–392.

Pedersen, J.S., Christensen, G., Otzen, D.E., 2004. Modulation of S6 fibrillation by unfolding rates and gatekeeper residues. J. Mol. Biol. 341, 575–588.

Pertinhez, T.A., Bouchard, M., Tomlinson, E.J., Wain, R., Ferguson, S.J., Dobson, C.M., et al., 2001. Amyloid fibril formation by a helical cytochrome. FEBS Lett. 495, 184–186.

Pervushin, K., Vamvaca, K., Vogeli, B., Hilvert, D., 2007. Structure and dynamics of a molten globular enzyme. Nat. Struct. Mol. Biol. 14, 1202–1206.

Petty, S.A., Decatur, S.M., 2005. Experimental evidence for the reorganization of beta-strands within aggregates of the Abeta(16–22) peptide. J. Am. Chem. Soc. 127, 13488–13489.

Plakoutsi, G., Taddei, N., Stefani, M., Chiti, F., 2004. Aggregation of the Acylphosphatase from Sulfolobus solfataricus: the folded and partially unfolded states can both be precursors for amyloid formation. J. Biol. Chem. 279, 14111–14119.

Plakoutsi, G., Bemporad, F., Calamai, M., Taddei, N., Dobson, C.M., Chiti, F., 2005. Evidence for a mechanism of amyloid formation involving molecular reorganisation within native-like precursor aggregates. J. Mol. Biol. 351, 910–922.

Plakoutsi, G., Bemporad, F., Monti, M., Pagnozzi, D., Pucci, P., Chiti, F., 2006. Exploring the mechanism of formation of native-like and precursor amyloid oligomers for the native acylphosphatase from Sulfolobus solfataricus. Structure 14, 993–1001.

Platt, G.W., Routledge, K.E., Homans, S.W., Radford, S.E., 2008. Fibril growth kinetics reveal a region of beta2-microglobulin important for nucleation and elongation of aggregation. J. Mol. Biol. 378, 251–263.

Poirier, M.A., Jiang, H., Ross, C.A., 2005. A structure-based analysis of huntingtin mutant polyglutamine aggregation and toxicity: evidence for a compact beta-sheet structure. Hum. Mol. Genet. 14, 765–774.

Ramazzotti, M., Parrini, C., Stefani, M., Manao, G., Degl'Innocenti, D., 2006. The intrachain disulfide bridge is responsible of the unusual stability properties of novel acylphosphatase from Escherichia coli. FEBS Lett. 580, 6763–6768.

Ramazzotti, M., Monsellier, E., Kamoun, C., Degl'Innocenti, D., Melki, R., 2012. Polyglutamine repeats are associated to specific sequence biases that are conserved among eukaryotes. PLoS One 7, e30824.

Rangarajan, N., Kulkarni, P., Hannenhalli, S., 2015. Evolutionarily conserved network properties of intrinsically disordered proteins. PLoS One 10, e0126729.

Redler, R.L., Shirvanyants, D., Dagliyan, O., Ding, F., Kim, D.N., Kota, P., et al., 2014. Computational approaches to understanding protein aggregation in neurodegeneration. J. Mol. Cell Biol. 6, 104–115.

Remaut, H., Waksman, G., 2006. Protein-protein interaction through beta-strand addition. Trends Biochem. Sci. 31, 436–444.

Richardson, J.S., Richardson, D.C., 2002. Natural beta-sheet proteins use negative design to avoid edge-to-edge aggregation. Proc. Natl. Acad. Sci. U.S.A. 99, 2754–2759.

Rodriguez, F., Arsene-Ploetze, F., Rist, W., Rudiger, S., Schneider-Mergener, J., Mayer, M.P., et al., 2008. Molecular basis for regulation of the heat shock transcription factor sigma32 by the DnaK and DnaJ chaperones. Mol. Cell 32, 347–358.

Romer, L., Scheibel, T., 2008. The elaborate structure of spider silk: structure and function of a natural high performance fiber. Prion 2, 154–161.

Romero, D., Kolter, R., 2014. Functional amyloids in bacteria. Int. Microbiol. 17, 65–73.

Ross, J.M., Olson, L., Coppotelli, G., 2015. Mitochondrial and ubiquitin proteasome system dysfunction in ageing and disease: two sides of the same coin? Int. J. Mol. Sci. 16, 19458–19476.

Rousseau, F., Schymkowitz, J., Serrano, L., 2006a. Protein aggregation and amyloidosis: confusion of the kinds? Curr. Opin. Struct. Biol. 16, 118–126.

Rousseau, F., Serrano, L., Schymkowitz, J.W., 2006b. How evolutionary pressure against protein aggregation shaped chaperone specificity. J. Mol. Biol. 355, 1037–1047.
Rudiger, S., Schneider-Mergener, J., Bukau, B., 2001. Its substrate specificity characterizes the DnaJ co-chaperone as a scanning factor for the DnaK chaperone. EMBO J. 20, 1042–1050.
Rujano, M.A., Kampinga, H.H., Salomons, F.A., 2007. Modulation of polyglutamine inclusion formation by the Hsp70 chaperone machine. Exp. Cell Res. 313, 3568–3578.
Ryadnov, M.G., Papapostolou, D., Woolfson, D.N., 2008. The leucine zipper as a building block for self-assembled protein fibers. Methods Mol. Biol. 474, 35–51.
Sablon-Carrazana, M., Fernandez, I., Bencomo, A., Lara-Martinez, R., Rivera-Marrero, S., Dominguez, G., et al., 2015. Drug development in conformational diseases: a novel family of chemical chaperones that bind and stabilise several polymorphic amyloid structures. PLoS One 10, e0135292.
Sanchez de Groot, N., Pallares, I., Aviles, F.X., Vendrell, J., Ventura, S., 2005. Prediction of "hot spots" of aggregation in disease-linked polypeptides. BMC Struct. Biol. 5, 18.
Sandal, M., Valle, F., Tessari, I., Mammi, S., Bergantino, E., Musiani, F., et al., 2008. Conformational equilibria in monomeric alpha-synuclein at the single-molecule level. PLoS Biol. 6, e6.
Sant'Anna, R., Braga, C., Varejao, N., Pimenta, K.M., Grana-Montes, R., Alves, A., et al., 2014. The importance of a gatekeeper residue on the aggregation of transthyretin: implications for transthyretin-related amyloidoses. J. Biol. Chem. 289, 28324–28337.
Sawaya, M.R., Sambashivan, S., Nelson, R., Ivanova, M.I., Sievers, S.A., Apostol, M.I., et al., 2007. Atomic structures of amyloid cross-beta spines reveal varied steric zippers. Nature 447, 453–457.
Scheibel, T., Lindquist, S.L., 2001. The role of conformational flexibility in prion propagation and maintenance for Sup35p. Nat. Struct. Biol. 8, 958–962.
Schmittschmitt, J.P., Scholtz, J.M., 2003. The role of protein stability, solubility, and net charge in amyloid fibril formation. Protein Sci. 12, 2374–2378.
Serio, T.R., Cashikar, A.G., Kowal, A.S., Sawicki, G.J., Moslehi, J.J., Serpell, L., et al., 2000. Nucleated conformational conversion and the replication of conformational information by a prion determinant. Science 289, 1317–1321.
Sharma, R., Raduly, Z., Miskei, M., Fuxreiter, M., 2015. Fuzzy complexes: specific binding without complete folding. FEBS Lett. 589, 2533–2542.
Shaykhalishahi, H., Gauhar, A., Wordehoff, M.M., Gruning, C.S., Klein, A.N., Bannach, O., et al., 2015. Contact between the beta1 and beta2 segments of alpha-synuclein that inhibits amyloid formation. Angew. Chem. Int. Ed. Engl. 54, 8837–8840.
Shorter, J., Lindquist, S., 2006. Destruction or potentiation of different prions catalyzed by similar Hsp104 remodeling activities. Mol. Cell 23, 425–438.
Smirnova, E., Safenkova, I., Stein-Margolina, V., Shubin, V., Polshakov, V., Gurvits, B., 2015. pH-responsive modulation of insulin aggregation and structural transformation of the aggregates. Biochimie 109, 49–59.
Snider, J., Kotlyar, M., Saraon, P., Yao, Z., Jurisica, I., Stagljar, I., 2015. Fundamentals of protein interaction network mapping. Mol. Syst. Biol. 11, 848.
Soldi, G., Bemporad, F., Torrassa, S., Relini, A., Ramazzotti, M., Taddei, N., et al., 2005. Amyloid formation of a protein in the absence of initial unfolding and destabilization of the native state. Biophys. J. 89, 4234–4244.
Soldi, G., Plakoutsi, G., Taddei, N., Chiti, F., 2006. Stabilization of a native protein mediated by ligand binding inhibits amyloid formation independently of the aggregation pathway. J. Med. Chem. 49, 6057–6064.
Soldi, G., Bemporad, F., Chiti, F., 2008. The degree of structural protection at the edge beta-strands determines the pathway of amyloid formation in globular proteins. J. Am. Chem. Soc. 130, 4295–4302.

Sormanni, P., Aprile, F.A., Vendruscolo, M., 2015. The CamSol method of rational design of protein mutants with enhanced solubility. J. Mol. Biol. 427, 478–490.
Stanislawski, J., Kotulska, M., Unold, O., 2013. Machine learning methods can replace 3D profile method in classification of amyloidogenic hexapeptides. BMC Bioinformatics 14, 21.
Stefani, M., Taddei, N., Ramponi, G., 1997. Insights into acylphosphatase structure and catalytic mechanism. Cell. Mol. Life Sci. 53, 141–151.
Sterner, R., Liebl, W., 2001. Thermophilic adaptation of proteins. Crit. Rev. Biochem. Mol. Biol. 36, 39–106.
Steward, A., Adhya, S., Clarke, J., 2002. Sequence conservation in Ig-like domains: the role of highly conserved proline residues in the fibronectin type III superfamily. J. Mol. Biol. 318, 935–940.
Stoycheva, A.D., Brooks 3rd, C.L., Onuchic, J.N., 2004. Gatekeepers in the ribosomal protein s6: thermodynamics, kinetics, and folding pathways revealed by a minimalist protein model. J. Mol. Biol. 340, 571–585.
Street, A.G., Mayo, S.L., 1999. Intrinsic beta-sheet propensities result from van der Waals interactions between side chains and the local backbone. Proc. Natl. Acad. Sci. U.S.A. 96, 9074–9076.
Sultan, A., Raman, B., Rao, Ch.M., Tangirala, R., 2013. The extracellular chaperone haptoglobin prevents serum fatty acid-promoted amyloid fibril formation of beta2-microglobulin, resistance to lysosomal degradation, and cytotoxicity. J. Biol. Chem. 288, 32326–32342.
Tartaglia, G.G., Vendruscolo, M., 2008. The Zyggregator method for predicting protein aggregation propensities. Chem. Soc. Rev. 37, 1395–1401.
Tartaglia, G.G., Vendruscolo, M., 2010. Proteome-level interplay between folding and aggregation propensities of proteins. J. Mol. Biol. 402, 919–928.
Tartaglia, G.G., Cavalli, A., Pellarin, R., Caflisch, A., 2004. The role of aromaticity, exposed surface, and dipole moment in determining protein aggregation rates. Protein Sci. 13, 1939–1941.
Tartaglia, G.G., Cavalli, A., Pellarin, R., Caflisch, A., 2005. Prediction of aggregation rate and aggregation-prone segments in polypeptide sequences. Protein Sci. 14, 2723–2734.
Tartaglia, G.G., Cavalli, A., Vendruscolo, M., 2007a. Prediction of local structural stabilities of proteins from their amino acid sequences. Structure 15, 139–143.
Tartaglia, G.G., Pechmann, S., Dobson, C.M., Vendruscolo, M., 2007b. Life on the edge: a link between gene expression levels and aggregation rates of human proteins. Trends Biochem. Sci. 32, 204–206.
Tartaglia, G.G., Pechmann, S., Dobson, C.M., Vendruscolo, M., 2009. A relationship between mRNA expression levels and protein solubility in E. coli. J. Mol. Biol. 388, 381–389.
Thakur, A.K., Jayaraman, M., Mishra, R., Thakur, M., Chellgren, V.M., Byeon, I.J., et al., 2009. Polyglutamine disruption of the huntingtin exon 1 N terminus triggers a complex aggregation mechanism. Nat. Struct. Mol. Biol. 16, 380–389.
Thanaraj, T.A., Argos, P., 1996. Protein secondary structural types are differentially coded on messenger RNA. Protein Sci. 5, 1973–1983.
Thompson, M.J., Sievers, S.A., Karanicolas, J., Ivanova, M.I., Baker, D., Eisenberg, D., 2006. The 3D profile method for identifying fibril forming segments of proteins. Proc. Natl. Acad. Sci. U.S.A. 103, 4074–4078.
Thunnissen, M.M., Taddei, N., Liguri, G., Ramponi, G., Nordlund, P., 1997. Crystal structure of common type acylphosphatase from bovine testis. Structure 5, 69–79.
Tjernberg, L., Hosia, W., Bark, N., Thyberg, J., Johansson, J., 2002. Charge attraction and beta propensity are necessary for amyloid fibril formation from tetrapeptides. J. Biol. Chem. 277, 43243–43246.

Toombs, J.A., McCarty, B.R., Ross, E.D., 2010. Compositional determinants of prion formation in yeast. Mol. Cell. Biol. 30, 319–332.
Toombs, J.A., Petri, M., Paul, K.R., Kan, G.Y., Ben-Hur, A., Ross, E.D., 2012. De novo design of synthetic prion domains. Proc. Natl. Acad. Sci. U.S.A. 109, 6519–6524.
Trovato, A., Chiti, F., Maritan, A., Seno, F., 2006. Insight into the structure of amyloid fibrils from the analysis of globular proteins. PLoS Comput. Biol. 2, e170.
Trovato, A., Seno, F., Tosatto, S.C., 2007. The PASTA server for protein aggregation prediction. Protein Eng. Des. Sel. 20, 521–523.
Tsolis, A.C., Papandreou, N.C., Iconomidou, V.A., Hamodrakas, S.J., 2013. A consensus method for the prediction of 'aggregation-prone' peptides in globular proteins. PLoS One 8, e54175.
Udgaonkar, J.B., 2013. Polypeptide chain collapse and protein folding. Arch. Biochem. Biophys. 531, 24–33.
Unsworth, L.D., van der Oost, J., Koutsopoulos, S., 2007. Hyperthermophilic enzymes—stability, activity and implementation strategies for high temperature applications. FEBS J. 274, 4044–4056.
Uversky, V.N., 2005. A GLYmmer of insight into fibril formation. Structure 13, 1090–1092.
Uversky, V.N., Dunker, A.K., 2010. Understanding protein non-folding. Biochim. Biophys. Acta 1804, 1231–1264.
Valleix, S., Gillmore, J.D., Bridoux, F., Mangione, P.P., Dogan, A., Nedelec, B., et al., 2012. Hereditary systemic amyloidosis due to Asp76Asn variant beta2-microglobulin. N. Engl. J. Med. 366, 2276–2283.
Vendruscolo, M., 2010. Enzymatic activity in disordered states of proteins. Curr. Opin. Chem. Biol. 14, 671–675.
Vendruscolo, M., Dobson, C.M., 2006. Structural biology. Dynamic visions of enzymatic reactions. Science 313, 1586–1587.
Ventura, S., Zurdo, J., Narayanan, S., Parreno, M., Mangues, R., Reif, B., et al., 2004. Short amino acid stretches can mediate amyloid formation in globular proteins: the Src homology 3 (SH3) case. Proc. Natl. Acad. Sci. U.S.A. 101, 7258–7263.
Walsh, I., Seno, F., Tosatto, S.C., Trovato, A., 2014. PASTA 2.0: an improved server for protein aggregation prediction. Nucleic Acids Res. 42, W301–W307.
Wang, W., Hecht, M.H., 2002. Rationally designed mutations convert de novo amyloid-like fibrils into monomeric beta-sheet proteins. Proc. Natl. Acad. Sci. U.S.A. 99, 2760–2765.
Wang, X., Zhou, Y., Ren, J.J., Hammer, N.D., Chapman, M.R., 2010. Gatekeeper residues in the major curlin subunit modulate bacterial amyloid fiber biogenesis. Proc. Natl. Acad. Sci. U.S.A. 107, 163–168.
Wei, L., Jiang, P., Xu, W., Li, H., Zhang, H., Yan, L., et al., 2011. The molecular basis of distinct aggregation pathways of islet amyloid polypeptide. J. Biol. Chem. 286, 6291–6300.
West, M.W., Hecht, M.H., 1995. Binary patterning of polar and nonpolar amino acids in the sequences and structures of native proteins. Protein Sci. 4, 2032–2039.
White Jr., F.H., 1961. Regeneration of native secondary and tertiary structures by air oxidation of reduced ribonuclease. J. Biol. Chem. 236, 1353–1360.
Wickner, R.B., Shewmaker, F.P., Bateman, D.A., Edskes, H.K., Gorkovskiy, A., Dayani, Y., et al., 2015. Yeast prions: structure, biology, and prion-handling systems. Microbiol. Mol. Biol. Rev. 79, 1–17.
Xiong, H., Buckwalter, B.L., Shieh, H.M., Hecht, M.H., 1995. Periodicity of polar and nonpolar amino acids is the major determinant of secondary structure in self-assembling oligomeric peptides. Proc. Natl. Acad. Sci. U.S.A. 92, 6349–6353.
Yen, H.C., Xu, Q., Chou, D.M., Zhao, Z., Elledge, S.J., 2008. Global protein stability profiling in mammalian cells. Science 322, 918–923.

Yoshimura, Y., Lin, Y., Yagi, H., Lee, Y.H., Kitayama, H., Sakurai, K., et al., 2012. Distinguishing crystal-like amyloid fibrils and glass-like amorphous aggregates from their kinetics of formation. Proc. Natl. Acad. Sci. U.S.A. 109, 14446–14451.
Zimmerman, S.B., Trach, S.O., 1991. Estimation of macromolecule concentrations and excluded volume effects for the cytoplasm of Escherichia coli. J. Mol. Biol. 222, 599–620.
Zuccotti, S., Rosano, C., Ramazzotti, M., Degl'Innocenti, D., Stefani, M., Manao, G., et al., 2004. Three-dimensional structural characterization of a novel *Drosophila melanogaster* acylphosphatase. Acta Crystallogr. D Biol. Crystallogr. 60, 1177–1179.
Zuccotti, S., Rosano, C., Bemporad, F., Stefani, M., Bolognesi, M., 2005. Preliminary characterization of two different crystal forms of acylphosphatase from the hyperthermophile archaeon Sulfolobus solfataricus. Acta Crystallogr. Sect. F: Struct. Biol. Cryst. Commun. 61, 144–146.

CHAPTER TWO

Protein Aggregation and Molecular Crowding: Perspectives From Multiscale Simulations

F. Musiani*,[1], A. Giorgetti†,[1]
*Laboratory of Bioinorganic Chemistry, University of Bologna, Bologna, Italy
†Applied Bioinformatics Group, University of Verona, Verona, Italy
[1]Corresponding authors: e-mail address: francesco.musiani@unibo.it; alejandro.giorgetti@univr.it

Contents

Abstract

Cells are extremely crowded environments, thus the use of diluted salted aqueous solutions containing a single protein is too simplistic to mimic the real situation. Macromolecular crowding might affect protein structure, folding, shape, conformational stability, binding of small molecules, enzymatic activity, interactions with cognate biomolecules, and pathological aggregation. The latter phenomenon typically leads to the formation of amyloid fibrils that are linked to several lethal neurodegenerative diseases, but that can also play a functional role in certain organisms. The majority of molecular simulations performed before the last few years were conducted in diluted solutions and were restricted both in the timescales and in the system dimensions by the available computational resources. In recent years, several computational solutions were developed to get close to physiological conditions. In this review we summarize the main computational techniques used to tackle the issue of protein aggregation both in a diluted and in a crowded environment.

International Review of Cell and Molecular Biology, Volume 329
ISSN 1937-6448
http://dx.doi.org/10.1016/bs.ircmb.2016.08.009

1. INTRODUCTION

1.1 Soluble Structured and Unstructured Proteins

Proteins attend a large number of functions within cells. Some are involved in structural support and movement, others in enzymatic activity, and still others in interactions with the outside world. Indeed, the functions of individual proteins are as varied as their unique amino acid sequences and complex three-dimensional structures. Proteins can be ascribed to different classes on the basis of their structural determinants (Andreeva et al., 2014; Fox et al., 2014; Greene et al., 2007; Hubbard et al., 1997). Among soluble proteins, globular proteins fold to a specific three-dimensional structure (also called native state), either on their own (Englander and Mayne, 2014) or with the help of chaperone molecules (Saibil, 2013). This native state generally corresponds to the functional state. The mechanism by which proteins fold to a specific structure involves a complex stochastic search of the large number of conformations accessible to the polypeptide chain itself (Dobson, 2004). In few words, the intrinsic fluctuations in the conformation of an incompletely folded polypeptide enable the contact between residues located in different positions of the amino acid sequence. Assuming that the establishment of correct interactions between different residues are on average more stable than the incorrect ones, such a search mechanism is in principle able to find the lowest-energy structure (Dinner et al., 2000). On the opposite side of globular proteins, the intrinsically disordered proteins (IDPs) (Habchi et al., 2014) are able to assume different folds and only populate a functional state once bound to a partner molecule or in specific conditions. IDPs break the classical protein sequence–structure–function paradigm by their inherent flexibility and peculiar features. In particular, IDPs do not possess a well-defined 3D structure, but they may be functionally active while adopting an ensemble of conformations in solution. Moreover, IDPs are able to assume different conformations as a result of changes of the chemical nature of the surrounding environment. This "ambivalent" behavior has been addressed using two concepts: chameleon sequences (Guo et al., 2007) and dual-personality (DP) sequences (Zhang et al., 2007). Chameleon sequences are those that fold in a template-dependent manner and could change conformation within the same protein as a result of point mutations, ligand binding, or a change in pH. DP fragments, on the other hand, are close to the order/disorder boundary (Habchi et al., 2014) and can cause conformational transitions depending on the solvation conditions and functional state (Tompa, 2010). In addition, some IDPs have the capacity to adapt their fold to the structure of distinct partners, consequently carrying out different functions, a phenomenon denoted as "moonlighting" (Hazy and Tompa, 2009). Initially, the concept of IDPs as functional biological units encountered skepticism. For a long time, disorder had no place in the perception of orchestrated molecular events controlling cell biology (Theillet et al., 2014). Over the past several years, however, this situation has changed and disorder has become an integral part of modern protein biochemistry (Theillet et al., 2014), with many in vitro studies that have identified and described IDP structural ensembles (Jensen et al., 2014; Uversky, 2011, 2013).

1.2 Protein Aggregation

Proteins can act alone or are able to form macromolecular complexes composed from few to large number of monomers. Protein oligomerization may be an advantageous feature for several reason: (i) scaffolds composed by more than one component may better support function and can form new active sites at the interface between subunits; (ii) oligomeric proteins can be allosterically regulated and can be thus controlled at different levels; (iii) large proteins composed of small subunits are more likely to be synthesized without errors with respect to a single-chain protein of similar size; (iv) where the monomer and oligomer feature different activity, regulation of the conditions of oligomerization by the cell can add regulatory flexibility; (v) deleterious mutations may have larger effects, and thus the mutated protein can be removed sooner from the gene pool; and finally (vi) larger proteins are more resistant to degradation and denaturation (Ali and Imperiali, 2005). Changes in cellular condition (pH, temperature, stress) or changes in the protein (mutation, posttranslational modification, overexpression) can lead to misfolding or partial unfolding of a protein and subsequent self-assembly into aggregate structures (Chiti and Dobson, 2006; Fink, 1998). In several cases protein aggregation does not cause harm to the cell and several organisms use it for functional purposes (Knowles and Buehler, 2011). In globular proteins this is, for example, the case of actin, a protein necessary to control the mobility and shape of the cells that is able to perform a transition between a globular monomeric form and a filamentous state able to self-aggregate (Dominguez and Holmes, 2011); or the enzyme glutamate dehydrogenase that forms inactive hexamers in equilibrium with polymeric forms with molecular weights of up to 2×10^6 Da (Fisher, 1973).

In pathological cases, protein aggregation reduces the effective number of active proteins and causes the formation of toxic aggregate species that are harmful to the cell. The typical example of pathologic protein (mis) aggregation for globular proteins is constituted by the single mutation in the β-globulin chain of human hemoglobin that promotes the polymerization of the deoxygenated form of the protein and causes the sickle cell disease (Pace et al., 2012). It also appears clear that the aggregates formed by human superoxide dismutase 1 (SOD1) mutants are a pathological feature of familial amyotrophic lateral sclerosis, a fatal neurodegenerative disease that affects motor neurons (Prudencio et al., 2009). In the case of IDP, the typical final product of aggregation, observed in both pathological and physiological cases, is the amyloid fibril, highly enriched in β content, with a cross-β structure. The latter is characterized by an arrangement of β-sheets running parallel to the fibril axis, with perpendicular hydrogen bonds (Fitzpatrick et al., 2013; Sunde et al., 1997).

Protein aggregation occurs through several steps (Fig. 1A) where the starting point is the monomeric form of the protein itself. The monomeric "inactive" protein can be "activated" as discussed earlier and can lead to the formation of oligomers that can precipitate as insoluble amorphous aggregates or can nucleate into larger oligomers. In the cases of amyloid fibrils, these larger oligomers constitute the basis for the formation of protofilament and eventually fibrils (Morris et al., 2009). The entire process can occur in a timescale range that goes from the order of seconds to years, depending on several factors such as the primary sequence, the dimension of the protein, and the cellular or extracellular conditions (Kashchiev et al., 2013; Luiken and Bolhuis, 2015) (Fig. 1B).

1.3 Protein Aggregation in a Crowded Environment

Another element of interest that may give rise to nonspecific protein aggregation events is due to the fact that in the cell interior biomolecules function in a densely crowded and highly heterogeneous environment, which is filled up to a volume of 40% with macromolecules (Zimmerman and Trach, 1991). Specifically, the high total concentration of macromolecules (reduced available space) of the interior of the cell inevitably influences the probability of intermolecular encounters and shifts the binding equilibria. Thus, crowding and the other factors regulate short-lived associations and tune the balance between functional and unintended "random" interactions. Such interactions are expected to introduce significant alterations in

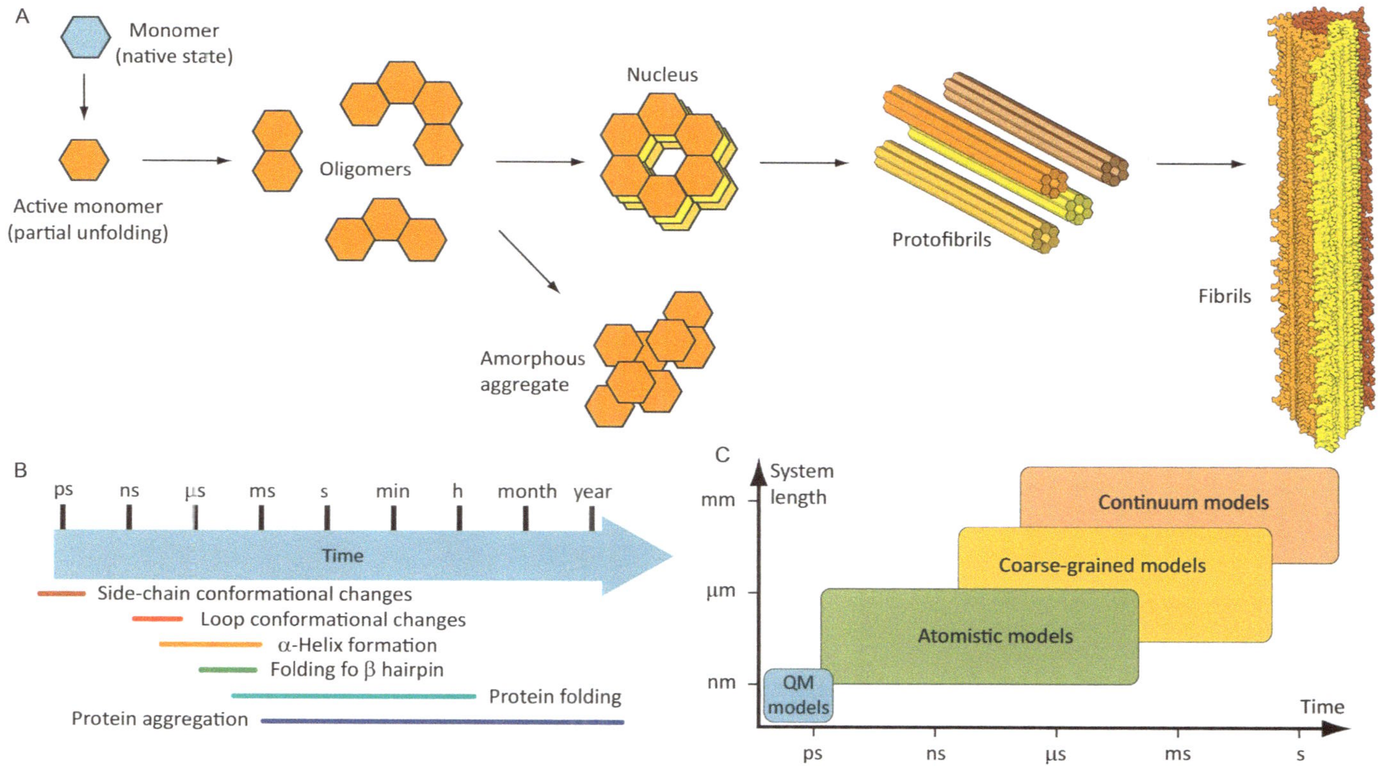

Fig. 1 (A) Schematic representation of a general pathway for the formation of protein fibrils or amorphous protein aggregates. (B) Timescales of protein motions, folding, and aggregation. (C) Representation of the time and length scales involved in different classes of molecular simulations.

the thermodynamics and kinetics of the processes they are involved in Abriata et al. (2013). Indeed, nonspecific attractive interactions between a protein and crowding agents (crowders) sometimes lead to partial destabilization, aggregation, or misfolding (Kuznetsova et al., 2014; Sanfelice et al., 2012; Van Den Berg et al., 1999; Wang et al., 2012). Furthermore, transient interactions with cytoplasmic components significantly and differently affect protein mobility. Protein aggregation and crowding effects have been studied both experimentally (Batra et al., 2009; Benton et al., 2012; Ellis and Minton, 2003; Gnutt and Ebbinghaus, 2016; Kuznetsova et al., 2014; Morris et al., 2009; Mourao et al., 2014; Munishkina et al., 2008; Pastore and Temussi, 2012; Politou and Temussi, 2015; Sanfelice et al., 2012; Van Den Berg et al., 1999; Wang et al., 2012; Zhou et al., 2008) and by using computational methods (Baaden and Marrink, 2013; Doig and Derreumaux, 2015; Erlkamp et al., 2015; Latshaw et al., 2014; Li and Mehler, 2006; Morriss-Andrews and Shea, 2015; Nasica-Labouze et al., 2015; Phillip and Schreiber, 2013; Qin and Zhou, 2009). Computer simulations on biological systems have been propelled in recent years by the availability of growing computational resources, the application of enhanced sampling algorithms (Bernardi et al., 2015), and the development of novel coarse-grained (CG) models (Morriss-Andrews and Shea, 2015). These approaches, when combined with experimental data, are the key for gaining insights into the mechanisms, and processes that occur when proteins function and/or aggregate in a crowded environment. At our advice, the field is mature enough to produce this quality jump in the understanding of biological systems by getting closer and closer to physiological conditions. In this review we describe the recent advances in the computational models and methodologies used to study the process of protein aggregation together with the techniques used to include the effect of molecular crowding in the simulation environment.

2. COMPUTATIONAL TECHNIQUES

Atomistic molecular dynamics (MD) simulations conducted at constant temperature is one of the most used computational methods to study proteins' behavior in solution. Such simulations are limited in the number of atoms of the system and in the simulated time by the computational cost of such simulations in both terms of execution time and data storage (Fig. 1C). Folded proteins tend to have energy landscape characterized by deep wells separated by high barriers. Thus, the configurations that are sampled using conventional MD simulations tend to get trapped into one of the energy wells available to the protein in solution. Moreover, simulations aimed to study protein aggregation require multiple proteins in the same simulation box that can assemble over a large range of length scales (up to fractions of micrometers) and timescales larger than the second. The same complications arise in simulations that mimic a crowded environment, where several macromolecules of different sizes and natures can be present at the same time. The typical solutions adopted to tackle these kinds of problems are the use of algorithms able to accelerate the sampling of the energy landscape and climb over energy barriers, or a reduction of the number of atoms involved in the simulations through the adoption of CG molecular models. Both atomistic and CG simulations combine well with enhanced sampling methods.

2.1 Enhanced Sampling Algorithms

The entity of energy barriers between the different conformations of a protein mean that transitions between them can configure as rare events not observable in affordable simulation times. A large number of methods have been developed in the last decades to overcome the computational bottleneck of molecular simulations. These methods are designed to accelerate the search in the configuration and trajectory space and thus allow to achieve fast thermodynamics and/or kinetics calculations (see Abrams and Bussi, 2014; Bernardi et al., 2015; Yang et al., 2015, for recent exhaustive overviews of these methods). If it is possible to identify predefined collective coordinates (i.e., one-dimensional coordinates which represent progress along a pathway on the free energy surface), the accelerated exploration of the free energy surface can be guided using methods such as umbrella sampling (Torrie and Valleau, 1977) or metadynamics (Bussi et al., 2006) (Fig. 2A). Alternatively, generalized ensemble methods can be used to generate a uniform distribution of energy (or temperature) in an MD simulation. The latter methods include, among others, replica-exchange molecular dynamics (REMD) (Sugita and Okamoto, 1999) (Fig. 2B).

2.1.1 Umbrella Sampling

In umbrella sampling, a defined collective variable is held by a potential well at a selected target value. Several simulations, differing only for their target values, are then run such that the statistics of neighboring "umbrellas" overlap. Umbrella sampling forces the exploration of regions of state space that would otherwise have insufficient sampling. The statistics of the unbiased

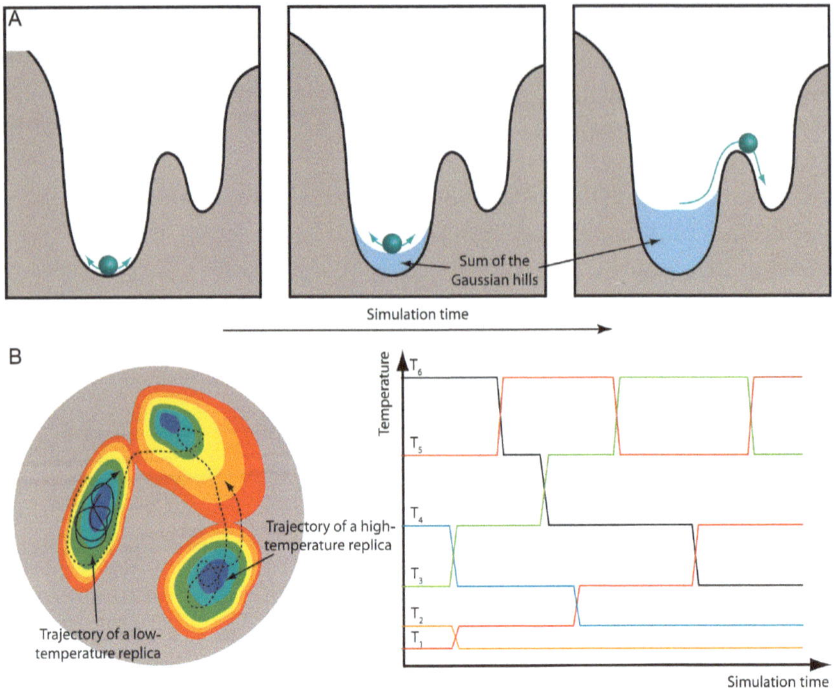

Fig. 2 (A) Schematic representation of the well-filling process by Gaussian hills occurring during metadynamics. (B) 2D representation of phase space (*left*) sampled by a low temperature replica (*continuous line*) and by a high temperature replica (*dashed line*); and schematic representation of parallel tempering swaps between adjacent replicas at different temperatures (*right*).

potential can be recovered from the biased statistics obtained from the simulations because the bias potential is known. In the study of protein aggregation, this can be applied to the determination of the energy of the separation of monomers, as the distance between monomers can be easily defined as a collective variable (Rivera et al., 2009), or to the study of the effect of membrane binding to the conformational space of the amyloid-β (Aβ) peptide (Davis and Berkowitz, 2009).

2.1.2 *Metadynamics*

Metadynamics involves enhanced sampling over collective variables using a biased potential to force the system to leave local minima and thus sample low-probability states (Laio and Parrinello, 2002). Metadynamics is an adaptive method, automatically biasing configurations away from those most visited. Also in this case, the correct unbiased statistics of the system are derived by subtracting the applied bias. To escape local minima, the method periodically adds a small Gaussian hill to the potential energy of the current region of state space. The local minimum is slowly filled by the addition of several Gaussian hills, and the system is thus forced to explore different configurations (Fig. 2A). If the simulation is prolonged for a sufficient time, the total potential energy will become flat as all the minima fill up with accumulated Gaussian hills. The parameters of this method (hill height, width, and frequency of the addition) should be chosen carefully in order to ensure that the statistics do not depend on them (see Baftizadeh Baghal et al., 2012; Barducci et al., 2013; Camilloni et al., 2012; Chiu et al., 2013; Rossetti et al., 2011; Wang et al., 2011a, for some recent applications of metadynamics to protein aggregation).

2.1.3 Replica-Exchange Methods

In replica-exchange enhances sampling, several parallel simulations are launched at the same time. Each simulation explores a specific point in the parameter space, and the definition of the specific parameter distinguishes the different methods. At predefined intervals, neighboring pairs of replicas are exchanged with a specific transition probability. Replica exchange can be used with both Monte Carlo and MD simulations (Earl and Deem, 2005). In most REMD studies the parameter of choice is temperature. A replica at lower temperatures can become trapped in a non-representative sample of the low free energy minima, while at higher temperatures, a simulation can sample more of phase space (Fig. 2B). Configuration swaps between the lower and higher temperature systems allow the lower temperature systems to escape from one region of phase space where they were effectively "stuck" and to sample a representative set of the low free energy minima. Efficient exchange between neighboring replicas requires the overlap of the potential energies and results in a high numbers of replicas when studying protein systems in explicit solvent (Rathore et al., 2005). In order to reduce the number of replicas used in temperature REMD, it is possible to simulate all the replicas at a constant temperature, while the force field or Hamiltonian of the system is used as the replica distinguishing parameter (Fukunishi et al., 2002). If the changes introduced in the different Hamiltonians are limited to only a subset of the degrees of freedom of the system (e.g., the solute in the case of replica exchange with solute-tempering MD (REST); Liu et al., 2005; Musiani et al., 2013; Rossetti et al., 2016; Wang et al., 2011b), the number of replicas needed to cover a given range in "effective temperature" can be greatly reduced compared to temperature REMD simulations. Recent REMD applications to protein aggregation can be found in Ball et al. (2013), Rosenman et al. (2013), Wu and Shea (2013), and Rossetti et al. (2016). A REMD tutorial related to the aggregation of the Tau peptide is provided in Shea and Levine (2016).

2.2 CG Simulations

Atomistic simulations are limited to timescales up to fractions of milliseconds even with the use of state-of-the-art, hardware-accelerated supercomputers (Lindorff-Larsen et al., 2011). These timescales are far shorter than that required for protein aggregation. Fortunately, aggregation propensity appears to be a sequence-independent property of all polypeptide chains (Dobson, 1999) and it is governed by hydrophobicity and electrostatics. CG simulations represent a valid choice for the modeling of physiologically relevant aggregation events (Morriss-Andrews and Shea, 2014, 2015; Wu and Shea, 2011). Coarse graining of a macromolecule is achieved by grouping multiple atoms into beads via effective force fields: for example, a single bead could represent an entire amino acid residue, the side chain of a residue, or a particular chemical group, like a methyl group. Despite the reduction in accuracy due to simplification, CG simulations can reach longer time and length scales thanks to the reduced number of particles included in the simulation that in turns reduces the number of calculations needed at each time step. Moreover, these techniques work well for studying the interaction between proteins and other biomolecules present in the cellular milieu, such as membranes (Ceccon et al., 2015; Li and Gorfe, 2013; Morriss-Andrews et al., 2014; Pannuzzo et al., 2013; Parton et al., 2011; Santo and Berkowitz, 2012; Simunovic et al., 2013).

A wide spectrum of CG models with different resolutions and parametrization schemes has been introduced in recent years. On one extreme there are several example of CG models representing each peptide as a single unit (Auer et al., 2008a,b; Irback et al., 2013; Vacha and Frenkel, 2011; Zhang and Muthukumar, 2009) (Fig. 3A). These models sacrifice sequence-level resolution but allow the simulation of a large number of peptides (up to 10^5). Others CG models involve one or more per residue. These models can be phenomenological or parametrized from atomistic or experimental data. Phenomenological models do not focus on the reproduction of a polypeptide sequence, but rather on the study of physical properties that guide protein aggregation (Abeln et al., 2014; Bellesia and Shea, 2007, 2009a,b;

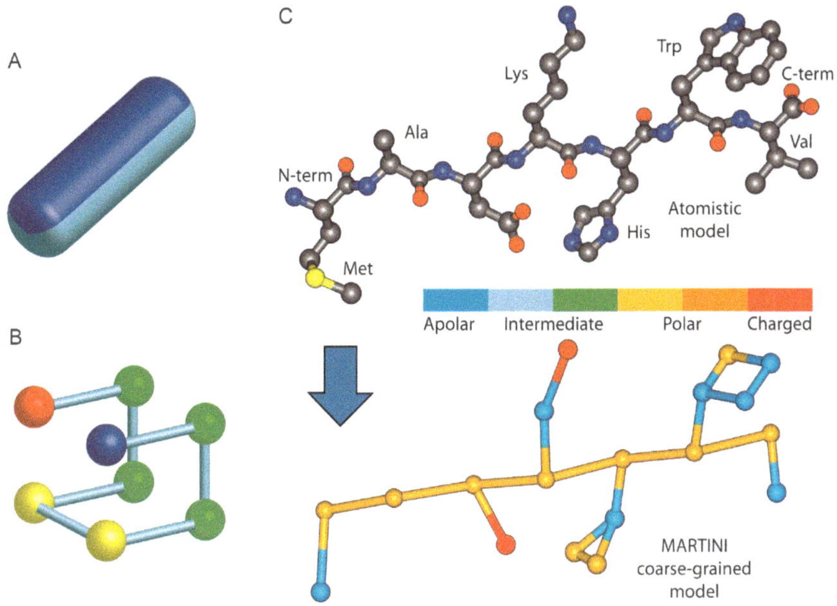

Fig. 3 (A) Example of a spherocylindrical simple CG model. (B) Structures of some of the conformations of the peptide used in a CG phenomenological lattice model (Li et al., 2008). Hydrophobic, polar, positively, and negatively charged beads are shown in *green*, *yellow*, *blue*, and *red*, respectively. (C) Example of the transformation from a peptide atomistic model to a MARTINI high-resolution CG model (Monticelli et al., 2008).

Li et al., 2008, 2010; Ni et al., 2013; Pellarin and Caflisch, 2006; Pellarin et al., 2007) (Fig. 3B). These CG models are designed to explore interaction space and do not provide sequence-dependent information. Nonphenomenological models allow to describe the properties of a given sequence. In the case of "systematic coarse graining," the aim is to obtain the optimal CG potential for a given protein by starting from atomistic simulations of experimental data (Carmichael and Shell, 2012; Izvekov and Voth, 2005; Reith et al., 2003; Shell, 2008; Wang and Voth, 2010). Systematic coarse-graining methods have the advantage of eliminate nonessential degrees of freedom. On the other hand, these potentials are specific for the studied system and any change in the protein require a complete reparameterization. High-resolution nonphenomenological CG models include the MARTINI model (Monticelli et al., 2008), the OPEP model (Chebaro et al., 2012), the PRIME model (Cheon et al., 2010), and the recent SIRAH model (Darre et al., 2015). The MARTINI model unites roughly four heavy atoms to a single bead (Fig. 3C) and is parametrized to reproduce experimental thermodynamic data. The OPEP model has an all-atom backbone with CG side chains and is parametrized from a library of test proteins to predict the lowest-energy conformations. The PRIME model is similar to the MARTINI model and is parametrized via learning algorithms to maximize the energetic gap between PDB structures and decoys. Finally, the SIRAH model was developed following a top-down approach fitting structural properties of macromolecules using a standard pairwise Hamiltonian common to most MD simulation packages.

2.2.1 CG Modeling of Water Molecules

CG simulations of biological systems in explicit solvent necessitate a computationally efficient and accurate CG model for solvent molecules. Water, for sure the most used solvent in simulations of biological interest (Hadley and Mccabe, 2012), has several peculiar properties which make its dynamical and physical properties harder to simulate with respect to other solvents (Guillot and Guissani, 2001). In some cases, water is modeled implicitly and the CG model of the solute is able to interact with other molecules as it would in the presence of water (Fennell and Dill, 2011; Vorobjev, 2011). In the case of explicit solvation, CG models have been developed by mapping one or more water molecules in a single CG bead (Hadley and Mccabe, 2012). In the case of one water molecule mapped to one CG bead, the CG modeling is usually implemented with center-of-mass-based techniques (Basdevant et al., 2004, 2006; Bedrov et al., 2006; Bizjak et al., 2009; De Oliveira et al., 2006a,b; Dias et al., 2009; Izvekov and Voth, 2005; Lyubartsev, 2005; Masella et al., 2011; Molinero and Moore, 2009; Wang et al., 2009), but only saves a marginal amount of simulation time. In the second and more common case, to each CG bead is assigned a fixed number of arbitrary water molecules and beads interact with each other through an analytical potential. In the latter case, the CG bead represents multiple waters as a function of its size, interaction strength, and other thermodynamic and structural parameters (Chiu et al., 2010; Darré et al., 2010; Hadley and Mccabe, 2010; Marrink et al., 2007; Riniker and Van Gunsteren, 2011; Shinoda et al., 2007; Van Hoof et al., 2011; Wu et al., 2010; Yesylevskyy et al., 2010).

2.2.2 Backmapping Algorithms

In several cases the CG detail is not enough and atomistic insight is required even if simulations are carried out at a coarser level. In this aim it should be useful to convert CG models into corresponding atomistic models, either for the analysis of the residues' interactions or for extracting some representative frames that can be simulated with a higher resolution. The retransformation of a CG model into its corresponding atomistic model is commonly termed backmapping, inverse mapping, or reverse transformation (Ingólfsson et al., 2014; Noid, 2013). This process generally consists of two stages: (i) first an atomistic starting structure is generated from CG coordinates and (ii) the atomistic structure is relaxed. The available methods for backmapping usually emphasize either the first (Feig et al., 2000; Heath et al., 2007; Shih et al., 2007; Stansfeld and Sansom, 2011) or the second (Hess et al., 2006; Machado and Pantano, 2016; Peter and Kremer, 2009; Rzepiela et al., 2010; Thøgersen et al., 2008; Wassenaar et al., 2014) stage.

2.3 Electrostatic Analysis

Intermolecular interactions are essential for nearly every cellular activity. The forces that underlie these interactions include van der Waals dispersion and repulsion, hydrogen bonding, and electrostatics. Electrostatic forces are especially important in biological systems because most biomolecules are charged or polar. The role of electrostatic interactions on the protein–protein association has been broadly studied, and it was shown that electrostatic interactions play a more important role in protein-binding mechanisms than they do in folding (see Hu et al., 2000; Kundrotas and Alexov, 2006, and references within). Thus, a deep characterization of electrostatics should be performed when dealing with protein aggregation. In general most of the polar and charged residues on the protein–protein interfaces are "hot spots," and thus, redesigning of charged interfacial residues leads complexes with better affinity (Ren et al., 2012). Charge complementarity is indeed also an important factor affecting the binding affinity. For an excellent review of computational methods for calculating electrostatics, see Ren et al. (2012).

Recently, Costabel and collaborators used electrostatic calculations to assess the interaction of proteins with the membrane (Zamarreno et al., 2012). They used a novel protocol in which the solutions of the Poisson–Boltzmann equation were used to calculate the total electrostatic free energy of the system (Holst and Saied, 1993). They showed that binding of fatty acid-binding proteins (FABPs) to membranes involves a significant electrostatic component that discriminates among possible membrane-bound mechanisms and classify a list of FABPs according to the mechanism involved in the interaction. The classification proposed and the results of this model are in agreement with experimental observations for FABP–membrane interaction (Zamarreno et al., 2012).

2.4 (Macro)molecular Crowding: Application Cases

In general, standard biochemical and biophysical studies use to consider proteins as highly purified entities that act in isolation, freely diffusing until finding their cognate partner to bind to. While in vitro experiments that reproduce these conditions largely remain the only way to investigate the intrinsic properties of molecules, this approach ignores an important factor: in their natural milieu, proteins are surrounded by several other molecules of different chemical nature, and this crowded environment can considerably modify their behavior. About 40% of the cellular volume on average is occupied by all sorts of molecules. Furthermore, biological macromolecules live and operate in an extremely structured and complex environment within the cell (endoplasmic reticulum, Golgi apparatus, cytoskeletal structures, etc.). Hence, to further complicate the picture, the interior of the cell is by no means a *simply crowded* medium, rather, a most crowded and *confining* one (Foffi et al., 2013).

Experimentally, artificial macromolecular crowding agents are used to mimic these conditions in vitro and the excluded volume theory is applied to explain the observed effects (Zhou et al., 2008). However, recent studies emphasize the role of further contributions aside from a pure volume effect including enthalpic and solvent effects (Gilman-Politi and Harries, 2011; Wang et al., 2012). Ebbinghaus and coworkers study cosolute effects at high-molecular and macromolecular concentrations by using a thermodynamic analysis of the thermal unfolding of ubiquitin in the presence of different concentrations of cosolutes (glucose, dextran, polyethylene glycol, potassium chloride) (Gnutt et al., 2015; Senske et al., 2014). In contrast to the excluded volume theory, they have observed enthalpic stabilization and entropic destabilization forces for all tested cosolutes. The enthalpic stabilization mechanism of ubiquitin in macromolecular polysaccharide solutions of dextran was thereby similar to the effects observed in monomeric glucose. In several cases, computational and experimental techniques were used in combination to study the interaction of proteins with crowding agents (Wang et al., 2012). Indeed, transient interactions with cytoplasmic components significantly and differently affect protein mobility (Crowley et al., 2011; Wang et al., 2010, 2011c). Unfortunately, in vitro experiments alone cannot offer detailed characterizations of the mechanisms underlying those processes. In this section, we review some successful cases, published in the last years, in which in silico approaches, sometimes in combination with in vitro experiments in crowding conditions, were able to unravel the effects of a crowded environment.

Diffusional movement of cytosolic proteins is also influenced by a confined environment: a variety of surface boundaries include the plasma membrane (c. 700 mm^2 in an average cell) and internal lipid membranes (c. 7000 mm^2) (Lodish et al., 2012). Thus, cytosolic proteins are potentially exposed to a significant concentration of membrane lipids. In a recent application, Ceccon and coworkers studied—using a combined experimental/theoretical approach—the propensity for weak chemical interactions between a cytosolic protein and cosolutes mimicking macromolecular and membrane cellular components (Ceccon et al., 2015). They focused on a test protein, the human ileal bile acid-binding protein (IBABP), a small (15 kDa) macromolecule of the family of intracellular lipid-binding proteins. For gaining structural insights into the modes of IBABP binding to anionic lipid membranes, they performed MD simulations. In particular they have used CG MD and the MARTINI force field for both the protein and the membrane (Marrink et al., 2007; Monticelli et al., 2008) to perform 20 4-μs MD simulations of the interaction. The calculations were in agreement with the experimental results and allowed to characterize the differences of the interactions in different ionic concentrations. Indeed, CG MD simulations support the finding (based on fluorescence spectroscopy) that the major docking site was formed by residues in the IBABP sheet domain rather than the helix-turn-helix motif. Then the interactions of IBABP with protein cosolutes, in particular, it was found using NMR and the Haddock knowledge-based protein–protein docking algorithm (De Vries et al., 2007) that lysozyme, another well-characterized crowder, binds specifically to two distinct surface epitopes on IBABP.

Dal Peraro and coworkers studied the macromolecular crowding effects by using MD simulation studies of the internal dynamics of proteins, at an all-atom description of the protein, the solvent, and a crowding element (Abriata et al., 2013). Their system focused on ubiquitin, a model protein that shows dynamical features closely related to its ability to bind to multiple partners, in a 325 g L^{-1} solution of glucose in water, a condition widely employed in in vitro studies of crowding effects. The authors showed that there is a small reduction in flexibility of loops, correlated with an important restriction of the conformational space explored during the c. 0.5 μs of the simulations (Abriata et al., 2013). Indeed they showed that a crowded environment contributes to the decrease of the collective motions and the rate of exploration of the conformational space. This effect is attributed, by the authors, to the formation of "stable" interactions between protein residues, on an extended surface region, and glucose. Kim and coworkers very recently have applied REMD simulations on the protein complex between phosphorylated kinase-inducible domain (pKID) of transcription factor CREB to the KIX domain of a CREB-binding protein (pKID−KIX), in order to assess the putative effects of a crowded environment on folding and binding events (Kim et al., 2014). They used a simplified residue-based protein model, able to capture the flexibility of the complex, while crowding macromolecules are represented as spherical particles. The interactions between crowders and protein residues can be either purely repulsive or a combination of short-range repulsion and intermediate-range attraction. The authors, by varying the size and strength of the crowders, found that the complex formation is stabilized by repulsive protein–crowder interactions and destabilized by sufficiently strong attractive protein–crowder interactions. The main conclusion is that the mechanism of coupled folding and binding is not affected in highly crowded conditions over a broad range of protein–crowder interaction strengths. Using a similar approach, Predeus et al. used REMD simulations to characterize the effect of crowders in another system (i.e., trp-cage and the melittin peptide) with the principal aim of characterizing peptide dynamics in a crowded environment (Predeus et al., 2012). In this case, the authors used a three-level multiscale modeling approach that includes atomistic representation for peptides, CG for crowders, and the use of implicit solvent. The multiscale simulations were compared with simulations of the same peptides in different dielectric media (Tanizaki et al., 2008). The authors found that the effect of crowders in the sampling resembles sampling with reduced dielectric constants. Furthermore, diverse conformational ensembles are generated in the presence of crowders including partially unfolded states for trp-cage. These findings emphasize the importance of enthalpic interactions over volume exclusion effects in describing the effects of cellular crowding. Soon after the work by Predeus et al., another study focused on trp-cage in the presence of crowders. Indeed, Bille and collaborators used Monte Carlo replica-exchange methods to explore the equilibrium behavior of the system, but this time to infer the effect on protein stability and using an atomistic representation of the crowder (Bille et al., 2015). Their observations suggest that the crowded environment hampered trp-cage from adopting its global native fold, while stabilizing its main secondary-structure element, an α-helix. They found specific interactions between the crowder, bovine pancreatic trypsin inhibitor, and the protein. The authors conclude that the use of an atomistic description of the crowder may have different effects than those of hard-sphere crowders.

Macdonald et al. recently used multicanonical MD and a CG protein model (Bereau and Deserno, 2009) to study the folding thermodynamics of α-helices in the presence of crowding agents (Macdonald et al., 2015). The calculations were carried out by using the REST MD approach (Church et al., 2012; Kim et al., 2012). They have shown that the stability of the test protein depends on the hydrophobicity of the crowding agents. Indeed, while for low hydrophobicity agents, the effect of the excluded volume is dominant, the α-helices are stabilized with respect to the dilute solution; for intermediate hydrophobicity values, the test protein is destabilized by favorable side chain–side chain interactions stabilizing the unfolded states. In the extreme case of a highly hydrophobic crowding agent, the native state is stabilized by the strong intermolecular attractions.

Qi and coworkers explored the effects of macromolecular crowding on the electrostatic desolvation and solvent-screened interaction components of protein–protein binding (Qi et al., 2014). They proposed a simple model: spherical, uncharged model crowders within an implicit solvent, and electrostatic free energies are computed via the Poisson/linearized Poisson–Boltzmann equation (Altman, 2006). This simplified model allows to run many calculations and thus to focus exclusively on the electrostatic effects of water depletion on protein binding due to crowding. The calculations were performed using the barnase–barstar complex as a model system and randomly placed, uncharged spheres within implicit solvent to model crowding in an aqueous environment. The authors found that the desolvation free energy penalties incurred by partners upon binding are lowered in a crowded environment and solvent-screened interactions are amplified.

2.5 Other Computational and Bioinformatics Approaches

In this section we review some recent works in which protein aggregation was studied using approaches not included in the previous sections. Recently, De Baets and coworkers have presented an in silico survey of the impact of human sequence variation on the aggregation propensity of human proteins (De Baets et al., 2015). The authors used the TANGO program (Fernandez-Escamilla et al., 2004) to analyze the aggregation-prone regions (APRs) in several human proteins involved in human diseases. They defined the intrinsic aggregation propensity of a protein as the propensity of an unfolded protein to aggregate. Independent grafting experiments have shown that the intrinsic aggregation propensity is related to the presence of short APRs that self-associate to form intermolecular β-structured assemblies. APRs are short hydrophobic sequence segments, spanning 5–15 amino acids. For assessing this, the authors have used TANGO, a statistical thermodynamics algorithm that identifies aggregation-nucleation sites by considering the factors described earlier and also the competition between β-sheet formation and other structured states. The authors found that disease-associated variations are statistically significantly enriched in mutations that increase the aggregation potential of human proteins when compared to neutral sequence variations. The authors also used the FoldX3b5 force field (Schymkowitz et al., 2005) to model the mutations and to calculate the effect of the mutation on protein stability. Other TANGO algorithm applications can be found in Chiti and Dobson (2006), Lashuel and Lansbury (2006), Winklhofer et al. (2008), Xu et al. (2011), and Siekierska et al. (2012).

In other works, the technique called discrete molecular dynamics (DMD) is used. It is a physics-based simulation technique that allows the acceleration of MD simulations as to perform microsecond timescale simulations of biomolecular systems on personal computers rather than supercomputers or specialized hardware (Ding et al., 2008; Dokholyan et al., 1998, 2000; Proctor et al., 2011; Shirvanyants et al., 2012). The idea behind the method is to use the physics of ballistic motion to generate trajectories of particles in space over time according to discrete, distance-based energetic potentials. Instead of integrating continuous energetic potentials at set time steps to determine forces that will impact new velocities and position, DMD assumes ballistic motion and assigns time step as the time until the next occurring interaction (or "event"), saving time and computational resources. Upon interaction, energy is assessed with a distance-based step function (Fig. 4), and velocity and position change instantaneously upon collision according to the conservation of momentum. DMD uses implicit solvent thus providing the opportunity for parameterization to taken into account the crowded cellular milieu, and the event-driven nature of the algorithms that advance the simulations are more appropriate to large, crowded systems than are continuous potentials with necessary integration over increasingly complex energetic potentials.

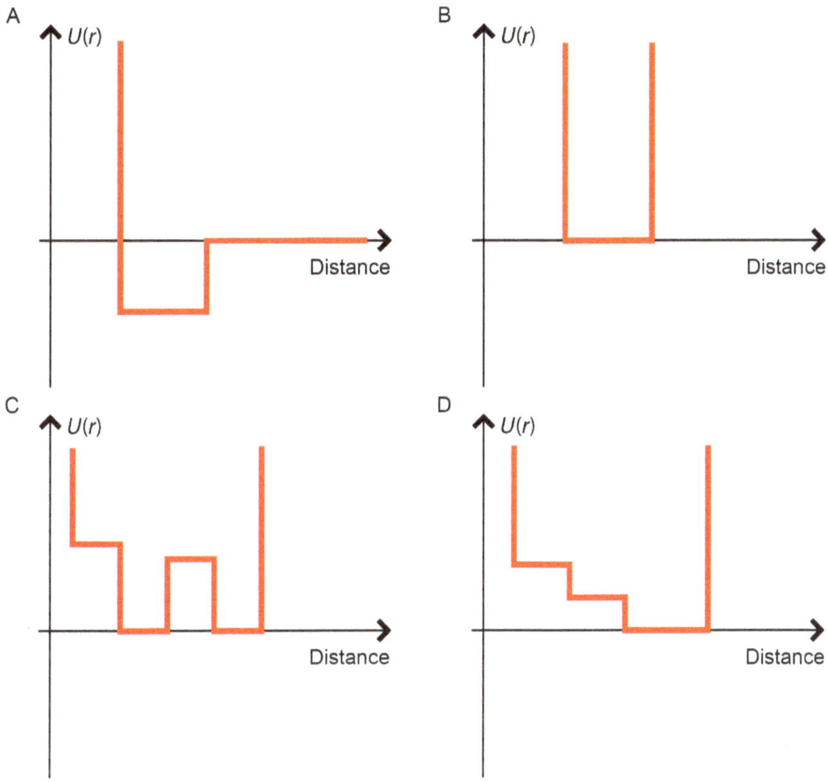

Fig. 4 Schematics of some of the step-function potentials used in DMD: hard-sphere interaction potential (A), single-infinite square well used for covalent bonds (B), dihedral constraint potential (C), and hydrogen-bonding auxiliary distance potential function (D) (Proctor et al., 2011).

Some recent application of the DMD method includes: Williams et al. utilized DMD simulations to identify a misfolded intermediate of the protein ApoE4 (Williams et al., 2015); Ding et al. combined a CG protein model with experimentally derived structural on ALS-associated SOD1 mutants (Ding et al., 2012); and Kimura et al. performed DMD simulations to observe the folding of HIV-1 protease monomers and their assembly into active dimers (Kimura et al., 2014).

3. PERSPECTIVES

The present-day challenge is now to understand *how* in vitro structural properties relate to the biological effects of protein *in cell* (Theillet et al.,

2014). The intracellular environment has ionic and metabolites composition, dielectric properties, viscosity, together with pH and crowding conditions (inside cells the concentration of macromolecules can reach up to 400 g/L; Zimmerman and Minton, 1993; Zimmerman and Trach, 1991) very different from the standard in vitro conditions used in experiments and simulations. In particular, in the case of IDPs and considering their proven conformational mutability, it is likely that this class of proteins may behave differently *in cell* as compared to in vitro. While from the experimental point of view many efforts are being undertaken to tackle this, a smaller number of works in the computational world introduced crowding conditions when trying to asses protein function at different levels in general and related to protein aggregation in particular. Here we have reviewed several ways of approaching protein aggregation and crowding from the computational point of view. Indeed, computational techniques are the key for gaining insights into the extremely complex mechanisms that take place in a densely crowded environment. Although, as reviewed here, studies still focus on model protein/systems, we consider that the field is mature enough, i.e., very accurate force fields and strong computational power that may allow, using theoretical approaches, to offer a deep and detailed characterization of the molecular determinants underlying the effect of protein aggregation, principally in a crowded environment. This is the direction that future approaches should go in order to be able to simulate events occurring in the cell.

REFERENCES

Abeln, S., Vendruscolo, M., Dobson, C.M., Frenkel, D., 2014. A simple lattice model that captures protein folding, aggregation and amyloid formation. PLoS One 9, e85185.

Abrams, C., Bussi, G., 2014. Enhanced sampling in molecular dynamics using metadynamics, replica-exchange, and temperature-acceleration. Entropy 16, 163–199.

Abriata, L.A., Spiga, E., Dal Peraro, M., 2013. All-atom simulations of crowding effects on ubiquitin dynamics. Phys. Biol. 10, 045006.

Ali, M.H., Imperiali, B., 2005. Protein oligomerization: how and why. Bioorg. Med. Chem. 13, 5013–5020.

Altman, M.D., 2006. Computational ligand design and analysis in protein complexes using inverse methods, combinatorial search, and accurate solvation modeling. Ph.D Massachusetts Institute of Technology, Cambridge, MA.

Andreeva, A., Howorth, D., Chothia, C., Kulesha, E., Murzin, A.G., 2014. SCOP2 prototype: a new approach to protein structure mining. Nucleic Acids Res. 42, D310–D314.

Auer, S., Dobson, C.M., Vendruscolo, M., Maritan, A., 2008a. Self-templated nucleation in peptide and protein aggregation. Phys. Rev. Lett. 101, 258101.

Auer, S., Meersman, F., Dobson, C.M., Vendruscolo, M., 2008b. A generic mechanism of emergence of amyloid protofilaments from disordered oligomeric aggregates. PLoS Comput. Biol. 4, e1000222.

Baaden, M., Marrink, S.J., 2013. Coarse-grain modelling of protein-protein interactions. Curr. Opin. Struct. Biol. 23, 878–886.

Baftizadeh Baghal, F., Biarnes, X., Pietrucci, F., Laio, A., Affinito, F., 2012. Simulation of amyloid nucleation with bias-exchange metadynamics. Biophys. J. 102, 242a.
Ball, K.A., Phillips, A.H., Wemmer, D.E., Head-Gordon, T., 2013. Differences in beta-strand populations of monomeric Abeta40 and Abeta42. Biophys. J. 104, 2714–2724.
Barducci, A., Bonomi, M., Prakash, M.K., Parrinello, M., 2013. Free-energy landscape of protein oligomerization from atomistic simulations. Proc. Natl. Acad. Sci. U.S.A. 110, E4708–E4713.
Basdevant, N., Borgis, D., Ha-Duong, T., 2004. A semi-implicit solvent model for the simulation of peptides and proteins. J. Comput. Chem. 25, 1015–1029.
Basdevant, N., Ha-Duong, T., Borgis, D., 2006. Particle-based implicit solvent model for biosimulations: application to proteins and nucleic acids hydration. J. Chem. Theory Comput. 2, 1646–1656.
Batra, J., Xu, K., Qin, S., Zhou, H.X., 2009. Effect of macromolecular crowding on protein binding stability: modest stabilization and significant biological consequences. Biophys. J. 97, 906–911.
Bedrov, D., Ayyagari, C., Smith, G.D., 2006. Multiscale modeling of poly(ethylene oxide)-poly(propylene oxide)-poly(ethylene oxide) triblock copolymer micelles in aqueous solution. J. Chem. Theory Comput. 2, 598–606.
Bellesia, G., Shea, J.-E., 2007. Self-assembly of β-sheet forming peptides into chiral fibrillar aggregates. J. Chem. Phys. 126, 245104.
Bellesia, G., Shea, J.E., 2009a. Diversity of kinetic pathways in amyloid fibril formation. J. Chem. Phys. 131, 111102.
Bellesia, G., Shea, J.E., 2009b. Effect of beta-sheet propensity on peptide aggregation. J. Chem. Phys. 130, 145103.
Benton, L.A., Smith, A.E., Young, G.B., Pielak, G.J., 2012. Unexpected effects of macromolecular crowding on protein stability. Biochemistry 51, 9773–9775.
Bereau, T., Deserno, M., 2009. Generic coarse-grained model for protein folding and aggregation. J. Chem. Phys. 130, 235106.
Bernardi, R.C., Melo, M.C., Schulten, K., 2015. Enhanced sampling techniques in molecular dynamics simulations of biological systems. Biochim. Biophys. Acta 1850, 872–877.
Bille, A., Linse, B., Mohanty, S., Irback, A., 2015. Equilibrium simulation of trp-cage in the presence of protein crowders. J. Chem. Phys. 143, 175102.
Bizjak, A., Urbic, T., Vlachy, V., Dill, K.A., 2009. Theory for the three-dimensional Mercedes-Benz model of water. J. Chem. Phys. 131, 194504.
Bussi, G., Laio, A., Parrinello, M., 2006. Equilibrium free energies from nonequilibrium metadynamics. Phys. Rev. Lett. 96, 090601.
Camilloni, C., Schaal, D., Schweimer, K., Schwarzinger, S., De Simone, A., 2012. Energy landscape of the prion protein helix 1 probed by metadynamics and NMR. Biophys. J. 102, 158–167.
Carmichael, S.P., Shell, M.S., 2012. A new multiscale algorithm and its application to coarse-grained peptide models for self-assembly. J. Phys. Chem. B 116, 8383–8393.
Ceccon, A., Busato, M., Perez Santero, S., D'onofrio, M., Musiani, F., Giorgetti, A., Assfalg, M., 2015. Transient interactions of a cytosolic protein with macromolecular and vesicular cosolutes: unspecific and specific effects. ChemBioChem 16, 2633–2645.
Chebaro, Y., Pasquali, S., Derreumaux, P., 2012. The coarse-grained OPEP force field for non-amyloid and amyloid proteins. J. Phys. Chem. B 116, 8741–8752.
Cheon, M., Chang, I., Hall, C.K., 2010. Extending the PRIME model for protein aggregation to all 20 amino acids. Proteins 78, 2950–2960.
Chiti, F., Dobson, C.M., 2006. Protein misfolding, functional amyloid, and human disease. Annu. Rev. Biochem. 75, 333–366.
Chiu, S.W., Scott, H.L., Jakobsson, E., 2010. A coarse-grained model based on morse potential for water and n-alkanes. J. Chem. Theory Comput. 6, 851–863.

Chiu, C.C., Singh, S., De Pablo, J.J., 2013. Effect of proline mutations on the monomer conformations of amylin. Biophys. J. 105, 1227–1235.
Church, M.S., Ferry, C.E., Van Giessen, A.E., 2012. Thermodynamics of peptide dimer formation. J. Chem. Phys. 136, 245102.
Crowley, P.B., Chow, E., Papkovskaia, T., 2011. Protein interactions in the Escherichia coli cytosol: an impediment to in-cell NMR spectroscopy. ChemBioChem 12, 1043–1048.
Darré, L., Machado, M.R., Dans, P.D., Herrera, F.E., Pantano, S., 2010. Another coarse grain model for aqueous solvation: WAT FOUR? J. Chem. Theory Comput. 6, 3793–3807.
Darre, L., Machado, M.R., Brandner, A.F., Gonzalez, H.C., Ferreira, S., Pantano, S., 2015. SIRAH: a structurally unbiased coarse-grained force field for proteins with aqueous solvation and long-range electrostatics. J. Chem. Theory Comput. 11, 723–739.
Davis, C.H., Berkowitz, M.L., 2009. Interaction between amyloid-beta (1-42) peptide and phospholipid bilayers: a molecular dynamics study. Biophys. J. 96, 785–797.
De Baets, G., Van Doorn, L., Rousseau, F., Schymkowitz, J., 2015. Increased aggregation is more frequently associated to human disease-associated mutations than to neutral polymorphisms. PLoS Comput. Biol. 11, e1004374.
De Oliveira, A.B., Netz, P.A., Colla, T., Barbosa, M.C., 2006a. Structural anomalies for a three dimensional isotropic core-softened potential. J. Chem. Phys. 125, 124503.
De Oliveira, A.B., Netz, P.A., Colla, T., Barbosa, M.C., 2006b. Thermodynamic and dynamic anomalies for a three-dimensional isotropic core-softened potential. J. Chem. Phys. 124, 084505.
De Vries, S.J., Van Dijk, A.D., Krzeminski, M., Van Dijk, M., Thureau, A., Hsu, V., Wassenaar, T., Bonvin, A.M., 2007. HADDOCK versus HADDOCK: new features and performance of HADDOCK2.0 on the CAPRI targets. Proteins 69, 726–733.
Dias, C.L., Ala-Nissila, T., Grant, M., Karttunen, M., 2009. Three-dimensional "Mercedes-Benz" model for water. J. Chem. Phys. 131, 054505.
Ding, F., Tsao, D., Nie, H., Dokholyan, N.V., 2008. Ab initio folding of proteins with all-atom discrete molecular dynamics. Structure 16, 1010–1018.
Ding, F., Furukawa, Y., Nukina, N., Dokholyan, N.V., 2012. Local unfolding of Cu, Zn superoxide dismutase monomer determines the morphology of fibrillar aggregates. J. Mol. Biol. 421, 548–560.
Dinner, A.R., Sali, A., Smith, L.J., Dobson, C.M., Karplus, M., 2000. Understanding protein folding via free-energy surfaces from theory and experiment. Trends Biochem. Sci. 25, 331–339.
Dobson, C.M., 1999. Protein misfolding, evolution and disease. Trends Biochem. Sci. 24, 329–332.
Dobson, C.M., 2004. Principles of protein folding, misfolding and aggregation. Semin. Cell Dev. Biol. 15, 3–16.
Doig, A.J., Derreumaux, P., 2015. Inhibition of protein aggregation and amyloid formation by small molecules. Curr. Opin. Struct. Biol. 30, 50–56.
Dokholyan, N.V., Buldyrev, S.V., Stanley, H.E., Shakhnovich, E.I., 1998. Discrete molecular dynamics studies of the folding of a protein-like model. Fold. Des. 3, 577–587.
Dokholyan, N.V., Buldyrev, S.V., Stanley, H.E., Shakhnovich, E.I., 2000. Identifying the protein folding nucleus using molecular dynamics. J. Mol. Biol. 296, 1183–1188.
Dominguez, R., Holmes, K.C., 2011. Actin structure and function. Annu. Rev. Biophys. 40, 169–186.
Earl, D.J., Deem, M.W., 2005. Parallel tempering: theory, applications, and new perspectives. Phys. Chem. Chem. Phys. 7, 3910–3916.
Ellis, R.J., Minton, A.P., 2003. Cell biology: join the crowd. Nature 425, 27–28.
Englander, S.W., Mayne, L., 2014. The nature of protein folding pathways. Proc. Natl. Acad. Sci. U.S.A. 111, 15873–15880.

Erlkamp, M., Marion, J., Martinez, N., Czeslik, C., Peters, J., Winter, R., 2015. Influence of pressure and crowding on the sub-nanosecond dynamics of globular proteins. J. Phys. Chem. B 119, 4842–4848.

Feig, M., Rotkiewicz, P., Kolinski, A., Skolnick, J., Brooks, C.L., 2000. Accurate reconstruction of all-atom protein representations from side-chain-based low-resolution models. Proteins 41, 86–97.

Fennell, C.J., Dill, K.A., 2011. Physical modeling of aqueous solvation. J. Stat. Phys. 145, 209–226.

Fernandez-Escamilla, A.M., Rousseau, F., Schymkowitz, J., Serrano, L., 2004. Prediction of sequence-dependent and mutational effects on the aggregation of peptides and proteins. Nat. Biotechnol. 22, 1302–1306.

Fink, A.L., 1998. Protein aggregation: folding aggregates, inclusion bodies and amyloid. Fold. Des. 3, R9–R23.

Fisher, H.F., 1973. Glutamate dehydrogenase-ligand complexes and their relationship to the mechanism of the reaction. In: Meister, A. (Ed.), Advances in Enzymology and Related Areas of Molecular Biology. John Wiley & Sons, Inc., New York, pp. 369–417.

Fitzpatrick, A.W., Debelouchina, G.T., Bayro, M.J., Clare, D.K., Caporini, M.A., Bajaj, V.S., Jaroniec, C.P., Wang, L., Ladizhansky, V., Muller, S.A., Macphee, C.E., Waudby, C.A., Mott, H.R., De Simone, A., Knowles, T.P., Saibil, H.R., Vendruscolo, M., Orlova, E.V., Griffin, R.G., Dobson, C.M., 2013. Atomic structure and hierarchical assembly of a cross-beta amyloid fibril. Proc. Natl. Acad. Sci. U.S.A. 110, 5468–5473.

Foffi, G., Pastore, A., Piazza, F., Temussi, P.A., 2013. Macromolecular crowding: chemistry and physics meet biology (Ascona, Switzerland, 10–14 June 2012). Phys. Biol. 10, 040301.

Fox, N.K., Brenner, S.E., Chandonia, J.M., 2014. SCOPe: structural classification of proteins—extended, integrating SCOP and ASTRAL data and classification of new structures. Nucleic Acids Res. 42, D304–D309.

Fukunishi, H., Watanabe, O., Takada, S., 2002. On the Hamiltonian replica exchange method for efficient sampling of biomolecular systems: application to protein structure prediction. J. Chem. Phys. 116, 9058–9067.

Gilman-Politi, R., Harries, D., 2011. Unraveling the molecular mechanism of enthalpy driven peptide folding by polyol osmolytes. J. Chem. Theory Comput. 7, 3816–3828.

Gnutt, D., Ebbinghaus, S., 2016. The macromolecular crowding effect—from in vitro into the cell. Biol. Chem. 397, 37–44.

Gnutt, D., Gao, M., Brylski, O., Heyden, M., Ebbinghaus, S., 2015. Excluded-volume effects in living cells. Angew. Chem. Int. Ed. 54, 2548–2551.

Greene, L.H., Lewis, T.E., Addou, S., Cuff, A., Dallman, T., Dibley, M., Redfern, O., Pearl, F., Nambudiry, R., Reid, A., Sillitoe, I., Yeats, C., Thornton, J.M., Orengo, C.A., 2007. The CATH domain structure database: new protocols and classification levels give a more comprehensive resource for exploring evolution. Nucleic Acids Res. 35, D291–D297.

Guillot, B., Guissani, Y., 2001. How to build a better pair potential for water. J. Chem. Phys. 114, 6720–6733.

Guo, J.-T., Jaromczyk, J.W., Xu, Y., 2007. Analysis of chameleon sequences and their implications in biological processes. Proteins 67, 548–558.

Habchi, J., Tompa, P., Longhi, S., Uversky, V.N., 2014. Introducing protein intrinsic disorder. Chem. Rev. 114, 6561–6588.

Hadley, K.R., Mccabe, C., 2010. On the investigation of coarse-grained models for water: balancing computational efficiency and the retention of structural properties. J. Phys. Chem. B 114, 4590–4599.

Hadley, K.R., Mccabe, C., 2012. Coarse-grained molecular models of water: a review. Mol. Simul. 38, 671–681.
Hazy, E., Tompa, P., 2009. Limitations of induced folding in molecular recognition by intrinsically disordered proteins. ChemPhysChem 10, 1415–1419.
Heath, A.P., Kavraki, L.E., Clementi, C., 2007. From coarse-grain to all-atom: toward multiscale analysis of protein landscapes. Proteins 68, 646–661.
Hess, B., Leon, S., Van Der Vegt, N., Kremer, K., 2006. Long time atomistic polymer trajectories from coarse grained simulations: bisphenol-A polycarbonate. Soft Matter 2, 409–414.
Holst, M., Saied, F., 1993. Multigrid solution of the Poisson–Boltzmann equation. J. Comput. Chem. 14, 105–113.
Hu, Z., Ma, B., Wolfson, H., Nussinov, R., 2000. Conservation of polar residues as hot spots at protein interfaces. Proteins 39, 331–342.
Hubbard, T.J., Murzin, A.G., Brenner, S.E., Chothia, C., 1997. SCOP: a structural classification of proteins database. Nucleic Acids Res. 25, 236–239.
Ingólfsson, H.I., Lopez, C.A., Uusitalo, J.J., De Jong, D.H., Gopal, S.M., Periole, X., Marrink, S.J., 2014. The power of coarse graining in biomolecular simulations. Wiley Interdiscip. Rev. Comput. Mol. Sci. 4, 225–248.
Irback, A., Jonsson, S.A.E., Linnemann, N., Linse, B., Wallin, S., 2013. Aggregate geometry in amyloid fibril nucleation. Phys. Rev. Lett. 110, 058101.
Izvekov, S., Voth, G.A., 2005. A multiscale coarse-graining method for biomolecular systems. J. Phys. Chem. B 109, 2469–2473.
Jensen, M.R., Zweckstetter, M., Huang, J.-R., Blackledge, M., 2014. Exploring free-energy landscapes of intrinsically disordered proteins at atomic resolution using NMR spectroscopy. Chem. Rev. 114, 6632–6660.
Kashchiev, D., Cabriolu, R., Auer, S., 2013. Confounding the paradigm: peculiarities of amyloid fibril nucleation. J. Am. Chem. Soc. 135, 1531–1539.
Kim, J., Straub, J.E., Keyes, T., 2012. Replica exchange statistical temperature molecular dynamics algorithm. J. Phys. Chem. B 116, 8646–8653.
Kim, Y.C., Bhattacharya, A., Mittal, J., 2014. Macromolecular crowding effects on coupled folding and binding. J. Phys. Chem. B 118, 12621–12629.
Kimura, S., Caldarini, M., Broglia, R.A., Dokholyan, N.V., Tiana, G., 2014. The maturation of HIV-1 protease precursor studied by discrete molecular dynamics. Proteins 82, 633–639.
Knowles, T.P., Buehler, M.J., 2011. Nanomechanics of functional and pathological amyloid materials. Nat. Nanotechnol. 6, 469–479.
Kundrotas, P.J., Alexov, E., 2006. Electrostatic properties of protein–protein complexes. Biophys. J. 91, 1724–1736.
Kuznetsova, I.M., Turoverov, K.K., Uversky, V.N., 2014. What macromolecular crowding can do to a protein. Int. J. Mol. Sci. 15, 23090–23140.
Laio, A., Parrinello, M., 2002. Escaping free-energy minima. Proc. Natl. Acad. Sci. U.S.A. 99, 12562–12566.
Lashuel, H.A., Lansbury Jr., P.T., 2006. Are amyloid diseases caused by protein aggregates that mimic bacterial pore-forming toxins? Q. Rev. Biophys. 39, 167–201.
Latshaw, D.C., Cheon, M., Hall, C.K., 2014. Effects of macromolecular crowding on amyloid beta (16-22) aggregation using coarse-grained simulations. J. Phys. Chem. B 118, 13513–13526.
Li, H., Gorfe, A.A., 2013. Aggregation of lipid-anchored full-length H-Ras in lipid bilayers: simulations with the MARTINI force field. PLoS One 8, e71018.
Li, X., Mehler, E.L., 2006. Simulation of molecular crowding effects on an Alzheimer's α-amyloid peptide. Cell Biochem. Biophys. 46, 123–141.

Li, M.S., Klimov, D.K., Straub, J.E., Thirumalai, D., 2008. Probing the mechanisms of fibril formation using lattice models. J. Chem. Phys. 129, 175101.
Li, M.S., Co, N.T., Reddy, G., Hu, C.K., Straub, J.E., Thirumalai, D., 2010. Factors governing fibrillogenesis of polypeptide chains revealed by lattice models. Phys. Rev. Lett. 105, 218101.
Lindorff-Larsen, K., Piana, S., Dror, R.O., Shaw, D.E., 2011. How fast-folding proteins fold. Science 334, 517–520.
Liu, P., Kim, B., Friesner, R.A., Berne, B.J., 2005. Replica exchange with solute tempering: a method for sampling biological systems in explicit water. Proc. Natl. Acad. Sci. U.S.A. 102, 13749–13754.
Lodish, H.F., Berk, A., Kaiser, C., Krieger, M., Bretscher, A., Ploegh, H., Amon, A., Scott, M., 2012. Biomembranes and cell architecture. In: Molecular Cell Biology, seventh ed. W.H. Freeman and Company, New York.
Luiken, J.A., Bolhuis, P.G., 2015. Primary nucleation kinetics of short fibril-forming amyloidogenic peptides. J. Phys. Chem. B 119, 12568–12579.
Lyubartsev, A.P., 2005. Multiscale modeling of lipids and lipid bilayers. Eur. Biophys. J. 35, 53–61.
Macdonald, B., Mccarley, S., Noeen, S., Van Giessen, A.E., 2015. Protein-protein interactions affect alpha helix stability in crowded environments. J. Phys. Chem. B 119, 2956–2967.
Machado, M.R., Pantano, S., 2016. SIRAH tools: mapping, backmapping and visualization of coarse-grained models. Bioinformatics 32, 1568–1570.
Marrink, S.J., Risselada, H.J., Yefimov, S., Tieleman, D.P., De Vries, A.H., 2007. The MARTINI force field: coarse grained model for biomolecular simulations. J. Phys. Chem. B 111, 7812–7824.
Masella, M., Borgis, D., Cuniasse, P., 2011. Combining a polarizable force-field and a coarse-grained polarizable solvent model. II. Accounting for hydrophobic effects. J. Comput. Chem. 32, 2664–2678.
Molinero, V., Moore, E.B., 2009. Water modeled as an intermediate element between carbon and silicon. J. Phys. Chem. B 113, 4008–4016.
Monticelli, L., Kandasamy, S.K., Periole, X., Larson, R.G., Tieleman, D.P., Marrink, S.J., 2008. The MARTINI coarse-grained force field: extension to proteins. J. Chem. Theory Comput. 4, 819–834.
Morris, A.M., Watzky, M.A., Finke, R.G., 2009. Protein aggregation kinetics, mechanism, and curve-fitting: a review of the literature. Biochim. Biophys. Acta 1794, 375–397.
Morriss-Andrews, A., Shea, J.E., 2014. Simulations of protein aggregation: insights from atomistic and coarse-grained models. J. Phys. Chem. Lett. 5, 1899–1908.
Morriss-Andrews, A., Shea, J.E., 2015. Computational studies of protein aggregation: methods and applications. Annu. Rev. Phys. Chem. 66, 643–666.
Morriss-Andrews, A., Brown, F.L., Shea, J.E., 2014. A coarse-grained model for peptide aggregation on a membrane surface. J. Phys. Chem. B 118, 8420–8432.
Mourao, M.A., Hakim, J.B., Schnell, S., 2014. Connecting the dots: the effects of macromolecular crowding on cell physiology. Biophys. J. 107, 2761–2766.
Munishkina, L.A., Ahmad, A., Fink, A.L., Uversky, V.N., 2008. Guiding protein aggregation with macromolecular crowding. Biochemistry 47, 8993–9006.
Musiani, F., Ippoliti, E., Micheletti, C., Carloni, P., Ciurli, S., 2013. Conformational fluctuations of UreG, an intrinsically disordered enzyme. Biochemistry 52, 2949–2954.
Nasica-Labouze, J., Nguyen, P.H., Sterpone, F., Berthoumieu, O., Buchete, N.V., Cote, S., De Simone, A., Doig, A.J., Faller, P., Garcia, A., Laio, A., Li, M.S., Melchionna, S., Mousseau, N., Mu, Y., Paravastu, A., Pasquali, S., Rosenman, D.J., Strodel, B., Tarus, B., Viles, J.H., Zhang, T., Wang, C., Derreumaux, P., 2015. Amyloid beta

protein and Alzheimer's disease: when computer simulations complement experimental studies. Chem. Rev. 115, 3518–8563.

Ni, R., Abeln, S., Schor, M., Cohen Stuart, M.A., Bolhuis, P.G., 2013. Interplay between folding and assembly of fibril-forming polypeptides. Phys. Rev. Lett. 111, 058101.

Noid, W.G., 2013. Perspective: coarse-grained models for biomolecular systems. J. Chem. Phys. 139, 090901.

Pace, B.S., Ofori-Acquah, S.F., Peterson, K.R., 2012. Sickle cell disease: genetics, cellular and molecular mechanisms, and therapies. Anemia 2012, 143594.

Pannuzzo, M., Milardi, D., Raudino, A., Karttunen, M., La Rosa, C., 2013. Analytical model and multiscale simulations of Abeta peptide aggregation in lipid membranes: towards a unifying description of conformational transitions, oligomerization and membrane damage. Phys. Chem. Chem. Phys. 15, 8940–8951.

Parton, D.L., Klingelhoefer, J.W., Sansom, M.S., 2011. Aggregation of model membrane proteins, modulated by hydrophobic mismatch, membrane curvature, and protein class. Biophys. J. 101, 691–699.

Pastore, A., Temussi, P.A., 2012. The two faces of Janus: functional interactions and protein aggregation. Curr. Opin. Struct. Biol. 22, 30–37.

Pellarin, R., Caflisch, A., 2006. Interpreting the aggregation kinetics of amyloid peptides. J. Mol. Biol. 360, 882–892.

Pellarin, R., Guarnera, E., Caflisch, A., 2007. Pathways and intermediates of amyloid fibril formation. J. Mol. Biol. 374, 917–924.

Peter, C., Kremer, K., 2009. Multiscale simulation of soft matter systems—from the atomistic to the coarse-grained level and back. Soft Matter 5, 4357–4366.

Phillip, Y., Schreiber, G., 2013. Formation of protein complexes in crowded environments—from in vitro to in vivo. FEBS Lett. 587, 1046–1052.

Politou, A., Temussi, P.A., 2015. Revisiting a dogma: the effect of volume exclusion in molecular crowding. Curr. Opin. Struct. Biol. 30, 1–6.

Predeus, A.V., Gul, S., Gopal, S.M., Feig, M., 2012. Conformational sampling of peptides in the presence of protein crowders from AA/CG-multiscale simulations. J. Phys. Chem. B 116, 8610–8620.

Proctor, E.A., Ding, F., Dokholyan, N.V., 2011. Discrete molecular dynamics. WIREs Comput. Mol. Sci. 1, 80–92.

Prudencio, M., Hart, P.J., Borchelt, D.R., Andersen, P.M., 2009. Variation in aggregation propensities among ALS-associated variants of SOD1: correlation to human disease. Hum. Mol. Genet. 18, 3217–3226.

Qi, H.W., Nakka, P., Chen, C., Radhakrishnan, M.L., 2014. The effect of macromolecular crowding on the electrostatic component of barnase-barstar binding: a computational, implicit solvent-based study. PLoS One 9, e98618.

Qin, S., Zhou, H.X., 2009. Atomistic modeling of macromolecular crowding predicts modest increases in protein folding and binding stability. Biophys. J. 97, 12–19.

Rathore, N., Chopra, M., De Pablo, J.J., 2005. Optimal allocation of replicas in parallel tempering simulations. J. Chem. Phys. 122, 024111.

Reith, D., Putz, M., Muller-Plathe, F., 2003. Deriving effective mesoscale potentials from atomistic simulations. J. Comput. Chem. 24, 1624–1636.

Ren, P., Chun, J., Thomas, D.G., Schnieders, M.J., Marucho, M., Zhang, J., Baker, N.A., 2012. Biomolecular electrostatics and solvation: a computational perspective. Q. Rev. Biophys. 45, 427–491.

Riniker, S., Van Gunsteren, W.F., 2011. A simple, efficient polarizable coarse-grained water model for molecular dynamics simulations. J. Chem. Phys. 134, 084110.

Rivera, E., Straub, J., Thirumalai, D., 2009. Sequence and crowding effects in the aggregation of a 10-residue fragment derived from islet amyloid polypeptide. Biophys. J. 96, 4552–4560.

Rosenman, D.J., Connors, C.R., Chen, W., Wang, C., Garcia, A.E., 2013. Abeta monomers transiently sample oligomer and fibril-like configurations: ensemble characterization using a combined MD/NMR approach. J. Mol. Biol. 425, 3338–3359.

Rossetti, G., Cossio, P., Laio, A., Carloni, P., 2011. Conformations of the Huntingtin N-term in aqueous solution from atomistic simulations. FEBS Lett. 585, 3086–3089.

Rossetti, G., Musiani, F., Abad, E., Dibenedetto, D., Mouhib, H., Fernandez, C.O., Carloni, P., 2016. Conformational ensemble of human alpha-synuclein physiological form predicted by molecular simulations. Phys. Chem. Chem. Phys. 18, 5702–5706.

Rzepiela, A.J., Schäfer, L.V., Goga, N., Risselada, H.J., De Vries, A.H., Marrink, S.J., 2010. Reconstruction of atomistic details from coarse-grained structures. J. Comput. Chem. 31, 1333–1343.

Saibil, H., 2013. Chaperone machines for protein folding, unfolding and disaggregation. Nat. Rev. Mol. Cell Biol. 14, 630–642.

Sanfelice, D., Adrover, M., Martorell, G., Pastore, A., Temussi, P.A., 2012. Crowding versus molecular seeding: NMR studies of protein aggregation in hen egg white. J. Phys. Condens. Matter 24, 244107.

Santo, K.P., Berkowitz, M.L., 2012. Difference between magainin-2 and melittin assemblies in phosphatidylcholine bilayers: results from coarse-grained simulations. J. Phys. Chem. B 116, 3021–3030.

Schymkowitz, J., Borg, J., Stricher, F., Nys, R., Rousseau, F., Serrano, L., 2005. The FoldX web server: an online force field. Nucleic Acids Res. 33, W382–W388.

Senske, M., Tork, L., Born, B., Havenith, M., Herrmann, C., Ebbinghaus, S., 2014. Protein stabilization by macromolecular crowding through enthalpy rather than entropy. J. Am. Chem. Soc. 136, 9036–9041.

Shea, J.E., Levine, Z.A., 2016. Studying the early stages of protein aggregation using replica exchange molecular dynamics simulations. Methods Mol. Biol. 1345, 225–250.

Shell, M.S., 2008. The relative entropy is fundamental to multiscale and inverse thermodynamic problems. J. Chem. Phys. 129, 144108.

Shih, A.Y., Freddolino, P.L., Sligar, S.G., Schulten, K., 2007. Disassembly of nanodiscs with cholate. Nano Lett. 7, 1692–1696.

Shinoda, W., Devane, R., Klein, M.L., 2007. Multi-property fitting and parameterization of a coarse grained model for aqueous surfactants. Mol. Simul. 33, 27–36.

Shirvanyants, D., Ding, F., Tsao, D., Ramachandran, S., Dokholyan, N.V., 2012. Discrete molecular dynamics: an efficient and versatile simulation method for fine protein characterization. J. Phys. Chem. B 116, 8375–8382.

Siekierska, A., De Baets, G., Reumers, J., Gallardo, R., Rudyak, S., Broersen, K., Couceiro, J., Van Durme, J., Schymkowitz, J., Rousseau, F., 2012. Alpha-galactosidase aggregation is a determinant of pharmacological chaperone efficacy on Fabry disease mutants. J. Biol. Chem. 287, 28386–28397.

Simunovic, M., Srivastava, A., Voth, G.A., 2013. Linear aggregation of proteins on the membrane as a prelude to membrane remodeling. Proc. Natl. Acad. Sci. U.S.A. 110, 20396–20401.

Stansfeld, P.J., Sansom, M.S.P., 2011. From coarse grained to atomistic: a serial multiscale approach to membrane protein simulations. J. Chem. Theory Comput. 7, 1157–1166.

Sugita, Y., Okamoto, Y., 1999. Replica-exchange molecular dynamics method for protein folding. Chem. Phys. Lett. 314, 141–151.

Sunde, M., Serpell, L.C., Bartlam, M., Fraser, P.E., Pepys, M.B., Blake, C.C., 1997. Common core structure of amyloid fibrils by synchrotron X-ray diffraction. J. Mol. Biol. 273, 729–739.

Tanizaki, S., Clifford, J., Connelly, B.D., Feig, M., 2008. Conformational sampling of peptides in cellular environments. Biophys. J. 94, 747–759.

Theillet, F.-X., Binolfi, A., Frembgen-Kesner, T., Hingorani, K., Sarkar, M., Kyne, C., Li, C., Crowley, P.B., Gierasch, L., Pielak, G.J., Elcock, A.H., Gershenson, A., Selenko, P., 2014. Physicochemical properties of cells and their effects on intrinsically disordered proteins (IDPs). Chem. Rev. 114, 6661–6714.
Thøgersen, L., Schiøtt, B., Vosegaard, T., Nielsen, N.C., Tajkhorshid, E., 2008. Peptide aggregation and pore formation in a lipid bilayer: a combined coarse-grained and all atom molecular dynamics study. Biophys. J. 95, 4337–4347.
Tompa, P., 2010. Structure and Function of Intrinsically Disordered Proteins. CRC Press, Taylor & Francis Group, Boca Raton, FL.
Torrie, G.M., Valleau, J.P., 1977. Nonphysical sampling distributions in Monte Carlo free-energy estimation: umbrella sampling. J. Comput. Phys. 23, 187–199.
Uversky, V.N., 2011. Intrinsically disordered proteins from A to Z. Int. J. Biochem. Cell Biol. 43, 1090–1103.
Uversky, V.N., 2013. A decade and a half of protein intrinsic disorder: biology still waits for physics. Protein Sci. 22, 693–724.
Vacha, R., Frenkel, D., 2011. Relation between molecular shape and the morphology of self-assembling aggregates: a simulation study. Biophys. J. 101, 1432–1439.
Van Den Berg, B., Ellis, R.J., Dobson, C.M., 1999. Effects of macromolecular crowding on protein folding and aggregation. EMBO J. 18, 6927–6933.
Van Hoof, B., Markvoort, A.J., Van Santen, R.A., Hilbers, P.A.J., 2011. The CUMULUS coarse graining method: transferable potentials for water and solutes. J. Phys. Chem. B 115, 10001–10012.
Vorobjev, Y.N., 2011. Advances in implicit models of water solvent to compute conformational free energy and molecular dynamics of proteins at constant pH. Adv. Protein Chem. Struct. Biol. 85, 281–322.
Wang, Y., Voth, G.A., 2010. Molecular dynamics simulations of polyglutamine aggregation using solvent-free multiscale coarse-grained models. J. Phys. Chem. B 114, 8735–8743.
Wang, H., Junghans, C., Kremer, K., 2009. Comparative atomistic and coarse-grained study of water: what do we lose by coarse-graining? Eur. Phys. J. E. Soft Matter 28, 221–229.
Wang, Y., Li, C., Pielak, G.J., 2010. Effects of proteins on protein diffusion. J. Am. Chem. Soc. 132, 9392–9397.
Wang, H., Barreyro, L., Provasi, D., Djemil, I., Torres-Arancivia, C., Filizola, M., Ubarretxena-Belandia, I., 2011a. Molecular determinants and thermodynamics of the amyloid precursor protein transmembrane domain implicated in Alzheimer's disease. J. Mol. Biol. 408, 879–895.
Wang, L., Friesner, R.A., Berne, B.J., 2011b. Replica exchange with solute scaling: a more efficient version of replica exchange with solute tempering (REST2). J. Phys. Chem. B 115, 9431–9438.
Wang, Q., Zhuravleva, A., Gierasch, L.M., 2011c. Exploring weak, transient protein–protein interactions in crowded in vivo environments by in-cell nuclear magnetic resonance spectroscopy. Biochemistry 50, 9225–9236.
Wang, Y., Sarkar, M., Smith, A.E., Krois, A.S., Pielak, G.J., 2012. Macromolecular crowding and protein stability. J. Am. Chem. Soc. 134, 16614–16618.
Wassenaar, T.A., Pluhackova, K., Böckmann, R.A., Marrink, S.J., Tieleman, D.P., 2014. Going backward: a flexible geometric approach to reverse transformation from coarse grained to atomistic models. J. Chem. Theory Comput. 10, 676–690.
Williams 2nd, B., Convertino, M., Das, J., Dokholyan, N.V., 2015. ApoE4-specific misfolded intermediate identified by molecular dynamics simulations. PLoS Comput. Biol. 11, e1004359.
Winklhofer, K.F., Tatzelt, J., Haass, C., 2008. The two faces of protein misfolding: gain- and loss-of-function in neurodegenerative diseases. EMBO J. 27, 336–349.

Wu, C., Shea, J.E., 2011. Coarse-grained models for protein aggregation. Curr. Opin. Struct. Biol. 21, 209–220.

Wu, C., Shea, J.E., 2013. Structural similarities and differences between amyloidogenic and non-amyloidogenic islet amyloid polypeptide (IAPP) sequences and implications for the dual physiological and pathological activities of these peptides. PLoS Comput. Biol. 9, e1003211.

Wu, Z., Cui, Q., Yethiraj, A., 2010. A new coarse-grained model for water: the importance of electrostatic interactions. J. Phys. Chem. B 114, 10524–10529.

Xu, J., Reumers, J., Couceiro, J.R., De Smet, F., Gallardo, R., Rudyak, S., Cornelis, A., Rozenski, J., Zwolinska, A., Marine, J.C., Lambrechts, D., Suh, Y.A., Rousseau, F., Schymkowitz, J., 2011. Gain of function of mutant p53 by coaggregation with multiple tumor suppressors. Nat. Chem. Biol. 7, 285–295.

Yang, L., Liu, C.W., Shao, Q., Zhang, J., Gao, Y.Q., 2015. From thermodynamics to kinetics: enhanced sampling of rare events. Acc. Chem. Res. 48, 947–955.

Yesylevskyy, S.O., Schäfer, L.V., Sengupta, D., Marrink, S.J., 2010. Polarizable water model for the coarse-grained MARTINI force field. PLoS Comput. Biol. 6, e1000810.

Zamarreno, F., Herrera, F.E., Corsico, B., Costabel, M.D., 2012. Similar structures but different mechanisms: prediction of FABPs-membrane interaction by electrostatic calculation. Biochim. Biophys. Acta 1818, 1691–1697.

Zhang, J., Muthukumar, M., 2009. Simulations of nucleation and elongation of amyloid fibrils. J. Chem. Phys. 130, 035102.

Zhang, Y., Stec, B., Godzik, A., 2007. Between order and disorder in protein structures: analysis of "dual personality" fragments in proteins. Structure 15, 1141–1147.

Zhou, H.X., Rivas, G., Minton, A.P., 2008. Macromolecular crowding and confinement: biochemical, biophysical, and potential physiological consequences. Annu. Rev. Biophys. 37, 375–397.

Zimmerman, S.B., Minton, A.P., 1993. Macromolecular crowding: biochemical, biophysical, and physiological consequences. Annu. Rev. Biophys. Biomol. Struct. 22, 27–65.

Zimmerman, S.B., Trach, S.O., 1991. Estimation of macromolecule concentrations and excluded volume effects for the cytoplasm of *Escherichia coli*. J. Mol. Biol. 222, 599–620.

CHAPTER THREE

Structural Characteristics of α-Synuclein Oligomers

N. Cremades*,[1], S.W. Chen†, C.M. Dobson†
*Biocomputation and Complex Systems Physics Institute (BIFI)-Joint Unit BIFI-IQFR (CSIC), Universidad de Zaragoza, Zaragoza, Spain
†University of Cambridge, Cambridge, United Kingdom
[1]Corresponding author: e-mail address: ncc@unizar.es

Contents

Abstract

Oligomeric forms of amyloid aggregates have been detected in the brains and tissues of patients suffering from neurodegenerative disorders such as Parkinson's disease, and it is widely thought that such species are key pathogenic agents in the development and spreading of the disease; however, the study of these species has been proven to be extremely challenging, primarily as a result of their intrinsically transient nature and high levels of heterogeneity. Identifying the structural nature and the details of the mechanisms of formation and interconversion of individual oligomeric species,

International Review of Cell and Molecular Biology, Volume 329
ISSN 1937-6448
http://dx.doi.org/10.1016/bs.ircmb.2016.08.010

particularly those with high toxicity, is of fundamental importance not only for understanding the mechanisms of protein misfolding and amyloid aggregation but also for the identification of diagnostic and therapeutic targets. In this review, we will focus on the current knowledge of the multitude of oligomeric forms of α-synuclein that have been reported to date, with particular emphasis on their structural features and possible relationship to other amyloid species, in order to build a clearer understanding of the types of oligomeric species that accumulate during the aggregation of α-synuclein and to develop a comprehensive picture of the misfolding behavior of the protein.

1. INTRODUCTION

The misfolding and aggregation of otherwise soluble proteins into fibrillar amyloid deposits is linked with the pathogenesis of approximately 50 human neurological and systemic disorders. These disorders have a multitude of disparate symptoms and some are among the most common and debilitating medical conditions in the modern world, including Alzheimer's disease and Parkinson's disease (PD). Growing evidence suggests that certain oligomeric forms, typically accumulated during the early events in the amyloid self-assembly process, are more inherently toxic than their fibrillar counterparts. Indeed, they are thought to be the culprits of the pathology associated with at least many of the amyloid diseases via a gain-of-toxic function mechanism, although the nature of the most pathogenic amyloid assemblies is still a subject of intense debate.

The process of amyloid formation observed in vivo can be reproduced in vitro, and the fibrils generated are morphologically and tinctorially similar to those extracted from patients, offering unique opportunities to characterize the details of the self-assembly processes including the nature of oligomeric forms, at a molecular level. Indeed, defining the molecular mechanisms of amyloid formation and the intermediate species and their relationships with mature amyloid fibrils is a vital objective in the field. It is, however, challenging because of the inherent difficulties in studying these soluble, transient, and highly heterogeneous protein species. A range of different methods have been used to stabilize and trap oligomeric species in order to obtain information about their nature, and a wide variety of structural and biophysical studies have been carried out in attempts to define connections between the different types of aggregated species. This review will examine our current knowledge of the structural nature of the aggregates formed by α-synuclein (αS), the protein whose deposition in the form of amyloid fibrils is linked with a family of neurodegenerative diseases including PD, with the aim of identifying common features between the multitude of intermediates that have been reported, and establishing the mechanistic relationships between them. As a result of the combined efforts of the scientific community to understand the structural nature of these complex protein species, important steps toward the elucidation of the underlying mechanisms of the conversion of proteins into the amyloid state have been made.

2. AMYLOID FORMATION

Amyloid formation refers to the process by which soluble, typically monomeric, peptides, and proteins misfold and self-assemble into the form of stable and highly ordered β-sheet-rich aggregates, of which the archetypal examples are amyloid fibrils. Such fibrils are thread-like structures typically composed of several protofilaments that twist around each other. The core of each protofilament adopts a common and amyloid-specific fold, termed a cross-β structure, in which β-strands align perpendicularly to the protofilament axis generating arrays of β-sheets that are oriented parallel to the fibrillar axis (Eisenberg and Jucker, 2012; Fitzpatrick et al., 2013; Sawaya et al., 2007; Sunde et al., 1997). The molecular mechanisms by which proteins adopt the amyloid state is of unquestionable interest, and much progress has been made recently through the development of new experimental approaches, such as single-molecule microfluidic techniques, and by combining experimental and theoretical methods using the formalism of chemical kinetics (Cohen et al., 2012; Knowles et al., 2009). There are important questions to clarify, for example, how and why a specific protein starts to self-assemble, how the aggregation and the acquisition of the cross-β structure occur, and how this process induces pathological behavior.

One of the most remarkable features of protein aggregation and fibril formation is its universal nature, a characteristic found originally in the late 1990s (Dobson, 1999; Guijarro et al., 1998). Since then, a large and increasing number of peptides and proteins, whether or not related to disease, have been shown to be able to self-assemble in vitro first into the form of low-molecular-weight prefibrillar aggregates (i.e., oligomers) and eventually into insoluble fibrillar aggregates (i.e., amyloid fibrils), adopting the distinctive cross-β configuration (Chiti and Dobson, 2006; Dobson, 2003; Eisenberg and Jucker, 2012; Knowles et al., 2014). Such structure can be acquired by a polypeptide chain in an aggregated state regardless of its amino acid sequence, the original topology of the native state of the protein, and the structure adopted by the protein under physiological conditions; indeed, both fully folded and intrinsically disordered proteins, including homopolymeric systems, have been shown to be able to self-assemble into cross-β fibrillar structures under appropriate conditions both in vitro and in vivo (Chiti and Dobson, 2006; Fandrich and Dobson, 2002).

These findings have led to the proposition that the amyloid state is a generic protein conformation (Chiti and Dobson, 2006; Dobson, 1999, 2003; Eisenberg and Jucker, 2012), and that it is the most favorable thermodynamic state under at least some physiological conditions (Baldwin et al., 2011; Gazit, 2002). The spontaneous adoption of the amyloid conformation by a large number of proteins challenges various previously accepted scientific paradigms including the assumption that the native functional state of a protein represents the minimum energy state and, therefore, the most thermodynamically favorable conformation of a protein. Interestingly, these findings highlight the fundamental role that both kinetic and thermodynamic factors play in controlling the balance between protein folding and misfolding in vivo (Dobson, 2003) and indicate the metastability of the functional proteome (Baldwin et al., 2011).

2.1 Mechanisms of Formation of Amyloid Aggregates

The transition of a protein from its functional soluble state to the amyloid state is a highly complex process that depends on both the intrinsic characteristics of the protein and the environmental conditions. In some cases, it has been shown to be favored by the presence of specific interfaces including lipid membranes (Campioni et al., 2014; Galvagnion et al., 2015). The early analysis of the kinetics of formation of amyloid fibrils demonstrated that the overall process follows a nucleation–polymerization model (Jarrett and Lansbury, 1992), where soluble species undergo a nucleation step that results in oligomeric species that are then able to grow through further monomer addition generating protofilaments and eventually mature fibrils. The growth profile is characteristically sigmoidal, reflecting the greater ease of addition of monomers onto existing aggregates compared to the novo formation of new oligomers directly from monomers. Amyloid formation can also be seeded by the addition of preformed fibrils, a phenomenon analogous to that observed in crystallization, in which the nucleation step is bypassed, and the structural features of the seeds are reproduced in the growing fibrils, a phenomenon also called templating. More recently, it has become evident that processes other than primary nucleation and elongation are important, including fibril fragmentation and surface-catalyzed nucleation. These secondary processes can dominate the kinetics of fibril growth under many circumstances, increasing the multiplicity of steps in the formation process (Cohen et al., 2011a; Knowles et al., 2009). An important additional observation is that secondary processes, particularly surface-catalyzed nucleation, can be a dominant factor in the generation of toxic oligomers (Cohen et al., 2013).

The detailed analysis of experimental kinetic data determined for the fibril formation process, typically obtained by means of fluorescence dyes such as Thioflavin T, has significantly advanced our understanding of the mechanisms of the fibril formation process. Nevertheless, characterization of the nucleation events themselves is challenging, particularly as a consequence of the often rapid elongation rates of the transient intermediate species and their highly heterogeneous nature. In order to overcome these difficulties, a number of techniques have been developed, including methods that enable the direct observation and characterization of individual molecular species populated during the aggregation reaction (Cremades et al., 2012; Ding et al., 2009; Horrocks et al., 2016; Kostka et al., 2008; Orte et al., 2008; Pitschke et al., 1998). Single-molecule fluorescence methods in which intermolecular Förster resonance energy transfer effects can be monitored have been shown to be extremely useful for identifying specific types of oligomeric species and following their formation and evolution during the early stages of protein aggregation (Cremades et al., 2012). Investigations of the primary nucleation process of fibril formation by the protein αS using this methodology revealed the existence of two important groups of αS oligomers. The αS monomers were found to self-assemble initially into relatively disordered oligomers that very slowly convert into compact and more stable oligomers, a key step that leads not only to fibril formation but also to toxicity and resistance to degradation by the cellular proteostasis machinery (Cremades et al., 2012).

2.2 Generic Features of the Structure of Amyloid Fibrils

Amyloid fibrils are typically long unbranched thread-like structures, just a few nanometers in diameter, and composed of several filaments that twist around each other (Eisenberg and Jucker, 2012; Fitzpatrick et al., 2013; Sawaya et al., 2007; Sunde et al., 1997). The individual protofilaments appear to be generally composed of a double layer of β-sheets whose strands are perpendicular to the fibril axis to form the classic cross-β structure (Fig. 1). In contrast to the globular native states of proteins, which generally possess highly diverse and extremely intricate folding topologies, the amyloid state possesses a generic architecture that is rich in β-sheet structure and shares distinctive structural features that are largely independent of the protein sequence (Dobson, 1999, 2003). The determination of the details at atomic level of the structure of fibrils formed by specific proteins has proven to be challenging, but recent technical developments, particularly in cryo-electron microscopy (cryo-EM) and solid-state nuclear magnetic resonance (NMR) spectroscopy, have greatly enhanced our understanding of the different levels of structural complexity inherent in the amyloid cross-β structure (Fitzpatrick et al., 2013; Zhang et al., 2009).

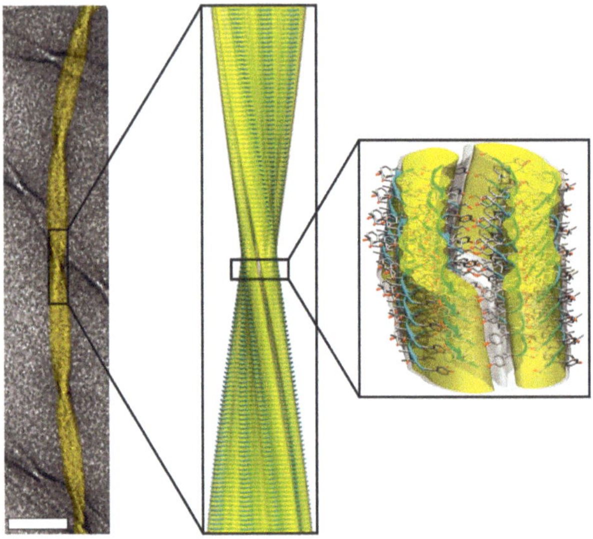

Fig. 1 Hierarchical assembly of the structure of amyloid fibrils. The structures of several polymorphisms of the amyloid fibrils formed by an 11-residue fragment of transthyretin at atomic resolution were obtained by the combination of cryo-EM imaging with solid-state NMR analysis (Fitzpatrick et al., 2013). In this figure, the structural architecture of one of the polymorphs is shown. These fibrils are composed of three filaments, each of them in turn formed by pairs of cross-β protofilaments, which are each composed of pairs of β-sheets that interact with other protofilaments through specific water-mediated interactions established between the side chains of the residues of the protein that form the core of the fibril. The scale bar in the cryo-EM image on the left is 50 nm in length. *Reprinted with permission from Fitzpatrick, A.W.P., Debelouchina, G.T., Bayro, M.J., Clare, D.K., Caporini, M.A., Bajaj, V.S., Jaroniec, C.P., Wang, L., Ladizhansky, V., Muller, S.A., et al., 2013. Atomic structure and hierarchical assembly of a cross-β amyloid fibril. Proc. Natl. Acad. Sci. U.S.A. 110, 5468–5473.*

Such studies, together with the X-ray diffraction analysis of peptide micro/nanocrystals (Nelson et al., 2005; Rodriguez et al., 2015; Sawaya et al., 2007), have confirmed the universality of the amyloid architecture. The origins of this architecture can be attributed to the nature of the intramolecular and intermolecular interactions within the β-sheets, which are dominated by the hydrogen bonds between the main-chain atoms of the polypeptide chain that are common to all peptide and protein molecules (Fandrich and Dobson, 2002). The structural variations between amyloid fibrils generated from different polypeptide chains arise from the manner in which the various side chains are incorporated into the structure, resulting in many cases in the existence of water-filled interfaces between protofilaments (Fitzpatrick et al., 2013; Jimenez et al., 1999; Serpell and Smith, 2000; Serpell et al., 1995, 2000b; Zhang et al., 2009). Moreover, except for short peptides, only specific segments of the chain are incorporated into the cross-β structure, with the remaining of the chain being external to the core elements of the structure.

The universal propensity of the polypeptide backbone to form hydrogen bonds is the main driving force for the formation and the stability of the cross-β structure of the protofilaments (Knowles et al., 2007). The lateral packing of the β-sheets, however, relies on the specific patterns of interactions between the side chains of the protein, which therefore depends on its specific amino acid sequence, explaining at least partially why mixed fibrils composed of several types of polypeptide chains are generally unstable. Another consequence of the specific features of the amyloid structure is the regular and generic spacing between polypeptide chains along the fibril axis as it is fixed by the intermain chain hydrogen bonding constrains and, therefore, independent of the amino acid sequence of the polypeptide. By contrast, the intersheet spacing (the spacing between β-sheets in the direction perpendicular to the fibril axis) is variable and highly dependent on the nature of the side chains (Fandrich and Dobson, 2002).

Another remarkable feature of the amyloid state is its extremely high stability that is primarily a consequence of the continuous array of hydrogen bonds along the direction of the fibril axis that is characteristic of the cross-β structure (Knowles et al., 2007). Indeed, increasing evidence suggests that, for many proteins, the amyloid state might be thermodynamically more stable than the native functional state even under physiological conditions (Baldwin et al., 2011; Gazit, 2002). This conclusion challenges one of the most fundamental assumptions in the field of protein folding and indicates that many proteins are metastable in their native state. In order to incorporate this characteristic in the energy landscape diagrams of proteins, a new dimension, the concentration of protein, needs to be considered. As the conversion of proteins from their soluble to the amyloid state involves the formation of intermolecular interactions, the thermodynamic stability of the amyloid state increases with the concentration of the protein. There will be, therefore, a protein concentration at which the stability of the amyloid state is the same as that of the native state, and above this critical protein concentration the protein will adopt the amyloid conformation unless there are high kinetic barriers for the structural conversion; it is, under such conditions, the functional native state becomes supersaturated (Ciryam et al., 2013). For proteins whose concentration in vivo is close to the critical concentration, favoring the conversion into the amyloid state, small perturbations in protein expression levels can lead to amyloid fibril formation.

A recent study (Baldwin et al., 2011) compared the experimental free energies per residue in the native and the amyloid states for a set of peptides and proteins as a function of the length of the polypeptide chain. Two main conclusions were observed. First, for short peptides, there is a high free energy difference between the two states, an observation that provides an explanation as to why pathogenic amyloid fibrils in vivo are formed by small peptides or proteins or fragments of proteins. Second, for polypeptide chains longer than c. 150 residues, there is virtually no free energy difference between the two states probably because of the topological constraints associated with the incorporation of long polypeptide chains into the amyloid state, and suggesting a reason why biology has a tendency to generate large proteins that have a natural low tendency to form fibrils. Interestingly, αS, having 140 residues, is one of the large proteins associated with amyloid diseases. In this study, the standard free energy per residue in the amyloid state for the different peptides and proteins was, however, calculated under seeding conditions and therefore effects of cooperativity and templating are implicit in the experimental values. It would be very interesting, although much more challenging experimentally, to establish analogous comparisons between the free energy of the native and the amyloid state for the oligomeric species with an early cross-β structure.

2.3 Generic Features of the Structure of Amyloid Oligomers

Aggregation of amyloidogenic systems produces an array of oligomeric species that will vary in size, β-sheet content, and stability, at least during the early stages of the aggregation reaction (notably the lag phase). In addition, it is becoming increasingly evident that amyloid formation can take place through a multiplicity of routes, each of them generating a multiplicity of oligomeric species (Bemporad and Chiti, 2012; Chen et al., 2015; Jimenez et al., 2002; Kad et al., 2003). In some cases, different mechanisms of aggregation will result from differences in solution conditions or by introducing mutations (Bader et al., 2006; Bitan et al., 2003; Gosal et al., 2005; Kumar and Udgaonkar, 2009), and in others, several pathways appear to be of similar significance and compete between each other in the aggregating sample (Jain and Udgaonkar, 2011; Kaylor et al., 2005). The extremely high levels of heterogeneity of the oligomers both in terms of size (from dimers to higher order multimers) and structure (from essentially random coil to in principle a similar degree of β-sheet content to that observed in the fibrillar species), together with their often transient nature and generally low population levels throughout the aggregation reaction, have hampered their direct identification, quantification, and detailed structural characterization. Developing a detailed structural understanding of the major types of oligomers that a polypeptide chain can generate at physiologically relevant conditions during its amyloid aggregation is, therefore, a high priority. A significant advance in this respect has been the development of single-molecule fluorescence techniques for the study of oligomeric complexes in highly heterogeneous samples, which have provided unprecedented information on the existence and nature of different types of oligomers formed during the early stages of aggregation (Cremades et al., 2012; Orte et al., 2008). These and other studies (Auer et al., 2008; Bleiholder et al., 2011; Carulla et al., 2009; Lee et al., 2011; Qi et al., 2008) are in agreement with a nucleation–conversion–polymerization model for the formation of amyloid fibrils at least for the intrinsically disordered proteins and peptides that have been studied with particular detail, in which monomers initially assemble into relatively disordered oligomers with little or no stable β-sheet structure that subsequently convert into more highly ordered and compact oligomers with increased stability, which can ultimately form amyloid fibrils after further growth and rearrangement. The nucleation–conversion–polymerization model appears also to be relevant for the aggregation of some globular proteins where the process is initiated from native-like states. In these cases, the polypeptide chains may maintain the native structure of the monomer in the initially formed oligomers, which, once formed, acquire β-sheet structure at later stages in the self-assembly process (Bouchard et al., 2000; Olofsson et al., 2004; Pagano et al., 2010; Plakoutsi et al., 2005).

As an alternative approach to reduce the extremely high heterogeneity of protein species in the sample, and in this way to extract more conclusive and detailed information on certain types of stable oligomers, experimental protocols to isolate specific species or to enrich the sample in certain types of oligomers have been developed for a wide variety of amyloidogenic peptides and proteins. A very interesting example is the discovery of two structurally different types of oligomers of HypF-N, a nondisease-related protein from *Escherichia coli*, that are formed and accumulate under different solution conditions and display different toxicities (Campioni et al., 2010). These two oligomeric species are similar in structure, with apparently identical secondary structure content, but show significantly different compactness; the more toxic oligomers display a higher degree of flexibility and a higher content of solvent-exposed hydrophobic patches on the surface than the less toxic ones. This amyloidogenic system has been characterized in great detail from the structural and biological point of view and it is contributing significantly to elucidating the structural determinants of oligomer cytotoxicity (Campioni et al., 2010; Mannini et al., 2014; Zampagni et al., 2011).

From the considerable number of studies of amyloid oligomers reported to date, and despite the variability of properties and proposed structures, several features appear to be common to the different types of oligomers; these include a content of β-sheet structure that is intermediate between that present in the monomeric and fibrillar forms, a high content of solvent-exposed hydrophobic regions and a globular or a tubular morphology. Interestingly, doughnut-shaped oligomers have been observed by atomic force microscopy (AFM) for a substantial number of amyloidogenic peptides and proteins such as Aβ, αS, amylin, and serum amyloid A (Lashuel et al., 2002a; Quist et al., 2005), which are widely recognized by a conformational-specific antibody, irrespective of the primary sequence of peptide/protein from which they are generated (Kayed et al., 2003). In most cases, oligomers capable of binding to this conformational-specific antibody were shown to be cytotoxic, which led to a hypothesis that there is a specific structure common to a variety of toxic amyloid oligomers that differs significantly from the monomeric and the fibrillar structures, as the latter have not been found to be recognized by the antibody (Kayed et al., 2003). Interestingly, this and other related antibodies have been widely used to monitor the presence of these in vitro-observed oligomers in Alzheimer's disease patients and the overall results indicate that such species are likely to exist in vivo and are related to disease conditions (Hillen et al., 2010; Kayed et al., 2003,

2007; Lacor et al., 2004; Lasagna-Reeves et al., 2012; Noguchi et al., 2009). Despite major efforts in obtaining more structural details of these and other common types of amyloid oligomers, little progress has so far been made, at least in part because of the inherent heterogeneity of the oligomeric species. Even if a specific type of oligomer can be isolated, the degree of variability in size and structure often remains significant. It is therefore of the utmost importance to control as much as possible this heterogeneity of the oligomeric samples to understand its origins, if meaningful structural models are to be obtained and related to toxicity.

2.4 Role in Disease

Amyloid formation has been recognized as a defining feature of a number of medical disorders, particularly neurodegenerative diseases, since the early 20th century. Currently, approximately 50 distinct human diseases and disorders, some of them among the most common and debilitating medical conditions in the modern world, including Alzheimer's disease and PD, have been linked with the accumulation of misfolded and abnormally aggregated proteins in the form of amyloid fibrils, primarily composed of one main type of peptide or protein depending on the specific disorder; indeed, these pathological inclusions represent the hallmark of this type of disease, collectively referred to as amyloid diseases, amyloidoses, or protein misfolding diseases (Chiti and Dobson, 2006). Despite the fact that the close relationship between the appearance of amyloid deposits and the onset of pathology in these conditions has been clear for many years, much remains to be understood about the specific mechanisms underlying the toxicity associated with the process of amyloid aggregation.

Amyloid fibrils are associated with a loss of function of the proteins that aggregate and are involved in the disease, but more generally with a gain-of-toxic function through the generation and accumulation of misfolded protein species during the process of self-assembly. The localization of the protein inclusions in the damaged regions of patients suffering from amyloid diseases led initially to the hypothesis that the inclusions themselves, or their fibrillar components, were responsible for the pathogenicity in these diseases. This may be the case for at least some systemic amyloidoses, in which large quantities of intractable amyloid aggregates are deposited in vital organs (Chiti and Dobson, 2006). In neurodegenerative disorders, however, the correlations between the quantities of protein inclusions and pathogenicity are not always highly correlated, suggesting that other forms of the proteins may be contributing to neurodegeneration (Kalia et al., 2013; Ross and Poirier, 2005).

Oligomeric forms of amyloid aggregates have been detected in the brains and tissues of patients suffering from neurodegenerative disorders and growing experimental evidence suggests that certain oligomeric forms, such as those that accumulate during the early events of the amyloid self-assembly process, are inherently more damaging than their fibrillar counterparts and are believed to be the primary origins of the toxicity associated with amyloid diseases (Arrasate et al., 2004; Bucciantini et al., 2002; Haass and Selkoe, 2007; Winner et al., 2011). The nature of the most toxic forms of amyloid assemblies is, however, still a subject of intense debate in the field, especially in the recent years (Luk et al., 2012; Peelaerts et al., 2015; Prusiner et al., 2015; Woerman et al., 2015). It is evident from in vitro and in vivo experiments, and from analysis of tissue samples and biofluids from human patients that oligomeric species play an important role in triggering pathology (for Aβ, Shankar et al., 2008; Townsend et al., 2006; tau, Lasagna-Reeves et al., 2012; Patterson et al., 2011; huntingtin, Nekooki-Machida et al., 2009; and αS, Horrocks et al., 2016; Kalia et al., 2013; Tsigelny et al., 2008; Winner et al., 2011). One of the possible mechanisms of toxicity that is emerging as a general primary step to neurodegeneration by amyloid oligomers is membrane perturbation (Kayed et al., 2004), although further validation of this conclusion is required. Other cytotoxic effects that have been associated with amyloid oligomers include the induction of oxidative stress (De Felice et al., 2007; Deas et al., 2016), dysregulation of calcium homeostasis (Danzer et al., 2007; Demuro et al., 2005), mitochondria dysfunction (Devi et al., 2008; Nakamura, 2013), impairment of the proteasome system (Xilouri et al., 2013), and blocking of trafficking in the ER–Golgi (Cooper et al., 2006).

The toxicity associated with oligomeric amyloid species has been long hypothesized to arise from a set of unique structural features characteristic of these species that are absent in the monomeric protein and in the fibrillar amyloid form. For the case of the widely observed doughnut-like oligomers described in Section 2.3, the presence of a central cavity in these oligomers led to the proposal that this specific "pore-like" structure was characteristic and unique of a certain type of toxic oligomers and a key determinant of their toxicity. In the light of such structure, and their ability to bind and disrupt lipid membranes, the mechanism of toxicity initially proposed was that these oligomers insert into membranes and create pores in the lipid bilayer, in an analogous manner to the behavior of well-known pore-forming peptides such as α-hemolysin (indeed, the oligomer conformation-specific antibody has also been reported to stain α-hemolysin pores; Yoshiike et al., 2007). The formation of pores in the cellular membranes would then result in dysregulation of signal transduction and changes in ion homeostasis (Arispe et al., 1993; Quist et al., 2005). The pore formation hypothesis has, however, been challenged by alternative scenarios in which the process of interaction of the oligomers with the lipid membranes has been proposed to be the main cause of membrane destabilization, without the need for specific pores in the membrane (Brender et al., 2012; Stockl et al., 2013; Walsh et al., 2014).

An alternative possibility is that oligomer toxicity arises simply from the inherently intermediate, partially folded nature of these species, that expose a range of side chains on their surface, particularly those of hydrophobic residues (Bolognesi et al., 2010; Chen et al., 2015; Mannini et al., 2014), that in the native state under normal physiological conditions, and also in highly compact fibrillar structures would not be accessible within the cellular environment. These groups could then establish inappropriate and probably highly promiscuous interactions with other proteins, lipid membranes, and additional cellular components that could disrupt their correct biological function. In addition, various types of oligomeric species are expected to be formed during aggregation, which may have the potential to induce different levels of toxicity, perhaps through different types of interactions, depending on the details of their particular structure and the nature of the cells in which they are formed or to which they are exposed (Campioni et al., 2010; Cremades et al., 2012).

3. α-SYNUCLEIN

αS is a 140-residue protein expressed abundantly in the brain (Bonini and Giasson, 2005). In neurons, αS is primarily located at the presynaptic termini (Iwai et al., 1995), either in the cytosol or, in some cases, with up to a third of the cellular αS pool, bound to synaptic membranes (Visanji et al., 2011). In the cytosol, αS is primarily monomeric and intrinsically disordered, with no persistent structure at physiological conditions, although it adopts an ensemble of relatively compact disordered conformations with an average hydrodynamic volume significantly smaller than that predicted for an extended random coil conformation (Bertoncini et al., 2005; Dedmon et al., 2005; Lendel et al., 2009). The sequence of αS is typically divided into three main regions: the N-terminal region (residues 1–60); the NAC (nonamyloid-β component) region (residues 61–95), which is mainly hydrophobic; and the C-terminal region (residues 96–140), which is highly acidic. The protein contains seven imperfect KTKEGV repeats spread between the N-terminal and the NAC region (Fig. 2) that generate a hydrophobic periodicity in this segment of the protein that resembles that found in the amphipathic lipid-binding α-helical domains of apolipoproteins (Clayton and George, 1998; George et al., 1995). The NAC region seems to be essential and sufficient for αS fibril formation (Giasson et al., 2001) and appears to remain protected in the monomeric form from aberrant interactions with the NAC region of other αS molecules in solution by establishing protective intramolecular long-range interactions with the highly acidic C-terminal region of the protein (Bertoncini et al., 2005; Dedmon et al., 2005). Due to the electrostatic character of these protective interactions, it is not surprising that αS aggregation can be strongly modulated by the ionic strength or pH of the solution (Buell et al., 2014).

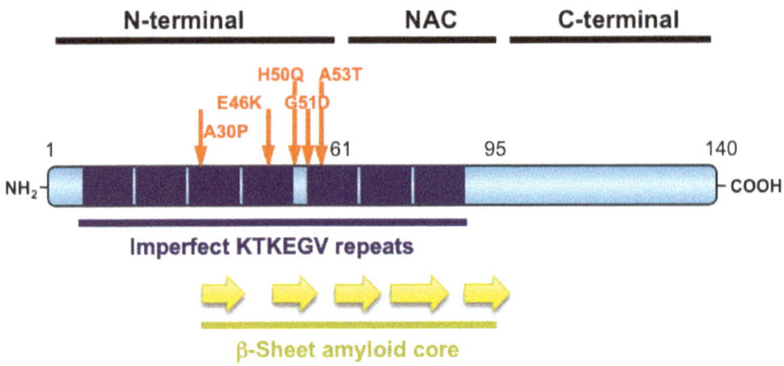

Fig. 2 Schematic representation of the sequence-dependent features of αS. The primary sequence of αS can be divided in three regions: the N-terminal, the non-β amyloid component (NAC), and the C-terminal region. Although the NAC region has been demonstrated to be essential and sufficient for the formation of amyloid fibrils, all the disease-associated point mutations reported so far lie in the N-terminus of the protein (their location is indicated by *orange arrows*). The protein contains seven imperfect apolipoprotein-like repeat motives in its first N-terminal 100 residues that result in the acquisition of an amphipathic α-helical structure in this region upon binding to lipid membranes. In addition, the protein is able to adopt β-sheet structure upon its self-assembly into amyloid fibrils. The folding core of αS in the fibrillar state has been reported to be composed of five β-strands that fall within residues 30–100 (Vilar et al., 2008). The approximate positions of the five strands that have been proposed to form the fibrillar core are indicated by *yellow arrows*.

Upon binding to lipid membranes, the N-terminal and most of the NAC region of the protein folds into an amphipathic α-helical structure, driven by the imperfect KTKEGV repeats, with two main differentiated regions: one involving the first 25 residues of the N-terminal region that acts as an anchor to the lipid bilayer, and the other involving residues 26–98 that is able to modulate the affinity of the protein for the bilayer in a manner that depends on the lipid composition of the membrane (Fusco et al., 2014). Upon membrane binding, the monomeric protein has been reported to trigger a modest amount of membrane remodeling (Ouberai et al., 2013), with no signs of major disruption or destabilization. Indeed, the ability of the protein to interact with lipid membranes has been associated with its physiological functions, although they are not yet well understood. Recent evidence points to a possible role of αS in regulating synaptic trafficking, homeostasis, and neurotransmitter release (Burre et al., 2010; Nemani et al., 2010). Interestingly, it has been proposed that αS could be a key player in these neuronal processes not only by interacting with synaptic vesicles but also through interactions with synaptic proteins such as phospholipase D2 (Payton et al., 2004), various members of the family of RAB small GTPases (Dalfo and Ferrer, 2005), and also SNARE complexes (Burre et al., 2010). In addition, αS has been proposed to be involved in other cellular processes including signal transduction, the functioning of mitochondria, and the regulation of oxidative stress (Bonini and Giasson, 2005).

αS has long been considered as an intracellular protein but the presence of the protein in biological fluids such as cerebrospinal fluid and blood plasma of both patients and normal subjects (El-Agnaf et al., 2003; Horrocks et al., 2016; Lee et al., 2006; Tokuda et al., 2006), together with evidence for specific mechanisms of αS release (Lee et al., 2005), suggests that a significant fraction of αS could be extracellular and that the protein may be secreted and transmitted between neuronal cells (Desplats et al., 2009; Hansen et al., 2011; Lee et al., 2008). Moreover, there is evidence suggesting that the secretion of αS is sensitive to its folding state and is stimulated by conditions of cellular stress (Jang et al., 2010).

3.1 Association With PD and Other Neurodegenerative Disorders

αS has been implicated in PD as well as in other neurodegenerative disorders including dementia with Lewy bodies (DLB) and multiple system atrophy (MSA), collectively referred to as synucleinopathies (McCann et al., 2014). These diseases share a common characteristic process, that is indeed the hallmark of these pathologies, namely the accumulation of αS in the form of amyloid fibrils into intractable intracellular inclusions (Baba et al., 1998; Spillantini et al., 1997, 1998). In PD and DLB, these inclusions are referred to as Lewy bodies (LBs) and Lewy neurites and are preferentially found inside neurons, while in MSA, αS accumulates into more diffuse inclusions of poorly organized bundles of αS fibrils that localize primarily in the cytoplasm of glial cells and, therefore, are called glial cytoplasmatic inclusions (Papp et al., 1989; Spillantini et al., 1998). Additionally, αS has also been identified as a component of amyloid inclusions from brain tissue of Alzheimer's disease patients (Ueda et al., 1993).

Several rare single-point mutations in the gene encoding αS (SNCA) have been identified in familial cases of PD (A53T, A30P, E46K, and more recently H50Q and G51D; Appel-Cresswell et al., 2013; Lesage et al., 2013; Polymeropoulos et al., 1997) and SNCA gene duplications and triplications lead to other types of familial PD with an early age of onset of the disease (Chartier-Harlin et al., 2004; Singleton et al., 2003). These familial cases account for only a small fraction of PD in the general population, but the LBs and Lewy neuritis observed both in familial and idiopathic PD stain strongly for αS (Galvin et al., 1999; Spillantini et al., 1997, 1998). Indeed, immunostaining of LBs and Lewy neurites for αS (Giasson et al., 2000; Jellinger, 2011) indicates that, although other proteins may also accumulate in the inclusions of PD, αS predominates. In addition, a genomewide association study has shown that individuals with certain polymorphisms of the SNCA gene have a higher risk of developing sporadic PD (Seidel et al., 2010), and some of these polymorphisms have been associated with higher expression of αS in neurons (Maraganore et al., 2006). Also, two single-point substitutions (A18T and A29S) that potentially add new phosphorylation sites in the protein have recently been suggested to be associated with sporadic PD, although the pathogenicity of these two variants needs to be established (Hoffman-Zacharska et al., 2013).

Taken together, the genetic evidence, along with the neuropathologic evidence for accumulation in essentially all patients with PD, indicate a central role for αS in the pathogenesis of both the inherited and sporadic forms of PD. The contribution of αS to PD and other synucleinopathies could in principle result from a loss or perturbation of the normal function of αS, from a toxic gain of function caused by its aggregation, or from a combination of the two. It is, however, becoming evident from both genetic and biochemical studies that there is a mechanistic link between an abnormal excess of αS concentration within neurons and the induction of neurodegeneration. Although the normal biological function of αS remains elusive, despite of its importance in the context of neurodegeneration, significant evidence points toward a gain-of-toxic function being a dominant factor in the pathology of PD and related conditions.

Recently, misfolded forms of αS, and other proteins associated with neurodegenerative disorders, have been shown to self-propagate and spread, sometimes described as in a "prion-like" manner, between interconnected regions of the central nervous system (Brundin et al., 2010; Danzer et al., 2009; Jucker and Walker, 2013). Indeed, cell-to-cell transmission of αS aggregates has been experimentally observed (Danzer et al., 2011; Desplats et al., 2009), pointing to a key role for αS in the stepwise spreading of LB pathology and the progressive nature of PD and other synucleinopathies. Interestingly, like prions, different types of strains or fibril polymorphs of αS have been identified, which have been proposed to present different degrees of infectivity and induce variable neuronal vulnerability and pathology (Peelaerts et al., 2015; Prusiner et al., 2015; Woerman et al., 2015), providing some insights into why fibrillar αS inclusions are associated with, and indeed constitute a hallmark of, distinct types of neurodegenerative disorders. Such "prion-like" behavior is indeed an intrinsic characteristic of the fundamental physicochemical principles of the self-assembly process of amyloid fibrils. Indeed, secondary nucleation mechanisms and seeding processes have been shown to be important catalytic processes in the aggregation of both αS in PD and Aβ42 in Alzheimer's disease (Buell et al., 2014; Cohen et al., 2011b; Knowles et al., 2014).

3.2 Structural Characteristics of the Fibrillar State

The fibrils generated in vitro are morphologically and tinctorially indistinguishable from those extracted from patients (Crowther et al., 2000), showing the typical characteristics of the amyloid cross-β structure (Serpell et al., 2000a). Although the structure of αS fibrils has not been unambiguously determined at atomic resolution, in part due to fibril polymorphism (Heise et al., 2005), extensive structural characterization of a number of αS fibrils using primarily EM, ssNMR, H/D exchange mass spectrometry, and electron paramagnetic resonance has been performed (Chen et al., 2007; Comellas et al., 2011; Dearborn et al., 2016; Del Mar et al., 2005; Gath et al., 2012, 2014; Vilar et al., 2008). The core structure of αS fibrils is generally found to include residues 30–110 (Chen et al., 2007; Comellas et al., 2011; Der-Sarkissian et al., 2003; Gath et al., 2012, 2014; Miake et al., 2002;

Vilar et al., 2008) (Fig. 2), although the exact position of the residues that are incorporated into the cross-β structure appears to vary in different fibril polymorphs. Most experimental data are consistent with a model in which αS monomers adopt an antiparallel in-register β-sandwich fold (with variations in the proposed number of β-strands involved) with monomeric units stacked in a parallel arrangement forming the fibril protofilament, as described in the model proposed by Riek and collaborators (Vilar et al., 2008). In this structure, both the C-terminus of the protein (residues 100–140) and the most N-terminal region (residues 1–30) remain unstructured, although both regions have been proposed to be involved in interactions between protofilaments (Conway et al., 2000a; Guilliams et al., 2013; Vilar et al., 2008). More recently, a high-resolution structure of a fibrillar form of full-length αS has been obtained by solid-state NMR. This structure presents similar overall structural features to earliest predicted β-serpentine models (Vilar et al., 2008), although the folding of this fibrillar form is substantially more complex, presenting a new orthogonal Greek key topology (Tuttle et al., 2016). Both twisted and straight αS fibrils have been documented, as have different fibril polymorphs that appear to vary both in the number and disposition of the protofilaments in the mature fibril, and in the specific molecular organization of the protofilaments, where differences in the amino acid stretches involved in the β-sandwich fold constituting the core of the protofilament have been reported (Comellas et al., 2011; Gath et al., 2012, 2014; Tuttle et al., 2016; Vilar et al., 2008). Indeed, in fibrils from other amyloidogenic proteins, a remarkable variety of stable and self-propagating protofilament assemblies have been observed (Fitzpatrick et al., 2013; Jimenez et al., 1999; Serpell and Smith, 2000; Serpell et al., 1995, 2000b; Zhang et al., 2009). In a recent EM study of a particular αS fibril polymorph, a model where αS monomers are arranged in a superpleated β-structure has been proposed (Dearborn et al., 2016), although this fold seems to be consistent only with this specific polymorphism.

An important common feature of a number of fibrils formed by αS that arises from EM studies, independently of the particular fibril polymorphism adopted, is the presence of a low electron density region in the center of the fibrillar structure, with a diameter of c. 1.5–3.0 nm (Dearborn et al., 2016; Vilar et al., 2008), suggesting that, as found in amyloid fibrils from other amyloidogenic peptides and proteins (Fitzpatrick et al., 2013; Jimenez et al., 1999; Serpell and Smith, 2000; Serpell et al., 1995, 2000b; Zhang et al., 2009), at least some αS fibrils would be composed of a number of protofilaments that interact in a face-to-face manner through water-filled interfaces generating a tubular-like ultrastructure appearance in both twisted and straight fibrils (see Fig. 1).

A recent structure of an 11-residue segment of the NAC region of αS (residues 68–78) adopting a cross-β structure in nanocrystals was resolved by microelectron diffraction at 1.4 Å resolution (Rodriguez et al., 2015). The structure exhibits protofibrils built of pairs of face-to-face β-sheets composed of nearly fully extended β-strands (each strand corresponding to one peptide) stack in register in a similar way to that observed in the structure of an 11-residue segment of the protein transthyretin in an amyloid fibril (see Fig. 1; Fitzpatrick et al., 2013). The pairs of β-sheets are interdigitated to form a steric-zipper structure as previously observed in many amyloid-like microcrystals formed from short segments of fibril-forming proteins (Nelson et al., 2005). The absence of water-filled interfaces in the protofibrils generated by these short polypeptide segments observed in the micro/nanocrystals could be a consequence of a preferred stabilization of dry interfaces in the completely ordered three-dimensional cross-β-type crystals of short peptides, which cannot acquire the typical fibrillar twisted structure (Knowles et al., 2012). This structure does, however, provide details of the possible structure of a region of the core of αS fibrils, incorporating a segment of the NAC region of the protein (Nelson et al., 2005).

3.3 Conversion From the Monomeric to the Fibrillar State

Evidence is accumulating that monomeric αS, under in vitro physiological conditions and in the absence of amphipathic molecules or lipid vesicles, populates an ensemble of disordered conformations including both extended conformers and also structures that are more compact than expected for a completely unfolded chain (Bernado et al., 2005; Bertoncini et al., 2005; Dedmon et al., 2005; Lee et al., 2004). There is also direct evidence that a similar dynamic structure is present within living cells (Lashuel et al., 2013; Theillet et al., 2016; Waudby et al., 2013). By contrast, as discussed earlier, in the fibrillar form of αS approximately two-thirds of the protein sequence adopts a highly stable and compact cross-β structure (Chen et al., 2007; Comellas et al., 2011; Der-Sarkissian et al., 2003; Gath et al., 2012, 2014; Miake et al., 2002; Vilar et al., 2008). The transition from the natively unfolded monomeric state to the fibrillar state is therefore a process of acquiring persistent structure within much of the polypeptide sequence.

Two types of model for the acquisition of the amyloid structure by αS, and more generally by other amyloidogenic proteins and peptides that are intrinsically disordered in the native state, have been proposed, each of them supported by a range of theoretical and experimental evidence. In one model, the monomeric form is able to adopt, at least transiently, conformations with β-sheet structure that have a propensity to establish intermolecular hydrogen bonds through specific side chains, generating oligomers with a folding core composed of β-sheet-rich stretches of different polypeptide chains. In this model, low-molecular-weight β-sheet oligomers, as small as dimers or trimers, would be formed, the rate-limiting step being the encounter of two or more polypeptide chains with high β-sheet propensities. This structural model would be analogous to the nucleation–polymerization kinetic model where the nucleus is the monomer, or indeed a subset of monomeric β-sheet conformations. A limiting case of this model would be one in which the monomer adopts essentially the same β-sheet structure as that presented in the mature amyloid fibrils, which would then act as a template for self-assembly (Fig. 3A).

It is difficult, however, to imagine that the monomeric αS can adopt conformations with a similar population of molecules with a fully formed amyloid β-sheet structure within the intrinsically disordered ensemble of conformations. A more realistic scenario in this context would be a situation in which the monomer adopts conformations with partially formed β-sheet structures that can then trigger self-assembly such that the aggregated species would adopt the fully formed amyloid structure at a later stage, either through the gradual increase in structure as the size of the aggregates increases or through a more cooperative process that converts the intermediate into the fully formed β-sheet structure. This model is supported by theoretical studies that show some tendency in the amino acid sequence of αS, specifically in the region suggested to form the core of the fibrils, to acquire β-sheet structure in its monomeric form (Jonsson et al., 2012; Zibaee et al., 2007). Indeed, a number of studies have shown that the protein is able to adopt conformations with some β-sheet structure and fibril-like tertiary contacts already present in its monomeric state (Esteban-Martin et al., 2013; Jonsson et al., 2012; Sandal et al., 2008), which are favored under certain conditions known to promote the formation of amyloid fibrils (Sandal et al., 2008).

An alternative model that has been proposed for the acquisition of amyloid structure by intrinsically disordered polypeptides assumes that monomers initially coalesce, most likely around hydrophobic stretches of the sequence, into predominantly disordered oligomers lacking persistent

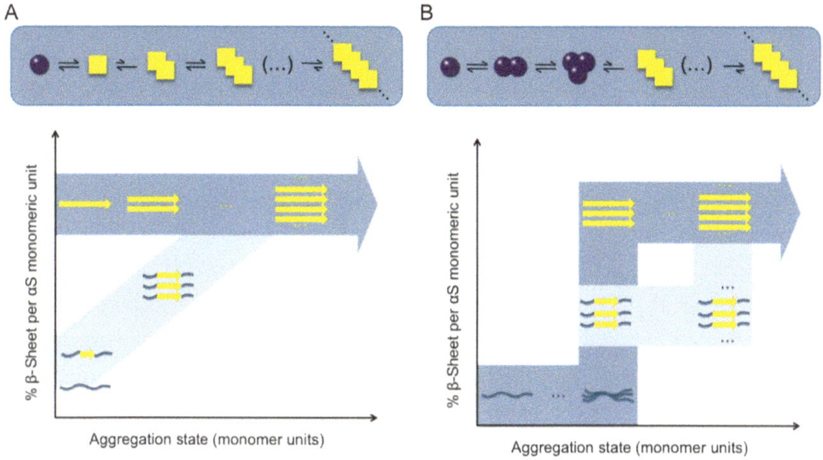

Fig. 3 Models proposed for the conversion of αS from its intrinsically disordered monomeric form to the cross-β structure characteristic of the amyloid form. Schematic representation of (A) a nucleation–polymerization model and (B) a nucleation–conversion–polymerization model. The intrinsically disordered nature of the polypeptide chain (or a region of the chain) is represented in *blue* and the fully formed β-sheet structure that the protein (or some regions of the protein) acquires in the fibrils is depicted in *yellow*. In the nucleation–polymerization model, the structural conversion from random coil to β-sheet structure is assumed to take place at the monomeric level. In the limiting case of the model, the monomeric protein is assumed to adopt the fully formed β-sheet structure, and the rate-limiting step is the encounter of several polypeptide chains with that conformation. A more likely scenario, however, is one in which the monomeric protein acquires only a partially formed β-sheet structure that acts as a template for the acquisition of more extensive structure upon self-assembly (shown in *light blue* on the graph). In the nucleation–conversion–polymerization model, the structural conversion occurs at the oligomeric level through a unimolecular reaction from disordered to β-sheet oligomers. Such β-sheet oligomers, in an extreme case, could have fully formed amyloid-like structure, or partially formed structure that later converts into the fully formed β-sheet structure in a subsequent step (shown in *light blue*).

structure, which subsequently convert slowly into oligomers with amyloid-like β-sheet structure that is sufficiently stable to promote the acquisition of similar structure by other molecules. The rate-limiting step in this model would, therefore, be the conversion from disordered to β-sheet-rich oligomers and, as a consequence of such slow conversion, the two structurally distinct types of oligomers would coexist at the early stages of the self-assembly process. In the limiting case of this model, the oligomers would possess a fully formed amyloid core immediately after the conversion process and would elongate by monomer addition through a templating or induced-fit mechanism at the oligomer ends. In a more likely scenario, the initially formed disordered oligomers would convert into oligomers with only partially formed amyloid-like β-sheet structures, which would then acquire the fully formed fibril conformation during subsequent structural conversions (Iljina et al., 2016) (Fig. 3B). This nucleation–conversion–polymerization model for the aggregation of αS and other amyloid systems is strongly supported by a number of experimental studies (Cremades et al., 2012; Iljina et al., 2016; Lee et al., 2011; Serio et al., 2000; Thakur et al., 2009; Wei et al., 2011), and indeed direct experimental evidence for the structural conversion from relatively disorganized oligomers to more compact structures with β-sheet structure that are able to grow into fibrillar species has recently been reported (Cremades et al., 2012; Iljina et al., 2016; Serio et al., 2000).

Interestingly, a comparative study of possible peptide self-assembly mechanisms on a series of short amyloid-forming peptides using ion mobility–mass spectrometry methods has revealed the importance of the specific properties of the peptide sequence in dictating the preference for a specific mechanism, with some peptides showing a preference for a nucleation–polymerization mechanism (such as VEALYL) and others (such as NNQQNY) for a nucleation–conversion–polymerization mechanism (Bleiholder et al., 2011). It is reasonable to assume, therefore, that changes in solution conditions could well alter the physicochemical properties of the polypeptide, and thus the preferential mechanism by which a given polypeptide self-assembles and forms amyloid fibrils (Auer et al., 2008), a situation likely to result in the formation of distinct polymorphs of both oligomers and fibrils.

4. OLIGOMERIC STATES OF αS

It is clear from the information already discussed in this chapter that developing a detailed understanding of the oligomeric species generated during protein amyloid aggregation, including their structural features, their mechanisms of formation, their relationships with other amyloid species, and their possible mechanisms of toxicity, is of fundamental importance. In the remainder of this review, we will focus on our current knowledge of the multitude of oligomeric forms of αS that have so far been reported, with special emphasis on their structural features and their possible relationships with other amyloid species, in order to develop a clearer understanding of the type of oligomeric conformations populated by αS during its aggregation. There are now large numbers of articles on these topics, and here we discuss those studies in which some degree of structural characterization has been reported. Indeed, an ultimate goal of this review is to decipher the structural similarities, and also the differences, between the multiplicity of αS oligomeric forms reported at the present time, and to explore the possible relationships between the dominant oligomeric and fibrillar forms of αS, in an attempt to suggest a structural classification of the oligomers and to identify some of the main routes that lead to amyloid forms of αS.

4.1 Types of Oligomers Identified During In Vitro Fibril Formation

αS oligomeric species with spherical and annular appearance have been observed during in vitro fibril formation by means of AFM (Conway et al., 1998, 2000a,b) and transmission electron microscopy (TEM; Lashuel et al., 2002a,b). In addition, further incubation of the spherical oligomers has been reported to result in circular and elliptical rings (Conway et al., 2000b) and incubation of these αS oligomers in the presence of excess monomeric αS has been found to result in the conversion of oligomers into fibrils within a few days at 37°C (Lashuel et al., 2002b). Moreover, incubation of A53T αS oligomers at 4°C for 2 weeks has been reported to produce long tubular structures (Lashuel et al., 2002b). We have recently used single-molecule techniques to observe the formation of two distinct types of oligomeric αS species; the initially formed species are nontoxic but convert to more compact and more stable proteinase K-resistant oligomers of high toxicity prior to the formation of fibrils (Cremades et al., 2012). In addition, studies of αS oligomerization in the presence of solid-state nanopores showed the presence of four types of oligomers and suggested that the formation of large oligomers resulted from the consumption of small oligomers (Hu et al., 2016). Both these studies, therefore, suggest the existence of different forms of oligomers whose interconversion could play a significant role in the aggregation of αS.

4.2 Stable In Vitro-Isolated Oligomers

In order to overcome the limitations imposed by studying the oligomeric species during the process of fibril formation, a variety of procedures have been used to trap or isolate some of these oligomeric forms. Some of these procedures involve a change in physicochemical conditions, whereas other approaches involve the addition of chemical compounds able to stabilize αS oligomers either by interacting preferentially with certain types of oligomers and inhibiting their elongation, such as epigallocatechin gallate (EGCG; Bieschke et al., 2010; Ehrnhoefer et al., 2008), selegiline (Braga et al., 2011), and a variety of metal ions (Cole et al., 2005), or by covalently modifying specific chemical groups in the protein to promote cross-linking, such as baicalein (Zhu et al., 2004), nitration agents (Souza et al., 2000; Uversky et al., 2005), glycation agents (Lee et al., 2009), 4-hydroxynonenal (HNE; Qin et al., 2007), or di-*tert*-butyl peroxide (DTBP; Cole et al., 2002). We summarize below the structural information reported for the various oligomeric forms of αS that have been stabilized through such procedures.

4.2.1 Stabilized Through Protein Lyophilization

One of the most commonly employed methods to enrich samples in oligomeric species involves the use of lyophilization. Lyophilization, in some cases in combination with agitation (Celej et al., 2012; Chen et al., 2015; Lorenzen et al., 2014a; Paslawski et al., 2014a,b; van Rooijen et al., 2009), has, for example, been used to generate oligomers at physiological pH and ionic strength without additives. While lyophilization is commonly used as a means of prolonging the storage of proteins, we (Chen et al., 2015) and other groups (Ding et al., 2002; Gallea and Celej, 2014; Volles et al., 2001) have shown that lyophilization can strongly promote the formation of oligomers, an effect that can be attributed to an increase in the access of protein molecules to solvent/air interfaces and in the intermolecular interactions during the vitrification process.

Oligomers formed in this way have been reported to bind to the oligomer-specific A11 antibody (Celej et al., 2012; Chen et al., 2015) and to contain largely antiparallel β-sheet structures (Celej et al., 2012; Chen et al., 2015; Gallea and Celej, 2014; Lorenzen et al., 2014a; Paslawski et al., 2014b). The isolated oligomers have been found to exist as a heterogenous population of species and to possess an average size of approximately 30 monomers (Chen et al., 2015; Lorenzen et al., 2014a; Zijlstra et al., 2012), although a wide range of sizes (from 10 mer to 90 mer) have been observed to coexist in samples in this type of oligomers (Chen et al., 2015). The oligomers are able to associate with lipid bilayers and induce membrane permeabilization (Chen et al., 2015; Giehm et al., 2011; Lorenzen et al., 2014a) and possess remarkable stability (Chen et al., 2015; Paslawski et al., 2014a). It has been suggested that the NAC region along with parts of the N-terminal region (residues 39–89) generates the core structure of the oligomers through well-defined intermolecular contacts (Gallea and Celej, 2014), whereas the rest of the protein remains in an essentially disordered state (Lorenzen et al., 2014a; Paslawski et al., 2014b; van Rooijen et al., 2009). The β-sheet core of these oligomers, therefore, appears to involve similar residues to that of the fibrillar core, although the latter involves a larger number of residues, but to have a significantly different β-sheet arrangement (Celej et al., 2012; Chen et al., 2015; Gallea and Celej, 2014). Importantly, these oligomers have been shown to be toxic to neuronal cells, as a result of the induction of an aberrant production of reactive oxygen species that eventually leads to caspase-3 activation and cell death (Deas et al., 2016). It has also been proposed that such oligomers are relevant in vivo, on the basis of the similarity of the cascade of cellular effects induced by these oligomers with that found at long term in a human iPS-derived αS triplication neuronal model (Deas et al., 2016).

Although most studies have been carried out with WT αS, recent work suggests that oligomers formed from disease-related mutants—A30P, A53T, E46K, H50Q, and G51D—are composed of similar numbers of monomers to those formed by the WT protein, although G51D oligomers appear to be unable to disrupt membranes unlike those of the other mutants (Stefanovic et al., 2015), and to have similar stabilities as those of WT αS oligomers (Paslawski et al., 2014a). Studies using HDX-MS have, however, suggested the presence of two different oligomeric types of species for WT αS and for three familial αS mutants (A30P, E46K, and A53T). "Type I" oligomers, while being more protected against HDX, were found to be transient in nature and be present only at low levels, dissociating rapidly to monomers and readily capable of forming mature fibrils. "Type II" oligomers, by contrast, were less protected against HDX and were reported not to be in rapid equilibrium with monomeric species or to be able to elongate to form fibrils, forming instead worm-like structures (Paslawski et al., 2014b). The presence of these two different oligomeric species with very different elongation capabilities may result from the specific protocol used to generate the oligomeric sample that combines protein lyophilization and incubation under strong agitation conditions. When no agitation is used during the incubation of the protein, only one type of oligomers is observed, which would correspond to the "type II" oligomers described by Paslawski et al. (Celej et al., 2012; Chen et al., 2015; Gallea and Celej, 2014). In general, oligomers formed by lyophilization appear unable to elongate and form fibrils, or to seed their formation at a significant rate (Celej et al., 2012; Chen et al., 2015; Giehm et al., 2011; Lorenzen et al., 2014a).

We have recently shown that the types of oligomers formed by lyophilization are remarkably stable and have suggested that they are kinetically trapped, do not elongate readily to form fibrils, and have very slow disaggregation kinetics, only slightly faster than those of the mature fibrils (Chen et al., 2015). By combining the information obtained by a wide range of biophysical and structural techniques, we were also able to define and minimize the degree of heterogeneity of the oligomers to such a degree that distinct subgroups of structures could be identified, enabling their overall structural properties and molecular architectures to be defined. Indeed, we were able to use cryo-EM image reconstruction techniques to obtain three-dimensional structural models for the different subgroups of amyloid oligomers, revealing the nature of the quaternary structure of these highly toxic amyloid oligomers. The 3D models reveal a common cylindrical architecture with a central cavity for the oligomers (Fig. 4), similar to those observed previously for a number of toxic oligomers formed from a range of amyloidogenic proteins and peptides, including some pathological variants of αS (Lashuel et al., 2002a,b). This cylindrical architecture has remarkable similarities to the structural models of amyloid fibrils formed from a variety of proteins, despite differences in their amino acid sequences. The similarities between the fibrillar and the oligomeric

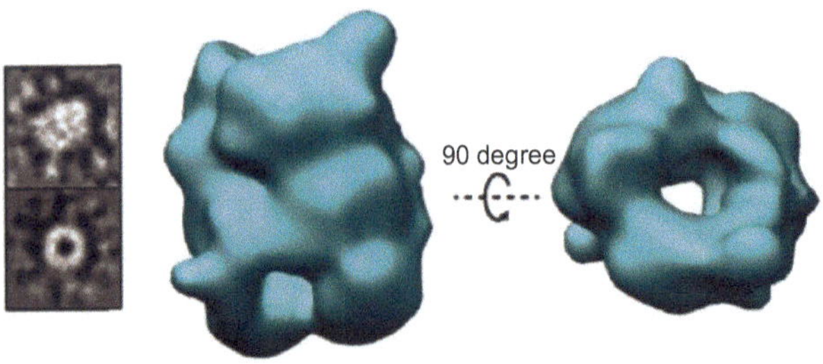

Fig. 4 Representative three-dimensional cryo-EM image reconstruction of the structural type of αS oligomers trapped by protein lyophilization. The images on the *left* are the representative of the typical side view (*top*) and end-on view (*bottom*) of this structural class of oligomers (Chen et al., 2015). The images on the *right* show the two orthogonal views, side (*left*) and end-on (*right*), of the 3D reconstruction, showing a cylindrical architecture of the type observed for αS fibrils despite having only c. 50% of the β-sheet content of the latter, and a different arrangement of the β-strands. *Reprinted with permission from Chen, S.W., Drakulic, S., Deas, E., Ouberai, M., Aprile, F.A., Arranz, R., Ness, S., Roodveldt, C., Guilliams, T., De-Genst, E.J., et al., 2015. Structural characterization of toxic oligomers that are kinetically trapped during α-synuclein fibril formation. Proc. Natl. Acad. Sci. U.S.A. 112, E1994–E2003.*

structural architectures have led us to suggest that similar types of interactions to those that stabilize amyloid fibrils are likely to be responsible for the initial acquisition of β-sheet structure, and indeed for the morphological characteristics of at least some forms of cytotoxic oligomers (Chen et al., 2015).

4.2.2 Stabilized by Chemical Compounds

The formation of stabilized forms of αS oligomers has also been observed through the addition of various organic molecules, often in an attempt to mimic physiological or pathological conditions.

4.2.2.1 Polyphenolic Compounds

Many polyphenolic compounds have been reported to be inhibitors of the formation of toxic amyloid oligomers and fibrils, such as curcumin (Singh et al., 2013), rosmarinic acid (Rao et al., 2008), nordihydroguaiaretic acid, and ferulic acid, among many others (Ono and Yamada, 2006). Two particularly well-characterized polyphenolic compounds are baicalein (Hong et al., 2008; Lu et al., 2011; Zhu et al., 2004) and EGCG (Bieschke et al., 2010; Ehrnhoefer et al., 2008), which have been reported to inhibit fibril formation by αS and also to destabilize preformed αS fibrils. Baicalein has also been shown to induce the formation of αS oligomers with a significant degree of β-sheet content, although these oligomers have been found to have little effect on membrane integrity; baicalein does, however, protect SH-SY5Y cells from αS oligomer-induced toxicity (Lu et al., 2011). In addition, EGCG has also been shown to actively remodel αS monomers, oligomers, and fibrils into nontoxic unstructured aggregates (Bieschke et al., 2010). Interestingly, preincubation of toxic oligomers, formed through lyophilization and agitation as described in Section 4.2.1, with EGCG was found to result in oligomers with reduced membrane binding and a decreased ability to disrupt and permeabilize DOPG vesicles. Importantly, the addition of EGCG to stable preformed αS oligomers inhibited their toxicity, as observed by the MTT assay and by nuclear staining using trypan, despite no evidences for major changes in their overall size, shape, or secondary structure content (Lorenzen et al., 2014b). The simplest anti-amyloid polyphenolic compound, gallic acid (GA), has also been shown to interact with monomeric αS, inhibiting the structural compaction associated with aggregation (Liu et al., 2014). The results demonstrate that GA not only inhibits αS aggregation and its associated toxicity, but it also disaggregates preformed αS fibrils to generate less toxic species, and has been suggested to bind to soluble, nontoxic, unstructured oligomers, and prevent their conformational transition into oligomers rich in β-sheet structure (Ardah et al., 2014).

4.2.2.2 Drug Molecules

The antibiotic, rifampicin, has been reported to stabilize αS oligomers, inhibit fibril formation, and disaggregate preformed fibrils (Li et al., 2005). Selegiline, a noncompetitive monoamino-oxidase B inhibitor, has been found to have neuroprotective effects when administered to PD patients as a monotherapy or in combination with L-dopa; studies have indicated that it promotes the formation of nontoxic amorphous aggregates of αS, including large annular species, which inhibit fibril formation (Braga et al., 2011). NMR experiments have indicated that selegiline does not interact with monomeric αS, but it could interfere with the process of nucleation since addition of fibrillar species abolished the inhibitory effect. Interestingly, when added in combination with dopamine, selegiline overrides the inhibitory effect of dopamine, favoring fibril formation and suppressing the formation of smaller toxic aggregates.

A small-molecule compound denoted FN075, which belongs to a class of compounds designed to mimic a small C-terminal peptide with an extended β-sheet conformation (Cegelski et al., 2009), has been reported to trigger the formation of αS oligomers that are able to form rapidly fibrils. FN075, which possess a bulky napthyl group and a CF3-phenyl substituent that mimic hydrophobic side chains, is designed to target hydrophobic β-sheet regions in proteins. The oligomers formed in the presence of FN075 have a radius of ~7 nm, with the C-terminal 40 residues remaining highly disordered (Horvath et al., 2012). The oligomers that are stabilized by FN075 have been shown to be able to disrupt lipid vesicles (Nors Perdersen et al., 2015) and their core structure and overall dimensions are similar to those previously reported for isolated oligomers generated upon agitation and lyophilization (Giehm et al., 2011), and the fibrils formed with and without FN075 are reported to be similar. The quantities of oligomers formed correlated with the quantity of fibrils formed at different FN075 concentrations, which suggests that the FN075 oligomers convert into fibrils. A systematic study of the manner in which different functional groups on FN075 affect oligomer and fibril formation show that specific modifications to the structure of FN075 led to the formation of relatively unstructured oligomers that inhibited fibril formation, pointing yet again to the multiplicity of misfolding pathways.

4.2.2.3 Metal Ions

An extensive AFM study of the effect of the addition of a wide range of metal ions to αS indicated that annular and spherical oligomeric αS species were formed prior to the fibrillar species (Lowe et al., 2004). Ca^{2+} was found to promote oligomer formation by αS in vitro (Nielsen et al., 2001), like Cu^{2+}, although in this later case the oligomers were found to have an amorphous "stellate" morphology and to be toxic when applied exogenously to cells (Nielsen et al., 2001; Wright et al., 2009). Interestingly, Cu^{2+} was also found to induce oligomer formation in cells, which ultimately lead to cell death (Wang et al., 2010). Treatment of αS with hydrogen peroxide in the presence of Fe^{2+} and Cu^{2+}, respectively, has also been reported to lead to the formation of oligomers (Hashimoto et al., 1999; Paik et al., 2000) and aggregation of αS in the presence of Fe^{3+} and DDT also resulted in the formation of oligomers, ranging from dimers to larger oligomers with annular worm-like structure (Cole et al., 2005). While these oligomers possess β-sheet structure, they inhibit the formation of fibrils and the formation of these oligomers was suppressed upon the addition of the iron chelator, deferoxamine.

4.2.2.4 Dopamine

The selective loss of dopamine-containing neurons in the substantia nigra is a key feature of PD, and as a result, extensive studies have been performed on αS in the presence of dopamine and its analogs. While there is some controversy as to whether or not it is dopamine or its reactive intermediates that interact with αS, as well as if covalent modifications of αS are involved in oligomer formation in the presence of dopamine, several studies show that dopamine and/or analogs of this compound promotes the formation of αS oligomers, inhibiting fibril formation, and disaggregates preformed fibrils to give disordered oligomeric aggregates (Conway et al., 2001; Li et al., 2004; Norris et al., 2005; Pham et al., 2009). Interestingly, while these oligomers were observed to inhibit fibril formation at pH values of 7 and above, the oligomers were found to associate further to form sheet-like assemblies at lower pH values (Pham et al., 2009). These oligomers were not, however, observed to bind to phospholipids or to affect membrane permeability (Pham and Cappai, 2013), although evidence was obtained for loss of function in the context of the chaperoning of the assembly of SNARE complexes (Choi et al., 2013). In addition, 3,4-dihydroxyphenylacetaldehyde (DOPAL), a monoamine oxidase metabolite of dopamine, has been observed to trigger αS aggregation in a cell-free system and in neuronal cell cultures, resulting in the formation of potentially toxic αS oligomers and aggregates (Burke et al., 2008). Subsequent injection of DOPAL into the substantia nigra of Sprague–Dawley rats resulted in the loss of dopaminergic neurons and the accumulation of high-molecular-weight αS oligomers (Burke et al., 2008).

4.2.3 Stabilized Through Chemical Modifications in the Protein

4.2.3.1 Oxidation and Nitration

There are four methionine residues in αS that are potential sites of oxidation. Oxidation of αS by hydrogen peroxide, which preferentially modifies Met 5 relative to Met 1, 116, and 127, was found to promote the formation of stable, nontoxic oligomers, but to delay the formation of fibrils (Zhou et al., 2010). The greater the degree of oxidation of αS, the slower the formation of fibrils, except at low pH (pH 3), where the ability of oxidized αS to form fibrils was unaltered relative to the nonoxidized protein; this latter finding suggests that oxidation of methionine could inhibit aggregation by strengthening the autoinhibitory effects of the negative charges in the C-terminal segment of αS, which contribute to the increased stability of the oxidized αS oligomers. When all four methionine residues were oxidized by H_2O_2, no oligomeric species were observed to form immediately following the modification reaction by size-exclusion chromatography (Xiang et al., 2013), although subsequent agitation stimulated the formation of a small number of soluble nontoxic oligomers, and fibril formation was dramatically retarded (Zhou et al., 2010).

There are four tyrosine residues in αS that are potential sites for nitration and treatment of αS with nitrating agents, peroxynitrite/CO_2 or myeloperoxidase/H_2O_2/nitrite, was found to result in the formation of oligomers via covalent cross-linking (Souza et al., 2000). It has also been reported that the nitration of αS by tetranitromethane leads to the formation of oligomeric αS species at neutral, but not acidic, pH values, and that the formation of these oligomers inhibited fibril formation (Uversky et al., 2005). At high protein concentrations (1 mg/mL), the predominant oligomeric species have been suggested to be octamers, and to be nontoxic and unable to seed fibril formation (Xiang et al., 2013).

4.2.3.2 Formation of Protein Adducts

αS, in LBs in the substantia nigra of PD brains and in neocortical and brainstem LBs in DLB, is often found to be enriched in the end product of lipid peroxidation, 4-hydroxy-2-nonenal (HNE; Castellani et al., 2002;

Dalfo and Ferrer, 2008; Yoritaka et al., 1996). It has been shown that the aldehyde HNE, and the chemically related reactive aldehyde 4-oxo-2-nonenal (ONE), can result in the in vitro formation of HNE/ONE-αS adducts, which appears to lead to the formation of αS oligomers (Bae et al., 2013; Nasstrom et al., 2009, 2011; Qin et al., 2007; Xiang et al., 2013). Some differences have been reported in the secondary structure content of the αS oligomers formed in the presence of HNE, with some studies proposing that they have a significant degree of secondary structure content with antiparallel β-sheet structure, which are incapable of forming fibrils (Cai et al., 2015; Nasstrom et al., 2009, 2011; Qin et al., 2007), and others reporting a predominantly disordered structure, which are able to elongate and trigger fibril formation (Bae et al., 2013). In addition, αS oligomers induced by HNE and ONE have been shown to be neurotoxic in vitro (Nasstrom et al., 2011; Qin et al., 2007; Xiang et al., 2013).

It has also been observed that advanced glycation end products (AGEs) are commonly found in autopsy samples of regions of the brain that contain LBs (Castellani et al., 1996), with AGEs and thus glycated αS colocalizing with LBs in PD brain samples (Munch et al., 2000) and in an MPTP mouse model of PD (Choi et al., 2011). In vitro glycation of αS by methylglyoxal (MGO) or glyoxal (GO) has been shown to inhibit αS aggregation (Lee et al., 2009), although nontoxic amorphous aggregates (Padmaraju et al., 2011) or β-sheet-rich oligomers were observed, which do not bind ThT efficiently (Lee et al., 2009).

4.3 Oligomers Generated Upon Fibril Disaggregation

The presence of spherical and annular oligomeric species has been observed following cold denaturation of amyloid fibrils using supercooling conditions (−15°C; Kim et al., 2008, 2009). These oligomers were found to have structures intermediate between that of monomers and fibrils and to be able to disrupt lipid membranes. Interestingly, these oligomers were able to elongate and form fibrils efficiently, a feature supported by later works where cold denatured fibrillar samples were able to reassemble to form fibrils upon an increase in temperature (Bousset et al., 2013; Ikenoue et al., 2014).

We have also observed the presence of monomeric and oligomeric species following disaggregation of low concentrations of preformed fibrillar samples kept at 37°C over an extended period of several weeks (Cremades et al., 2012). While the oligomers dissociate into monomers after longer incubation times, the oligomers formed at the beginning of the disaggregation process appear to be clearly similar to the toxic β-sheet oligomers identified during the formation of fibrils (Cremades et al., 2012).

4.4 Oligomers Generated Upon Binding to Lipid Membranes

Small oligomeric structures have been suggested to form on binding of αS to lipid membranes (Cole et al., 2002; Ding et al., 2002; Eichmann et al., 2016; Tsigelny et al., 2007, 2012). The formation of annular oligomers from spherical oligomeric species has been observed using TEM following incubation of αS with microsomal vesicles obtained from rat brains (Ding et al., 2002). Using a variety of chemical cross-linking methods, αS was found to form small oligomers within cells, mainly dimers and trimers, which preferentially bind to lipid droplets and cell membranes (Cole et al., 2002). A population of αS oligomeric species, otherwise known as αS lipoprotein (nano)particles to indicate their relationship to high-density lipoproteins formed by human apolipoproteins, was generated by removing the detergent from a preincubated mixture of αS and phosphatidylserine lipids dissolved in sodium cholate, followed by the isolation of the nanoparticles using size-exclusion chromatography (Eichmann et al., 2016). This stable and homogenous population of nanoparticles have been reported to possess largely helical secondary structure, and was found by means of chemical cross-linking to be composed of c. 8–10 αS monomers in addition to anionic phosphatidylserine lipids.

4.5 Features of αS Oligomers Identified In Vivo

4.5.1 Detection by Antibodies

Conformational antibodies that bind specifically to oligomers have been widely used to detect their presence in vivo. One of these, the A11 antibody, binds to amyloidogenic oligomers of many different proteins (Kayed et al., 2003), while others are specific of one protein, such as FILA-1, which interacts specifically with β-sheet-rich oligomers and fibrils of αS (Paleologou et al., 2009). The use of A11 and FILA-1 antibodies has, for example, led to the observation of αS oligomers accumulating in the neuronal ER of aged A53T αS transgenic mice (Colla et al., 2012). Overexpression of WT αS in the immortalized MN9D dopaminergic-like cell model has also been observed to result in the formation of toxic SDS-stable αS oligomers that can be detected by the A11 antibody (Feng et al., 2010). These oligomers appear to possess pore-like structures that accumulate at cell membranes, resulting in an increase in membrane conductance in cells overexpressing αS, an effect that could be blocked by the direct extracellular addition of the anti-αS antibody Syn-1. Nevertheless, further work is required to establish directly the link between specific oligomers and the observed cellular effects, as other species would undoubtedly be present in such heterogeneous samples.

4.5.2 Detection by Fluorescence Assays

The formation of EGFP-tagged oligomers consisting of less than 10 monomers has been observed in live SH-SY5Y cells overexpressing αS, using a method referred to as the "Number and Brightness Method" (Plotegher et al., 2014), which is based on fluorescence fluctuation analysis to determine the dimensions of the fluorescent molecules observed. These oligomers were found to be formed when the αS concentration exceeded a threshold concentration of c. 90 nM, and were observed to be partially sequestered by lysosomes and to alter mitochondria morphology (Plotegher et al., 2014). An alternative approach, referred to as "αS proximity ligation assay," has been reported to recognize specifically αS oligomers in vivo (Roberts et al., 2015). This assay is based on the use of a primary antibody recognizing the protein of interest, αS in this case, and a secondary antibody that is coupled with a short DNA strand, which can serve as a platform for DNA ligation and amplification using labeled complementary oligonucleotides when the proteins of interest are in close proximity. Using this assay, an extensive diffuse deposition of αS oligomers in postmortem brain tissue from patients with PD was often localize to neuroanatomical regions mildly affected in the brains of PD patients in the absence of LBs. Interestingly, these oligomers showed proteinase K resistance intermediate between that of the physiological, presynaptic αS species (proteinase K sensitive) and the highly organized αS fibrils within the LBs (proteinase K resistant), suggesting that this distinctive oligomeric conformer of αS is associated and specific of PD (Roberts et al., 2015).

4.5.3 Purification of Cell Lysates

αS aggregates can be isolated from postmortem brain tissue by means of detergent extraction. Using this method, samples of purified αS aggregates found in postmortem cortical tissue from patients with MSA were obtained and they were found to release annular particles when partially dissociated by treatment with mild concentrations of detergent (Pountney et al., 2004). In a separate study, soluble αS oligomers, which coeluted with the 26S proteasome in PC12 cells stably expressing αS, were found to contain

10–30 monomers based on native PAGE gel analysis. The presence of β-sheet structure in these oligomers was demonstrated by its disruption following the addition of Congo Red, hence, restoring proteasome activity (Emmanouilidou et al., 2010).

Although αS has long been thought to be present in solution in an intrinsically disordered monomeric state (Bertoncini et al., 2005; Dedmon et al., 2005; Fauvet et al., 2012; Lashuel et al., 2013), the possibility has been raised recently that it might occur in a tetrameric form in vivo from studies of endogenous αS purified and analyzed under nondenaturing conditions from a wide range of neuronal and nonneuronal cell lines, brain tissue, and living human cells (Bartels et al., 2011; Dettmer et al., 2015b). The results suggested that αS could exist in large part as a folded tetramer with α-helical structure and a greater lipid-binding capacity than recombinant monomeric αS, and that the tetramers were largely destabilized upon cell lysis, requiring them to be cross-linked in order to be observed and studied outside the intact cell (Dettmer et al., 2013). Tetrameric αS has also been reported to be stable outside cells when αS was N-terminally acetylated and purified using a mild protocol involving the use of the nonionic detergent BOG, although lipids or detergent molecules could remain associated with this material (Trexler and Rhoades, 2012), and by purification of αS from *E. coli* using nondenaturing conditions in the presence of 0.1% BOG and using cross-linking techniques (Wang et al., 2011).

Systematic mutagenesis was also carried out to probe the self-assembly of αS as tetrameric species, and showed unexpectedly that a series of 14 sequential 10-residue deletions in the 140-residue protein did not distort the formation of tetramers, although insertion of certain in-register missense mutations into the highly conserved repeat motifs (consensus: KTKEGV) as well as PD-associated point mutations appeared to suppress tetramer formation (Dettmer et al., 2015a,b), providing support for previous speculations on the existence of such species (Kara et al., 2013). This tetrameric structure displayed little or no tendency to aggregate, leading to the hypothesis that destabilization of the tetramer precedes αS misfolding in vivo. It has also been suggested that there are metastable tetrameric conformers as well as stable monomers present in the human brain, with up to 70% of native cytosolic αS present in the tetrameric α-helical form (Gould et al., 2014). Attempts to verify independently the existence of such tetrameric conformation have not, however, been successful and their existence is still a topic of intense debate in the field (Lashuel et al., 2013). NMR studies of αS in intact cells have revealed that αS remains largely disordered in its soluble state at normal conditions (Lashuel et al., 2013; Theillet et al., 2016; Waudby et al., 2013). This tetrameric form of αS has been associated with the possible functional activity of the protein and, not as part of the ensemble of oligomeric conformations of αS, generated during amyloid self-assembly.

5. RELATIONSHIPS BETWEEN DIFFERENT TYPES OF AMYLOID SPECIES

In order to understand the nature, origin, and relevance of the different types of amyloid species that can be formed by αS, it is necessary to identify the various different species present under different conditions, distinguish their common and distinctive features, and establish possible mechanistic links between them. With that aim, we have tried to gather together the key structural information on the different αS amyloid species, particularly the oligomeric forms, that has been published to date, and that we have summarized in Section 4. Because of the variety of conditions used to study αS oligomers, a multitude of apparently different species have been reported, making it extremely challenging to rationalize their intrinsic characteristics in the context of the differences in the solution conditions and protocols used to stabilize the various species, and of the scarce and diverse information provided about them. We have tried, however, to extract some general features between the various species, with the aim of classifying the major different types of species that exist and generating a global overview of αS oligomerization and fibril formation. Given the currently more detailed understanding for the structural features and mechanisms of formation of the amyloid species, both oligomeric and fibrillar forms, generated in the absence of any apparent modifying chemical compound or macromolecule, we have focused our discussion on these species, although an attempt has been made to establish relationships between the multitude of amyloid species generated under different conditions.

5.1 Relationships Between Different Oligomeric Species

From the information discussed in Section 4, it seems apparent that a variety of conditions can trigger the self-assembly of αS, with the result being an accumulation of different types of aggregates, some with fibrillar appearance but others more characteristic of soluble oligomeric species. It is particularly challenging to obtain a clear understanding of the relationship between the structure and properties of oligomers in the presence of additional molecules, since the binding of such molecules will undoubtedly modify to some degree the physicochemical properties of the various species and affect their ability to grow and elongate. In addition, the presence of additional molecules could affect currently available techniques to determine accurately the structural properties of the different aggregates. In Table 1, we summarize the major features of the various oligomeric species reported to date, and the key characteristics of the conditions under which they were generated and observed. Although, as we have explained earlier, it is difficult to establish commonalities and relationships between the different species reported to be stabilized in different ways, it seems clear that under some conditions, including the presence of dopamine, some drugs such as selegiline and certain polyphenolic compounds, act to stabilize disordered oligomers that appear to be nontoxic to neuronal cells. In addition, other conditions, such as the presence of certain metal ions and aldehydes, induce the formation and accumulation of oligomers with a significant degree of β-sheet structure (although with significant variations in their morphologies) that can induce significant level of toxicity in cells.

In the case of oligomeric species that are formed in the absence of any added compound or without any chemical modification, the majority of the species reported have been found to possess a significant degree of β-sheet structure and are typically toxic to cells. This β-sheet structure in the oligomers trapped through lyophilization appears to be largely antiparallel in nature (Celej et al., 2012; Chen et al., 2015; Gallea and Celej, 2014) and the oligomers are unable to seed fibril formation efficiently (Chen et al., 2015; Lorenzen et al., 2014a). By contrast, the oligomers generated from fibril disaggregation processes, including cold denaturation of fibrils, have been found to be able to reassemble and elongate to form parallel β-sheet fibrils (Kim et al., 2009). Both types of oligomers appear to have a content of β-sheet structure that is greater than the monomeric species but lower than the fibrillar species (Celej et al., 2012; Chen et al., 2015; Giehm et al., 2011; Kim et al., 2009; Lorenzen et al., 2014a) and to possess a degree of resistance to proteinase K degradation that is between that present in the monomeric and the fibrillar forms (Chen et al., 2015; Cremades et al., 2012). Both types also bind typically to the A11 antibody (Celej et al., 2012; Chen et al., 2015; Kim et al., 2009) as well as to lipid membranes; in the latter case, the binding causing major structural perturbations to the lipid bilayer and to induce membrane permeabilization (Chen et al., 2015; Giehm et al., 2011; Kim et al., 2009; Lorenzen et al., 2014a), and some have been reported to have an annular/cylindrical appearance (Chen et al., 2015; Lashuel et al., 2002b).

Table 1 Summary of the Properties of αS Oligomers

Morphology	Structure	Seeding Capability	Toxicity: Membrane	Toxicity: Cells	Refs.
Oligomers identified during in vitro fibril formation					
Spherical (AFM)	—	"Yes"	—	—	Conway et al. (1998, 2000a,b)
Annular (TEM)	β-Sheet (CD)	"Yes"	—	—	Lashuel et al. (2002a,b)
	Type A: disordered Type B: intermediate structure between monomers and fibrils (smFRET)	Yes	—	Type A: no Type B: yes (ROS)	Cremades et al. (2012)
—	—	Yes	—	—	Hu et al. (2016)
Oligomers stabilized through protein lyophilization					
Annular (TEM)	Antiparallel β-sheet (FTIR)	No	—	—	Celej et al. (2012)
Spherical (AFM); spherical + annular (TEM)	β-Sheet (CD)	—	Yes	—	van Rooijen et al. (2009)
Small: spherical + annular (TEM) Large: elongated short fibrils (TEM)	Antiparallel β-sheet (FTIR)	No	Yes	—	Lorenzen et al. (2014a)
Annular (TEM)	C-terminus remains disordered (NMR)	—	Yes	Yes (MTT, trypan)	Lorenzen et al. (2014b)
Closed wreath	—	No	Yes	—	Giehm et al. (2011)
—	—	—	—	—	Zijlstra et al. (2012)

Continued

Table 1 Summary of the Properties of αS Oligomers—cont'd

Morphology	Structure	Seeding Capability	Toxicity		Refs.
			Membrane	Cells	
Annular (TEM, cryo-EM)	Antiparallel β-sheet (FTIR)	No	Yes	Yes (ROS)	Chen et al. (2015)
Spherical (AFM)	β-Sheet (CD)	—	Yes	—	Volles et al. (2001)
Spherical + annular (TEM)	Antiparallel β-sheet (FTIR)	—	—	—	Gallea and Celej (2014)
Spherical (AFM); annular with membrane (AFM)	—	—	—	—	Ding et al. (2002)
~Residues 39–75 protected, C-terminus remains disordered (HDX)	—	—	—	—	Paslawski et al. (2014b)
Annular (TEM)	Antiparallel β-sheet (FTIR)	Yes (forms larger worm-like structure)	Yes	—	Paslawski et al. (2014a)
—		—	Yes	—	Stefanovic et al. (2015)
Oligomers stabilized by chemical compounds					
Polyphenolic compounds—Baicalein					
Spherical + annular (AFM)	—	No	—	—	Zhu et al. (2004)
Spherical (AFM)	Antiparallel β-sheet (FTIR)	No	No	—	Hong et al. (2008)

—	—	—	—	Reduces toxicity (MTT, LDH)	Lu et al. (2011)
Polyphenolic compounds—EGCG					
Spherical (TEM)	Disordered	No	—	No (MTT, LDH)	Bieschke et al. (2010) and Ehrnhoefer et al. (2008)
Effect of EGCG on lyophilized oligomer					
—	Less flexible C-terminus	—	Inhibits membrane permeabilization	Reduced toxicity (MTT, trypan)	Lorenzen et al. (2014b)
Polyphenolic compounds—Gallic acid					
—	—	No	—	No (MTT)	Ardah et al. (2014)
Drugs—Rifampicin					
Spherical/annular/short fib (TEM)	Disordered	No	—	No	Li et al. (2005)
Drugs—Selegiline					
Amorphous/annular species (TEM)	Disordered + some β-sheet content	No	—	No (DAPI)	Braga et al. (2011)
Drugs—FN075					
Spherical (AFM/TEM)	β-Sheet rich; C-terminus disordered	Yes	Yes	—	Horvath et al. (2012), Cegelski et al. (2009), and Nors Perdersen et al. (2015)
Metals					
(Ca^{2+}) Aggregates detected in gel	—	—	—	—	Nielsen et al. (2001)

Continued

Table 1 Summary of the Properties of αS Oligomers—cont'd

Morphology	Structure	Seeding Capability	Toxicity		Refs.
			Membrane	Cells	
(Range of metal ions) Annular/spherical (AFM)	—	Yes	—	—	Lowe et al. (2004)
(Cu^{2+})$^{-}$	—	—	—	Yes (MTT)	Wang et al. (2010)
(Cu^{2+}) Amorphous "stellate" (TEM)	—	—	—	Yes (MTT)	Wright et al. (2009)
(Fe^{2+}, Cu^{2+} with H_2O_2) Aggregates detected in gel	—	—	—	—	Hashimoto et al. (1999) and Paik et al. (2000)
(Fe^{2+}, DTT) Annular, worm-like structure (TEM)	β-Sheet (CD)	No	—	—	Cole et al. (2005)
Dopamine					
Spherical (AFM)	Disordered (CD, FTIR)	No	—	—	Li et al. (2004)
Spherical (AFM)	Largely disordered with some β-sheet (CD, FTIR)	No	—	—	Norris et al. (2005)
Annular, rod shaped (TEM)	Largely disordered (CD)	No	—	—	Cappai et al. (2005)
—	—	—	—	Yes (DA neuron loss)	Burke et al. (2008)
—	Largely disordered (CD)	No, $\geq$ pH 7 Yes, pH 4	No	—	Pham and Cappai (2013) and Pham et al. (2009)

Rod shaped (TEM)	—	—	—	Yes (inhibit SNARE-mediated vesicle lipid mixing)	Choi et al. (2013)
Oligomers stabilized through chemical modifications in protein					
Oxidation					
Annular (AFM, TEM)	—	No	—	No (dopamine uptake, GABA uptake)	Zhou et al. (2010)
Spherical (AFM)	—	No	—	No (MTS)	Xiang et al. (2013)
Nitration					
Aggregates detected in gel	—	—	—	—	Souza et al. (2000)
Spherical (AFM); annular (TEM)	β-Sheet (CD)	No at pH 7.5; Yes at pH 3	—	—	Uversky et al. (2005)
Spherical (AFM)	—	No	—	No (MTS)	Xiang et al. (2013)
Formation of protein adducts					
(HNE/ONE)	Antiparallel β-sheet (FTIR)	—	—	—	Cai et al. (2015)
(ONE) Spherical (AFM)	β-Sheet (CD)	No	—	Yes (MTT)	Nasstrom et al. (2009, 2011)
(HNE) Curved protofibril like	β-Sheet (CD)	No	—		
(HNE) Spherical (AFM)	β-Sheet (CD) antiparallel β-sheet (FTIR)	No	—	Yes (decrease in dopamine/GABA uptake)	Qin et al. (2007)

Continued

Table 1 Summary of the Properties of αS Oligomers—cont'd

Morphology	Structure	Seeding Capability	Toxicity		Refs.
			Membrane	Cells	
(HNE) Spherical (AFM)	—	No	—	Yes (ROS, MTS, caspase-3)	Xiang et al. (2013)
(HNE) Spherical, short fibrils (AFM)	Disordered (CD)	Yes	—	—	Bae et al. (2013)
(MGO/GO) Spherical (AFM)	β-Sheet (FTIR/CD)	No	—	No (cell proliferation assay)	Lee et al. (2009)
(MGO/GO) Amorphous (TEM)	—	—	—	—	Padmaraju et al. (2011)
Oligomers generated upon fibril disaggregation					
Cold denaturation					
Spherical (AFM); annular (TEM)	Intermediate between monomer and fibrils (ssNMR)	Yes	Yes	—	Kim et al. (2008, 2009)
Annular (TEM)	—	Yes	—	—	Bousset et al. (2013)
—	—	Yes	—	—	Ikenoue et al. (2014)
Absence of added monomeric protein					
—	Similar FRET efficiencies as toxic type B oligomer	—	—	—	Cremades et al. (2012)
Oligomers generated upon binding to lipid membranes					
—	—	—	—	—	Cole et al. (2002)

Annular upon membrane binding (TEM)	—	—	—	—	Ding et al. (2002)
Annular	—	—	Yes	—	Tsigelny et al. (2007, 2012)
Spherical/ellipsoidal (cryo-EM) with low-density inner region	Helical (CD/ssNMR)	—	—	—	Eichmann et al. (2016)
Oligomers identified in vivo					
Antibody					
Proposed pore-like structure	—	—	Yes	Yes (MTT)	Feng et al. (2010)
—	—	—	—	Yes (ER stress)	Colla et al. (2012)
Fluorescence assays					
—	Intermediate PK resistance between monomer and fibrils	—	—	—	Roberts et al. (2015)
—	—	—	—	Yes	Plotegher et al. (2014)
Purification of cell lysates					
Annular	—	—	—	Found in MSA brain	Pountney et al. (2004)
—	β-Sheet (Congo Red disruption)	—	—	Yes	Emmanouilidou et al. (2010)

Although no high-resolution structural data have yet been obtained for αS oligomers formed in vivo, some of the characteristics reported so far are very similar to those of the β-sheet-rich oligomers formed in vitro, both during the fibril formation reaction and through the lyophilization process, notably a significant content of β-sheet structure and containing a relatively small number of αS molecules (10–30 molecules; Emmanouilidou et al., 2010) or less than 10 monomers (Plotegher et al., 2014), and a pore-like morphology (Feng et al., 2010; Pountney et al., 2004). The oligomers formed in vivo also accumulated on membranes and generated membrane permeability (Feng et al., 2010), bind to the A11 (Kayed et al., 2003) and the FILA-1 (Paleologou et al., 2009) antibodies (Colla et al., 2012), are detergent resistant (Campbell et al., 2001; Feng et al., 2010; Pountney et al., 2004), possess proteinase K resistance intermediate between that of the monomer and the fibrils (Roberts et al., 2015), and are ~20–60 nm in diameter and ~12 nm in height (Pountney et al., 2004).

Although some common features can be recognized between the different oligomeric species, significant structural, physicochemical, and pathological variations appear to exist between the various species and much more detailed study needs to be carried out in order to elucidate the similarities and differences between the oligomers formed in the various ways discussed earlier. The apparent variability in the type of oligomeric species generated by the different methods is highly to reflect the presence of multiple aggregation pathways, although the data collected so far on the various αS oligomeric species suggest that it is likely that only a small number of preferred self-assembly pathways exist and that some of the differences between oligomers observed could be related with the details of the interactions with any molecule that has been used to stabilize the oligomers.

5.2 Relationships Between Oligomeric and Fibrillar Aggregates

As discussed earlier in Section 4.3, fibril disaggregation, particularly through cold denaturation processes, can result in the accumulation of β-sheet-rich oligomeric species that can readily elongate under appropriate conditions to reform fibrils that have the typical parallel β-sheet structure (Kim et al., 2009), suggesting that the oligomeric species themselves have already a predominance of parallel β-sheet structure. By contrast, other conditions, such as lyophilization, result in the accumulation of similar oligomeric species to those observed during fibril formation and disaggregation (Cremades et al., 2012; Iljina et al., 2016), but with a preference for the antiparallel β-sheet geometry (Celej et al., 2012; Chen et al., 2015; Gallea and Celej, 2014), suggesting that these species are kinetically trapped and hence accumulate as a consequence of the inability of the antiparallel β-sheet geometry to elongate efficiently under the conditions where fibrils readily form in vitro (Chen et al., 2015).

At a fundamental level, the rapid elongation rate of small fibrils in the presence of monomeric protein hampers the detailed study of the oligomers that are transient intermediate species, some of them with parallel β-sheet structure. The ability to produce and isolate the trapped oligomers with antiparallel β-sheet structure has, however, provided an opportunity to gain detailed insights into the nature and structure of both species. We have recently carried out a detailed study of the structural properties of the major species present in enriched samples of oligomers generated using lyophilization. These oligomers contain antiparallel β-sheet structure, and are in slow exchange with any monomeric protein present in solution and have remarkably slow rates of elongation under normal conditions of fibril formation (Chen et al., 2015). We were therefore able to perform a detailed analysis of the sizes and structures of the ensemble of oligomeric species using a wide variety of biophysical techniques including analytical ultracentrifugation, cryo-EM, and circular dichroism, fluorescence and Fourier transform infrared spectroscopy. The results indicated that each sample contained a group of oligomeric species having the same structural architecture and differing primarily in their size (10–90 mers, with an average of c. 25–30 mers), and the degree of β-sheet content, which we observed to depend primarily on the size of the oligomer.

The 3D image reconstruction that we obtained from cryo-EM for these oligomers reveals a common cylindrical architecture with a central cavity, which has remarkable similarities to the structural models of amyloid fibrils of a variety of proteins, despite their differences in amino acid sequences (Chen et al., 2015; Chiti and Dobson, 2006). The similarities between the fibrillar and the oligomeric structural architectures suggest, therefore, that similar types of interactions to those that stabilize amyloid fibrils are likely to be responsible for the initial acquisition of β-sheet structure in the oligomeric species. This rudimentary element of cross-β structure that we have found in the trapped antiparallel β-sheet oligomers could be well adopted by parallel β-sheet oligomers generated during the self-assembly of αS into amyloid fibrils. Indeed, oligomeric cylindrical species formed during fibril formation observed by TEM with remarkable similarities to the 3D cryo-EM model of antiparallel β-sheet oligomers have been reported (Lashuel et al., 2002b; Quist et al., 2005), suggesting that parallel β-sheet oligomers could well have a similar architecture, and hence that the parallel and antiparallel β-sheet oligomers are indeed analogous in their global structure.

Analysis of the relationships between the size and β-sheet content of the trapped antiparallel β-sheet oligomers has revealed that there is a strong linear correlation between the two parameters: the larger the oligomers, the higher the β-sheet content (Fig. 5A) (Chen et al., 2015). In agreement with this correlation, that suggests a sequential acquisition of β-sheet structure upon oligomer growth, an earlier study of oligomers formed by Aβ40 showed that the initial low-molecular-weight oligomers of Aβ40 have a less extensive β-sheet structure, judging from H/D exchange data, than of the larger intermediate oligomers that form later in the aggregation process (Qi et al., 2008).

An important finding in our recent study (Chen et al., 2015) is that we were unable to detect oligomers smaller than 10 mers, even under disaggregating conditions involving different concentrations of urea, indicating that this type of oligomer is unstable below a specific size (Fig. 5A). Under strongly denaturing conditions, we observed the loss of structure in the oligomeric species in addition to their overall destabilization, with an apparent minimum β-sheet content of c. 10% in the smallest detectable oligomers, that in the absence of denaturant would be expected to have c. 25% of β-sheet structure (Fig. 5A). Taken together, these findings indicate that there is a minimum size and β-sheet content that is needed to stabilize this type of oligomers and, therefore, that the smallest oligomers at the early stages of fibril formation under physiological conditions, in the absence of added organic molecules or lipids, are likely to have a largely disordered structure. These results are in agreement with our earlier findings from single-molecule fluorescence experiments in which we were able to observe and characterize the structural conversion of the initially formed oligomers, when they reached a certain size, into more compact and stable structures during the early stages of αS amyloid aggregation (Cremades et al., 2012).

The general picture emerging from a significant number of studies on αS, therefore, is that its aggregation, in the absence of other molecules that could influence its self-assembly through specific interactions, typically proceeds through a nucleated conformational conversion process in which the initially formed, predominantly disordered, oligomers undergo subsequent conformational rearrangements in which the cross-β amyloid structure is acquired (Fig. 5B). In those cases in which αS aggregation is promoted

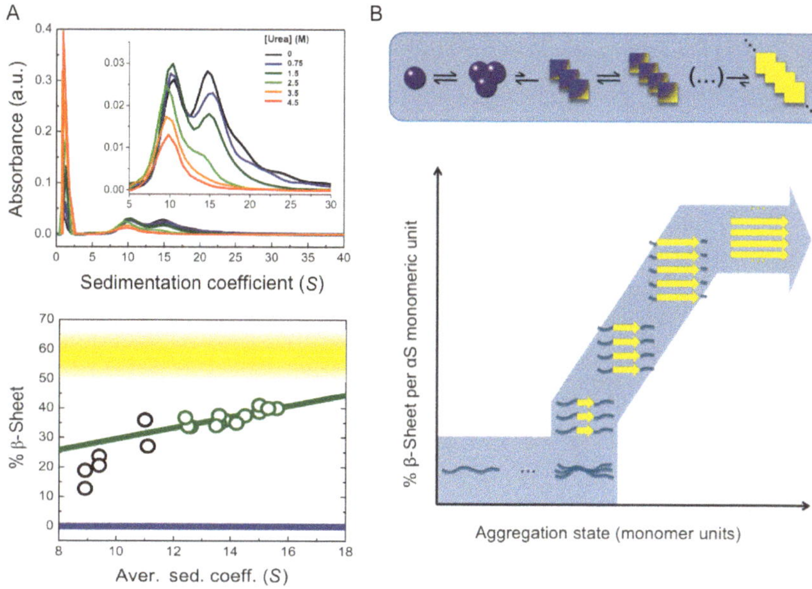

Fig. 5 A mechanism for the initial acquisition of the amyloid structure in αS in the absence of any modifying agent. Under physiologically relevant conditions and in the absence of any modifying molecule, αS has been shown to follow primarily a nucleation–conversion model, where the conversion from the initially formed oligomers to more compact and stable oligomers is remarkably slow (Cremades et al., 2012; Iljina et al., 2016), although the details of the structural conversions that take place during αS fibril formation remain to be established. The analogy between trapped antiparallel β-sheet oligomers that we have characterized recently in detail with parallel β-sheet oligomers formed during the direct formation of β-sheet fibrils suggests that the initially generated oligomers are essentially disordered. After increasing in size, however, they acquire β-sheet structure, which extends during oligomer growth, generating a continuous array of oligomeric species with differing sizes and β-sheet contents and geometries but with similar structural architecture. (A) Analysis of the dependency on size of the structure of the oligomers trapped during lyophilization, which have been demonstrated to be analogous to those found during fibril formation (Cremades et al., 2012; Iljina et al., 2016; Lashuel et al., 2002b; Quist et al., 2005), suggests sequential folding into the fully formed cross-β structure upon oligomer growth. In addition, this analysis suggests a threshold size of oligomers (c. 10 mers) in order to be able to acquire stable β-sheet structure, with a minimum β-sheet content of c. 10% (Chen et al., 2015). This information was obtained by varying the size of the ensemble of oligomers (depicted as a function of the sedimentation coefficient from analytical ultracentrifugation sedimentation velocity experiments) using different concentrations of urea as the denaturing agent (*upper graph* in A). The average sedimentation coefficient of the ensemble of oligomers for each urea concentration can then be correlated with the average percentage of β-sheet content displayed by the oligomers (*bottom graph* in A). The experimental data obtained under mildly denaturing conditions ([urea] < 1.5 M; shown by *green points*) were used to establish a structure–size correlation (*green line*).

(Continued)

by lipids or certain molecules, however, the predominant aggregation pathway could be significantly altered, leading to differences in the type of structural aggregates formed (Galvagnion et al., 2015).

5.3 Multiplicity of Misfolding Pathways

In addition to the existence of specific self-assembly pathways that vary with solution conditions and appear to generate structurally distinct types of aggregates, αS can form oligomeric species with structural features that are similar to each other and indeed even similar to that of the mature fibrils, but with significantly different arrangement of their corresponding β-strands. Thus, some species could have largely parallel β-strands within the β-sheets that can readily form fibrils, and others could have largely antiparallel β-strands such that they are kinetically trapped, as they grow into amyloid fibrils only very slowly. Such a situation reveals an inherent multiplicity in the process of protein misfolding and aggregation. In the light of our experimental findings, and those of others, we propose that the initial coalescence of several αS polypeptide chains is followed by the formation of intermolecular hydrogen bonds that result in the acquisition of β-sheet structure in specific regions of the protein. It is, however, likely that there will be differences in the regions of the protein included in the early β-sheet structure, the lengths of the β-strands of the folding cores, and the permutations of interstrand hydrogen bonding interactions, resulting in the formation of oligomers with a variety of differences in the details of the structure but with the same type of overall core architecture that are precursors of different structural forms or polymorphs of amyloid fibrils.

Fig. 5—Cont'd Under more strongly denaturing conditions, the high concentration of urea present in the sample starts to unfold the core of the oligomers, in addition to disaggregating them, and was therefore excluded from the correlation analysis. These data, however, indicate the apparent minimum size and stable β-sheet content that αS can acquire in this structural group of oligomers. Below this threshold size, these oligomers no longer appear to be stable, as the acquisition of β-sheet structure is no longer stabilized by the required number of intermolecular hydrogen bonds in the oligomer. These experimental data, together with those obtained from single-molecule fluorescence experiments, where a conversion from initially formed disordered oligomers of αS to more compact and stable oligomers at oligomer sizes of c. 5–15 mers was observed directly (Cremades et al., 2012), are in agreement with a model of nucleation–conversion–sequential folding upon polymerization (B). *The figures in Panel (A) have been adapted with permission from Chen, S.W., Drakulic, S., Deas, E., Ouberai, M., Aprile, F.A., Arranz, R., Ness, S., Roodveldt, C., Guilliams, T., De-Genst, E.J., et al., 2015. Structural characterization of toxic oligomers that are kinetically trapped during α-synuclein fibril formation. Proc. Natl. Acad. Sci. U.S.A. 112, E1994–E2003.*

Some such precursor species will, however, be able to rearrange as their size increases to generate the structure of lowest energy under given conditions (Petty and Decatur, 2005). Others, however, will lead to the formation of structurally distinct but energetically similar fibril polymorphs. Indeed, the comparison of a variety of such polymorphs of αS has been possible by ssNMR and the results show the existence of differences in the interstrand organization of the protein within the fibrillar cores (Gath et al., 2014). The formation of precursors with a high fraction of interactions that differ from those in the fibrillar species (for example, oligomeric species with an antiparallel β-structure when the fibrils have parallel β-sheet arrangement) can result in the accumulation of kinetically trapped amyloid oligomers that would require major structural rearrangements to enable them to elongate efficiently to generate the apparent fibrillar architecture (Chen et al., 2015). Indeed, interactions similar or dissimilar to those found in the fibrils appear to be present in the monomeric form of αS (Esteban-Martin et al., 2013), suggesting that both parallel and antiparallel structures are formed at least transiently in the early stages of αS self-assembly.

This multiplicity of misfolding pathways has interesting analogies to the multiplicity of folding pathways observed in studies of the conversion of unfolded states of proteins into compact globular structures. Such a picture suggests that a unifying framework exists to connect folding and misfolding processes and to reconcile apparently contradictory experimental data on the role that some oligomers play in the process of fibril formation. The concept of a multiplicity of assembly steps and associated pathways generates an ensemble of oligomers with a variety of β-sheet geometries and interstrand contacts that lead to differences in aggregation rates and associated toxicities. Such a picture also leads to the fascinating possibility that protein misfolding and aggregation within cells could generate species with differing pathological consequences depending on the nature of the stochastic processes occurring during the self-assembly process. Indeed, there is evidence that such processes are responsible for the variability in cellular vulnerability and pathology that are characteristic of the distinct types of synucleinopathies (Prusiner et al., 2015; Winner et al., 2011; Woerman et al., 2015). Overall, therefore, we can be optimistic that fundamental studies of the molecular origins of protein aggregation will invariably provide insights into the nature of its associated oligomeric states and potentially contribute to the development of new tools for the effective prevention or treatment of amyloid diseases.

REFERENCES

Appel-Cresswell, S., Vilarino-Guell, C., Encarnacion, M., Sherman, H., Yu, I., Shah, B., Weir, D., Thompson, C., Szu-Tu, C., Trinh, J., et al., 2013. Alpha-synuclein p.H50Q, a novel pathogenic mutation for Parkinson's disease. Mov. Disord. 28, 811–813.

Ardah, M.T., Paleologou, K.E., Lv, G., Abul Khair, S.B., Kazim, A.S., Minhas, S.T., Al-Tel, T.H., Al-Hayani, A.A., Haque, M.E., Eliezer, D., et al., 2014. Structure activity relationship of phenolic acid inhibitors of alpha-synuclein fibril formation and toxicity. Front. Aging Neurosci. 6, 197.

Arispe, N., Rojas, E., Pollard, H.B., 1993. Alzheimer disease amyloid beta protein forms calcium channels in bilayer membranes: blockade by tromethamine and aluminum. Proc. Natl. Acad. Sci. U.S.A. 90, 567–571.

Arrasate, M., Mitra, S., Schweitzer, E.S., Segal, M.R., Finkbeiner, S., 2004. Inclusion body formation reduces levels of mutant huntingtin and the risk of neuronal death. Nature 431, 805–810.

Auer, S., Meersman, F., Dobson, C.M., Vendruscolo, M., 2008. A generic mechanism of emergence of amyloid protofilaments from disordered oligomeric aggregates. PLoS Comput. Biol. 4, e1000222.

Baba, M., Nakajo, S., Tu, P.H., Tomita, T., Nakaya, K., Lee, V.M., Trojanowski, J.Q., Iwatsubo, T., 1998. Aggregation of alpha-synuclein in Lewy bodies of sporadic Parkinson's disease and dementia with Lewy bodies. Am. J. Pathol. 152, 879–884.

Bader, R., Bamford, R., Zurdo, J., Luisi, B.F., Dobson, C.M., 2006. Probing the mechanism of amyloidogenesis through a tandem repeat of the PI3-SH3 domain suggests a generic model for protein aggregation and fibril formation. J. Mol. Biol. 356, 189–208.

Bae, E.J., Ho, D.H., Park, E., Jung, J.W., Cho, K., Hong, J.H., Lee, H.J., Kim, K.P., Lee, S.J., 2013. Lipid peroxidation product 4-hydroxy-2-nonenal promotes seeding-capable oligomer formation and cell-to-cell transfer of alpha-synuclein. Antioxid. Redox Signal. 18, 770–783.

Baldwin, A.J., Knowles, T.P., Tartaglia, G.G., Fitzpatrick, A.W., Devlin, G.L., Shammas, S.L., Waudby, C.A., Mossuto, M.F., Meehan, S., Gras, S.L., et al., 2011. Metastability of native proteins and the phenomenon of amyloid formation. J. Am. Chem. Soc. 133, 14160–14163.

Bartels, T., Choi, J.G., Selkoe, D.J., 2011. alpha-Synuclein occurs physiologically as a helically folded tetramer that resists aggregation. Nature 477, 107–110.

Bemporad, F., Chiti, F., 2012. Protein misfolded oligomers: experimental approaches, mechanism of formation, and structure-toxicity relationships. Chem. Biol. 19, 315–327.

Bernado, P., Bertoncini, C.W., Griesinger, C., Zweckstetter, M., Blackledge, M., 2005. Defining long-range order and local disorder in native alpha-synuclein using residual dipolar couplings. J. Am. Chem. Soc. 127, 17968–17969.

Bertoncini, C.W., Jung, Y.S., Fernandez, C.O., Hoyer, W., Griesinger, C., Jovin, T.M., Zweckstetter, M., 2005. Release of long-range tertiary interactions potentiates aggregation of natively unstructured alpha-synuclein. Proc. Natl. Acad. Sci. U.S.A. 102, 1430–1435.

Bieschke, J., Russ, J., Friedrich, R.P., Ehrnhoefer, D.E., Wobst, H., Neugebauer, K., Wanker, E.E., 2010. EGCG remodels mature alpha-synuclein and amyloid-beta fibrils and reduces cellular toxicity. Proc. Natl. Acad. Sci. U.S.A. 107, 7710–7715.

Bitan, G., Kirkitadze, M.D., Lomakin, A., Vollers, S.S., Benedek, G.B., Teplow, D.B., 2003. Amyloid beta -protein (Abeta) assembly: Abeta 40 and Abeta 42 oligomerize through distinct pathways. Proc. Natl. Acad. Sci. U.S.A. 100, 330–335.

Bleiholder, C., Dupuis, N.F., Wyttenbach, T., Bowers, M.T., 2011. Ion mobility-mass spectrometry reveals a conformational conversion from random assembly to beta-sheet in amyloid fibril formation. Nat. Chem. 3, 172–177.

Bolognesi, B., Kumita, J.R., Barros, T.P., Esbjorner, E.K., Luheshi, L.M., Crowther, D.C., Wilson, M.R., Dobson, C.M., Favrin, G., Yerbury, J.J., 2010. ANS binding reveals common features of cytotoxic amyloid species. ACS Chem. Biol. 5, 735–740.

Bonini, N.M., Giasson, B.I., 2005. Snaring the function of alpha-synuclein. Cell 123, 359–361.

Bouchard, M., Zurdo, J., Nettleton, E.J., Dobson, C.M., Robinson, C.V., 2000. Formation of insulin amyloid fibrils followed by FTIR simultaneously with CD and electron microscopy. Protein Sci. 9, 1960–1967.

Bousset, L., Pieri, L., Ruiz-Arlandis, G., Gath, J., Jensen, P.H., Habenstein, B., Madiona, K., Olieric, V., Bockmann, A., Meier, B.H., et al., 2013. Structural and functional characterization of two alpha-synuclein strains. Nat. Commun. 4, 2575.

Braga, C.A., Follmer, C., Palhano, F.L., Khattar, E., Freitas, M.S., Romao, L., Di Giovanni, S., Lashuel, H.A., Silva, J.L., Foguel, D., 2011. The anti-Parkinsonian drug selegiline delays the nucleation phase of alpha-synuclein aggregation leading to the formation of nontoxic species. J. Mol. Biol. 405, 254–273.

Brender, J.R., Salamekh, S., Ramamoorthy, A., 2012. Membrane disruption and early events in the aggregation of the diabetes related peptide IAPP from a molecular perspective. Acc. Chem. Res. 45, 454–462.

Brundin, P., Melki, R., Kopito, R., 2010. Prion-like transmission of protein aggregates in neurodegenerative diseases. Nat. Rev. Mol. Cell Biol. 11, 301–307.

Bucciantini, M., Giannoni, E., Chiti, F., Baroni, F., Formigli, L., Zurdo, J., Taddei, N., Ramponi, G., Dobson, C.M., Stefani, M., 2002. Inherent toxicity of aggregates implies a common mechanism for protein misfolding diseases. Nature 416, 507–511.

Buell, A.K., Galvagnion, C., Gaspar, R., Sparr, E., Vendruscolo, M., Knowles, T.P., Linse, S., Dobson, C.M., 2014. Solution conditions determine the relative importance of nucleation and growth processes in alpha-synuclein aggregation. Proc. Natl. Acad. Sci. U.S.A. 111, 7671–7676.

Burke, W.J., Kumar, V.B., Pandey, N., Panneton, W.M., Gan, Q., Franko, M.W., O'Dell, M., Li, S.W., Pan, Y., Chung, H.D., et al., 2008. Aggregation of alpha-synuclein by DOPAL, the monoamine oxidase metabolite of dopamine. Acta Neuropathol. 115, 193–203.

Burre, J., Sharma, M., Tsetsenis, T., Buchman, V., Etherton, M.R., Sudhof, T.C., 2010. Alpha-synuclein promotes SNARE-complex assembly in vivo and in vitro. Science 329, 1663–1667.

Cai, Y., Lendel, C., Osterlund, L., Kasrayan, A., Lannfelt, L., Ingelsson, M., Nikolajeff, F., Karlsson, M., Bergstrom, J., 2015. Changes in secondary structure of alpha-synuclein during oligomerization induced by reactive aldehydes. Biochem. Biophys. Res. Commun. 464, 336–341.

Campbell, B.C., McLean, C.A., Culvenor, J.G., Gai, W.P., Blumbergs, P.C., Jakala, P., Beyreuther, K., Masters, C.L., Li, Q.X., 2001. The solubility of alpha-synuclein in multiple system atrophy differs from that of dementia with Lewy bodies and Parkinson's disease. J. Neurochem. 76, 87–96.

Campioni, S., Mannini, B., Zampagni, M., Pensalfini, A., Parrini, C., Evangelisti, E., Relini, A., Stefani, M., Dobson, C.M., Cecchi, C., et al., 2010. A causative link between the structure of aberrant protein oligomers and their toxicity. Nat. Chem. Biol. 6, 140–147.

Campioni, S., Carret, G., Jordens, S., Nicoud, L., Mezzenga, R., Riek, R., 2014. The presence of an air-water interface affects formation and elongation of alpha-synuclein fibrils. J. Am. Chem. Soc. 136, 2866–2875.

Cappai, R., Leck, S.L., Tew, D.J., Williamson, N.A., Smith, D.P., Galatis, D., Sharples, R.A., Curtain, C.C., Ali, F.E., Cherny, R.A., et al., 2005. Dopamine promotes alpha-synuclein aggregation into SDS-resistant soluble oligomers via a distinct folding pathway. FASEB J 19, 1377–1379.

Carulla, N., Zhou, M., Arimon, M., Gairi, M., Giralt, E., Robinson, C.V., Dobson, C.M., 2009. Experimental characterization of disordered and ordered aggregates populated during the process of amyloid fibril formation. Proc. Natl. Acad. Sci. U.S.A. 106, 7828–7833.

Castellani, R., Smith, M.A., Richey, P.L., Perry, G., 1996. Glycoxidation and oxidative stress in Parkinson disease and diffuse Lewy body disease. Brain Res. 737, 195–200.

Castellani, R.J., Perry, G., Siedlak, S.L., Nunomura, A., Shimohama, S., Zhang, J., Montine, T., Sayre, L.M., Smith, M.A., 2002. Hydroxynonenal adducts indicate a role for lipid peroxidation in neocortical and brainstem Lewy bodies in humans. Neurosci. Lett. 319, 25–28.

Cegelski, L., Pinkner, J.S., Hammer, N.D., Cusumano, C.K., Hung, C.S., Chorell, E., Aberg, V., Walker, J.N., Seed, P.C., Almqvist, F., et al., 2009. Small-molecule inhibitors target Escherichia coli amyloid biogenesis and biofilm formation. Nat. Chem. Biol. 5, 913–919.

Celej, M.S., Sarroukh, R., Goormaghtigh, E., Fidelio, G.D., Ruysschaert, J.M., Raussens, V., 2012. Toxic prefibrillar alpha-synuclein amyloid oligomers adopt a distinctive antiparallel beta-sheet structure. Biochem. J. 443, 719–726.

Chartier-Harlin, M.C., Kachergus, J., Roumier, C., Mouroux, V., Douay, X., Lincoln, S., Levecque, C., Larvor, L., Andrieux, J., Hulihan, M., et al., 2004. Alpha-synuclein locus duplication as a cause of familial Parkinson's disease. Lancet 364, 1167–1169.

Chen, M., Margittai, M., Chen, J., Langen, R., 2007. Investigation of alpha-synuclein fibril structure by site-directed spin labeling. J. Biol. Chem. 282, 24970–24979.

Chen, S.W., Drakulic, S., Deas, E., Ouberai, M., Aprile, F.A., Arranz, R., Ness, S., Roodveldt, C., Guilliams, T., De-Genst, E.J., et al., 2015. Structural characterization of toxic oligomers that are kinetically trapped during alpha-synuclein fibril formation. Proc. Natl. Acad. Sci. U.S.A. 112, E1994–E2003.

Chiti, F., Dobson, C.M., 2006. Protein misfolding, functional amyloid, and human disease. Annu. Rev. Biochem. 75, 333–366.

Choi, J.G., Kim, H.G., Kim, M.C., Yang, W.M., Huh, Y., Kim, S.Y., Oh, M.S., 2011. Polygalae radix inhibits toxin-induced neuronal death in the Parkinson's disease models. J. Ethnopharmacol. 134, 414–421.

Choi, B.K., Choi, M.G., Kim, J.Y., Yang, Y., Lai, Y., Kweon, D.H., Lee, N.K., Shin, Y.K., 2013. Large alpha-synuclein oligomers inhibit neuronal SNARE-mediated vesicle docking. Proc. Natl. Acad. Sci. U.S.A. 110, 4087–4092.

Ciryam, P., Tartaglia, G.G., Morimoto, R.I., Dobson, C.M., Vendruscolo, M., 2013. Widespread aggregation and neurodegenerative diseases are associated with supersaturated proteins. Cell Rep. 5, 781–790.

Clayton, D.F., George, J.M., 1998. The synucleins: a family of proteins involved in synaptic function, plasticity, neurodegeneration and disease. Trends Neurosci. 21, 249–254.

Cohen, S.I., Vendruscolo, M., Dobson, C.M., Knowles, T.P., 2011a. Nucleated polymerization with secondary pathways. II. Determination of self-consistent solutions to growth processes described by non-linear master equations. J. Chem. Phys. 135, 065106.

Cohen, S.I., Vendruscolo, M., Dobson, C.M., Knowles, T.P., 2011b. Nucleated polymerization with secondary pathways. III. Equilibrium behavior and oligomer populations. J. Chem. Phys. 135, 065107.

Cohen, S.I., Vendruscolo, M., Dobson, C.M., Knowles, T.P., 2012. From macroscopic measurements to microscopic mechanisms of protein aggregation. J. Mol. Biol. 421, 160–171.

Cohen, S.I., Linse, S., Luheshi, L.M., Hellstrand, E., White, D.A., Rajah, L., Otzen, D.E., Vendruscolo, M., Dobson, C.M., Knowles, T.P., 2013. Proliferation of amyloid-beta42 aggregates occurs through a secondary nucleation mechanism. Proc. Natl. Acad. Sci. U.S.A. 110, 9758–9763.

Cole, N.B., Murphy, D.D., Grider, T., Rueter, S., Brasaemle, D., Nussbaum, R.L., 2002. Lipid droplet binding and oligomerization properties of the Parkinson's disease protein alpha-synuclein. J. Biol. Chem. 277, 6344–6352.
Cole, N.B., Murphy, D.D., Lebowitz, J., Di Noto, L., Levine, R.L., Nussbaum, R.L., 2005. Metal-catalyzed oxidation of alpha-synuclein: helping to define the relationship between oligomers, protofibrils, and filaments. J. Biol. Chem. 280, 9678–9690.
Colla, E., Jensen, P.H., Pletnikova, O., Troncoso, J.C., Glabe, C., Lee, M.K., 2012. Accumulation of toxic alpha-synuclein oligomer within endoplasmic reticulum occurs in alpha-synucleinopathy in vivo. J. Neurosci. 32, 3301–3305.
Comellas, G., Lemkau, L.R., Nieuwkoop, A.J., Kloepper, K.D., Ladror, D.T., Ebisu, R., Woods, W.S., Lipton, A.S., George, J.M., Rienstra, C.M., 2011. Structured regions of alpha-synuclein fibrils include the early-onset Parkinson's disease mutation sites. J. Mol. Biol. 411, 881–895.
Conway, K.A., Harper, J.D., Lansbury, P.T., 1998. Accelerated in vitro fibril formation by a mutant alpha-synuclein linked to early-onset Parkinson disease. Nat. Med. 4, 1318–1320.
Conway, K.A., Harper, J.D., Lansbury Jr., P.T., 2000a. Fibrils formed in vitro from alpha-synuclein and two mutant forms linked to Parkinson's disease are typical amyloid. Biochemistry 39, 2552–2563.
Conway, K.A., Lee, S.J., Rochet, J.C., Ding, T.T., Williamson, R.E., Lansbury Jr., P.T., 2000b. Acceleration of oligomerization, not fibrillization, is a shared property of both alpha-synuclein mutations linked to early-onset Parkinson's disease: implications for pathogenesis and therapy. Proc. Natl. Acad. Sci. U.S.A. 97, 571–576.
Conway, K.A., Rochet, J.C., Bieganski, R.M., Lansbury Jr., P.T., 2001. Kinetic stabilization of the alpha-synuclein protofibril by a dopamine-alpha-synuclein adduct. Science 294, 1346–1349.
Cooper, A.A., Gitler, A.D., Cashikar, A., Haynes, C.M., Hill, K.J., Bhullar, B., Liu, K., Xu, K., Strathearn, K.E., Liu, F., et al., 2006. Alpha-synuclein blocks ER-Golgi traffic and Rab1 rescues neuron loss in Parkinson's models. Science 313, 324–328.
Cremades, N., Cohen, S.I., Deas, E., Abramov, A.Y., Chen, A.Y., Orte, A., Sandal, M., Clarke, R.W., Dunne, P., Aprile, F.A., et al., 2012. Direct observation of the interconversion of normal and toxic forms of alpha-synuclein. Cell 149, 1048–1059.
Crowther, R.A., Daniel, S.E., Goedert, M., 2000. Characterisation of isolated alpha-synuclein filaments from substantia nigra of Parkinson's disease brain. Neurosci. Lett. 292, 128–130.
Dalfo, E., Ferrer, I., 2005. Alpha-synuclein binding to rab3a in multiple system atrophy. Neurosci. Lett. 380, 170–175.
Dalfo, E., Ferrer, I., 2008. Early alpha-synuclein lipoxidation in neocortex in Lewy body diseases. Neurobiol. Aging 29, 408–417.
Danzer, K.M., Haasen, D., Karow, A.R., Moussaud, S., Habeck, M., Giese, A., Kretzschmar, H., Hengerer, B., Kostka, M., 2007. Different species of alpha-synuclein oligomers induce calcium influx and seeding. J. Neurosci. 27, 9220–9232.
Danzer, K.M., Krebs, S.K., Wolff, M., Birk, G., Hengerer, B., 2009. Seeding induced by alpha-synuclein oligomers provides evidence for spreading of alpha-synuclein pathology. J. Neurochem. 111, 192–203.
Danzer, K.M., Ruf, W.P., Putcha, P., Joyner, D., Hashimoto, T., Glabe, C., Hyman, B.T., McLean, P.J., 2011. Heat-shock protein 70 modulates toxic extracellular alpha-synuclein oligomers and rescues trans-synaptic toxicity. FASEB J. 25, 326–336.
De Felice, F.G., Velasco, P.T., Lambert, M.P., Viola, K., Fernandez, S.J., Ferreira, S.T., Klein, W.L., 2007. Abeta oligomers induce neuronal oxidative stress through an N-methyl-D-aspartate receptor-dependent mechanism that is blocked by the Alzheimer drug memantine. J. Biol. Chem. 282, 11590–11601.

Dearborn, A.D., Wall, J.S., Cheng, N., Heymann, J.B., Kajava, A.V., Varkey, J., Langen, R., Steven, A.C., 2016. alpha-Synuclein amyloid fibrils with two entwined, asymmetrically associated protofibrils. J. Biol. Chem. 291, 2310–2318.
Deas, E., Cremades, N., Angelova, P.R., Ludtmann, M.H., Yao, Z., Chen, S., Horrocks, M.H., Banushi, B., Little, D., Devine, M.J., et al., 2016. Alpha-synuclein oligomers interact with metal ions to induce oxidative stress and neuronal death in Parkinson's disease. Antioxid. Redox Signal. 24, 376–391.
Dedmon, M.M., Lindorff-Larsen, K., Christodoulou, J., Vendruscolo, M., Dobson, C.M., 2005. Mapping long-range interactions in alpha-synuclein using spin-label NMR and ensemble molecular dynamics simulations. J. Am. Chem. Soc. 127, 476–477.
Del Mar, C., Greenbaum, E.A., Mayne, L., Englander, S.W., Woods Jr., V.L., 2005. Structure and properties of alpha-synuclein and other amyloids determined at the amino acid level. Proc. Natl. Acad. Sci. U.S.A. 102, 15477–15482.
Demuro, A., Mina, E., Kayed, R., Milton, S.C., Parker, I., Glabe, C.G., 2005. Calcium dysregulation and membrane disruption as a ubiquitous neurotoxic mechanism of soluble amyloid oligomers. J. Biol. Chem. 280, 17294–17300.
Der-Sarkissian, A., Jao, C.C., Chen, J., Langen, R., 2003. Structural organization of alpha-synuclein fibrils studied by site-directed spin labeling. J. Biol. Chem. 278, 37530–37535.
Desplats, P., Lee, H.J., Bae, E.J., Patrick, C., Rockenstein, E., Crews, L., Spencer, B., Masliah, E., Lee, S.J., 2009. Inclusion formation and neuronal cell death through neuron-to-neuron transmission of alpha-synuclein. Proc. Natl. Acad. Sci. U.S.A. 106, 13010–13015.
Dettmer, U., Newman, A.J., Luth, E.S., Bartels, T., Selkoe, D., 2013. In vivo cross-linking reveals principally oligomeric forms of alpha-synuclein and beta-synuclein in neurons and non-neural cells. J. Biol. Chem. 288, 6371–6385.
Dettmer, U., Newman, A.J., Soldner, F., Luth, E.S., Kim, N.C., von Saucken, V.E., Sanderson, J.B., Jaenisch, R., Bartels, T., Selkoe, D., 2015a. Parkinson-causing alpha-synuclein missense mutations shift native tetramers to monomers as a mechanism for disease initiation. Nat. Commun. 6, 7314.
Dettmer, U., Newman, A.J., von Saucken, V.E., Bartels, T., Selkoe, D., 2015b. KTKEGV repeat motifs are key mediators of normal alpha-synuclein tetramerization: their mutation causes excess monomers and neurotoxicity. Proc. Natl. Acad. Sci. U.S.A. 112, 9596–9601.
Devi, L., Raghavendran, V., Prabhu, B.M., Avadhani, N.G., Anandatheerthavarada, H.K., 2008. Mitochondrial import and accumulation of alpha-synuclein impair complex I in human dopaminergic neuronal cultures and Parkinson disease brain. J. Biol. Chem. 283, 9089–9100.
Ding, T.T., Lee, S.J., Rochet, J.C., Lansbury Jr., P.T., 2002. Annular alpha-synuclein protofibrils are produced when spherical protofibrils are incubated in solution or bound to brain-derived membranes. Biochemistry 41, 10209–10217.
Ding, H., Wong, P.T., Lee, E.L., Gafni, A., Steel, D.G., 2009. Determination of the oligomer size of amyloidogenic protein beta-amyloid(1-40) by single-molecule spectroscopy. Biophys. J. 97, 912–921.
Dobson, C.M., 1999. Protein misfolding, evolution and disease. Trends Biochem. Sci. 24, 329–332.
Dobson, C.M., 2003. Protein folding and misfolding. Nature 426, 884–890.
Ehrnhoefer, D.E., Bieschke, J., Boeddrich, A., Herbst, M., Masino, L., Lurz, R., Engemann, S., Pastore, A., Wanker, E.E., 2008. EGCG redirects amyloidogenic polypeptides into unstructured, off-pathway oligomers. Nat. Struct. Mol. Biol. 15, 558–566.
Eichmann, C., Campioni, S., Kowal, J., Maslennikov, I., Gerez, J., Liu, X., Verasdonck, J., Nespovitaya, N., Choe, S., Meier, B., et al., 2016. Preparation and characterization of stable alpha-synuclein lipoprotein particles. J. Biol. Chem. 291, 8516–8527.

Eisenberg, D., Jucker, M., 2012. The amyloid state of proteins in human diseases. Cell 148, 1188–1203.
El-Agnaf, O.M., Salem, S.A., Paleologou, K.E., Cooper, L.J., Fullwood, N.J., Gibson, M.J., Curran, M.D., Court, J.A., Mann, D.M., Ikeda, S., et al., 2003. Alpha-synuclein implicated in Parkinson's disease is present in extracellular biological fluids, including human plasma. FASEB J. 17, 1945–1947.
Emmanouilidou, E., Stefanis, L., Vekrellis, K., 2010. Cell-produced alpha-synuclein oligomers are targeted to, and impair, the 26S proteasome. Neurobiol. Aging 31, 953–968.
Esteban-Martin, S., Silvestre-Ryan, J., Bertoncini, C.W., Salvatella, X., 2013. Identification of fibril-like tertiary contacts in soluble monomeric alpha-synuclein. Biophys. J. 105, 1192–1198.
Fandrich, M., Dobson, C.M., 2002. The behaviour of polyamino acids reveals an inverse side chain effect in amyloid structure formation. EMBO J. 21, 5682–5690.
Fauvet, B., Mbefo, M.K., Fares, M.B., Desobry, C., Michael, S., Ardah, M.T., Tsika, E., Coune, P., Prudent, M., Lion, N., et al., 2012. alpha-Synuclein in central nervous system and from erythrocytes, mammalian cells, and Escherichia coli exists predominantly as disordered monomer. J. Biol. Chem. 287, 15345–15364.
Feng, L.R., Federoff, H.J., Vicini, S., Maguire-Zeiss, K.A., 2010. Alpha-synuclein mediates alterations in membrane conductance: a potential role for alpha-synuclein oligomers in cell vulnerability. Eur. J. Neurosci. 32, 10–17.
Fitzpatrick, A.W., Debelouchina, G.T., Bayro, M.J., Clare, D.K., Caporini, M.A., Bajaj, V.S., Jaroniec, C.P., Wang, L., Ladizhansky, V., Muller, S.A., et al., 2013. Atomic structure and hierarchical assembly of a cross-beta amyloid fibril. Proc. Natl. Acad. Sci. U.S.A. 110, 5468–5473.
Fusco, G., De Simone, A., Gopinath, T., Vostrikov, V., Vendruscolo, M., Dobson, C.M., Veglia, G., 2014. Direct observation of the three regions in alpha-synuclein that determine its membrane-bound behaviour. Nat. Commun. 5, 3827.
Gallea, J.I., Celej, M.S., 2014. Structural insights into amyloid oligomers of the Parkinson disease-related protein alpha-synuclein. J. Biol. Chem. 289, 26733–26742.
Galvagnion, C., Buell, A.K., Meisl, G., Michaels, T.C., Vendruscolo, M., Knowles, T.P., Dobson, C.M., 2015. Lipid vesicles trigger alpha-synuclein aggregation by stimulating primary nucleation. Nat. Chem. Biol. 11, 229–234.
Galvin, J.E., Uryu, K., Lee, V.M., Trojanowski, J.Q., 1999. Axon pathology in Parkinson's disease and Lewy body dementia hippocampus contains alpha-, beta-, and gamma-synuclein. Proc. Natl. Acad. Sci. U.S.A. 96, 13450–13455.
Gath, J., Habenstein, B., Bousset, L., Melki, R., Meier, B.H., Bockmann, A., 2012. Solid-state NMR sequential assignments of alpha-synuclein. Biomol. NMR Assign. 6, 51–55.
Gath, J., Bousset, L., Habenstein, B., Melki, R., Meier, B.H., Bockmann, A., 2014. Yet another polymorph of alpha-synuclein: solid-state sequential assignments. Biomol. NMR Assign. 8, 395–404.
Gazit, E., 2002. The "correctly folded" state of proteins: is it a metastable state? Angew. Chem. Int. Ed. Engl. 41, 257–259.
George, J.M., Jin, H., Woods, W.S., Clayton, D.F., 1995. Characterization of a novel protein regulated during the critical period for song learning in the zebra finch. Neuron 15, 361–372.
Giasson, B.I., Jakes, R., Goedert, M., Duda, J.E., Leight, S., Trojanowski, J.Q., Lee, V.M., 2000. A panel of epitope-specific antibodies detects protein domains distributed throughout human alpha-synuclein in Lewy bodies of Parkinson's disease. J. Neurosci. Res. 59, 528–533.
Giasson, B.I., Murray, I.V., Trojanowski, J.Q., Lee, V.M., 2001. A hydrophobic stretch of 12 amino acid residues in the middle of alpha-synuclein is essential for filament assembly. J. Biol. Chem. 276, 2380–2386.

Giehm, L., Svergun, D.I., Otzen, D.E., Vestergaard, B., 2011. Low-resolution structure of a vesicle disrupting α-synuclein oligomer that accumulates during fibrillation. Proc. Natl. Acad. Sci. U.S.A. 108, 3246–3251.

Gosal, W.S., Morten, I.J., Hewitt, E.W., Smith, D.A., Thomson, N.H., Radford, S.E., 2005. Competing pathways determine fibril morphology in the self-assembly of beta2-microglobulin into amyloid. J. Mol. Biol. 351, 850–864.

Gould, N., Mor, D.E., Lightfoot, R., Malkus, K., Giasson, B., Ischiropoulos, H., 2014. Evidence of native alpha-synuclein conformers in the human brain. J. Biol. Chem. 289, 7929–7934.

Guijarro, J.I., Sunde, M., Jones, J.A., Campbell, I.D., Dobson, C.M., 1998. Amyloid fibril formation by an SH3 domain. Proc. Natl. Acad. Sci. U.S.A. 95, 4224–4228.

Guilliams, T., El-Turk, F., Buell, A.K., O'Day, E.M., Aprile, F.A., Esbjorner, E.K., Vendruscolo, M., Cremades, N., Pardon, E., Wyns, L., et al., 2013. Nanobodies raised against monomeric alpha-synuclein distinguish between fibrils at different maturation stages. J. Mol. Biol. 425, 2397–2411.

Haass, C., Selkoe, D.J., 2007. Soluble protein oligomers in neurodegeneration: lessons from the Alzheimer's amyloid beta-peptide. Nat. Rev. Mol. Cell Biol. 8, 101–112.

Hansen, C., Angot, E., Bergstrom, A.L., Steiner, J.A., Pieri, L., Paul, G., Outeiro, T.F., Melki, R., Kallunki, P., Fog, K., et al., 2011. alpha-Synuclein propagates from mouse brain to grafted dopaminergic neurons and seeds aggregation in cultured human cells. J. Clin. Invest. 121, 715–725.

Hashimoto, M., Hsu, L.J., Xia, Y., Takeda, A., Sisk, A., Sundsmo, M., Masliah, E., 1999. Oxidative stress induces amyloid-like aggregate formation of NACP/alpha-synuclein in vitro. Neuroreport 10, 717–721.

Heise, H., Hoyer, W., Becker, S., Andronesi, O.C., Riedel, D., Baldus, M., 2005. Molecular-level secondary structure, polymorphism, and dynamics of full-length alpha-synuclein fibrils studied by solid-state NMR. Proc. Natl. Acad. Sci. U.S.A. 102, 15871–15876.

Hillen, H., Barghorn, S., Striebinger, A., Labkovsky, B., Muller, R., Nimmrich, V., Nolte, M.W., Perez-Cruz, C., van der Auwera, I., van Leuven, F., et al., 2010. Generation and therapeutic efficacy of highly oligomer-specific beta-amyloid antibodies. J. Neurosci. 30, 10369–10379.

Hoffman-Zacharska, D., Koziorowski, D., Ross, O.A., Milewski, M., Poznanski, J., Jurek, M., Wszolek, Z.K., Soto-Ortolaza, A., Slawek, J., Janik, P., et al., 2013. Novel A18T and pA29S substitutions in alpha-synuclein may be associated with sporadic Parkinson's disease. Parkinsonism Relat. Disord. 19, 1057–1060.

Hong, D.P., Fink, A.L., Uversky, V.N., 2008. Structural characteristics of alpha-synuclein oligomers stabilized by the flavonoid baicalein. J. Mol. Biol. 383, 214–223.

Horrocks, M.H., Lee, S.F., Gandhi, S., Magdalinou, N.K., Chen, S.W., Devine, M.J., Tosatto, L., Kjaergaard, M., Beckwith, J.S., Zetterberg, H., et al., 2016. Single-molecule imaging of individual amyloid protein aggregates in human biofluids. ACS Chem. Neurosci. 7, 399–406.

Horvath, I., Weise, C.F., Andersson, E.K., Chorell, E., Sellstedt, M., Bengtsson, C., Olofsson, A., Hultgren, S.J., Chapman, M., Wolf-Watz, M., et al., 2012. Mechanisms of protein oligomerization: inhibitor of functional amyloids templates alpha-synuclein fibrillation. J. Am. Chem. Soc. 134, 3439–3444.

Hu, R., Diao, J., Li, J., Tang, Z., Li, X., Leitz, J., Long, J., Liu, J., Yu, D., Zhao, Q., 2016. Intrinsic and membrane-facilitated alpha-synuclein oligomerization revealed by label-free detection through solid-state nanopores. Sci. Rep. 6, 20776.

Ikenoue, T., Lee, Y.H., Kardos, J., Saiki, M., Yagi, H., Kawata, Y., Goto, Y., 2014. Cold denaturation of alpha-synuclein amyloid fibrils. Angew. Chem. Int. Ed. Engl. 53, 7799–7804.

Iljina, M., Garcia, G.A., Horrocks, M.H., Tosatto, L., Choi, M.L., Ganzinger, K.A., Abramov, A.Y., Gandhi, S., Wood, N.W., Cremades, N., et al., 2016. Kinetic model of the aggregation of alpha-synuclein provides insights into prion-like spreading. Proc. Natl. Acad. Sci. U.S.A. 113, E1206–E1215.

Iwai, A., Masliah, E., Yoshimoto, M., Ge, N., Flanagan, L., de Silva, H.A., Kittel, A., Saitoh, T., 1995. The precursor protein of non-A beta component of Alzheimer's disease amyloid is a presynaptic protein of the central nervous system. Neuron 14, 467–475.

Jain, S., Udgaonkar, J.B., 2011. Defining the pathway of worm-like amyloid fibril formation by the mouse prion protein by delineation of the productive and unproductive oligomerization reactions. Biochemistry 50, 1153–1161.

Jang, A., Lee, H.J., Suk, J.E., Jung, J.W., Kim, K.P., Lee, S.J., 2010. Non-classical exocytosis of alpha-synuclein is sensitive to folding states and promoted under stress conditions. J. Neurochem. 113, 1263–1274.

Jarrett, J.T., Lansbury Jr., P.T., 1992. Amyloid fibril formation requires a chemically discriminating nucleation event: studies of an amyloidogenic sequence from the bacterial protein OsmB. Biochemistry 31, 12345–12352.

Jellinger, K.A., 2011. Synuclein deposition and non-motor symptoms in Parkinson disease. J. Neurol. Sci. 310, 107–111.

Jimenez, J.L., Guijarro, J.I., Orlova, E., Zurdo, J., Dobson, C.M., Sunde, M., Saibil, H.R., 1999. Cryo-electron microscopy structure of an SH3 amyloid fibril and model of the molecular packing. EMBO J. 18, 815–821.

Jimenez, J.L., Nettleton, E.J., Bouchard, M., Robinson, C.V., Dobson, C.M., Saibil, H.R., 2002. The protofilament structure of insulin amyloid fibrils. Proc. Natl. Acad. Sci. U.S.A. 99, 9196–9201.

Jonsson, S.A., Mohanty, S., Irback, A., 2012. Distinct phases of free alpha-synuclein—a Monte Carlo study. Proteins 80, 2169–2177.

Jucker, M., Walker, L.C., 2013. Self-propagation of pathogenic protein aggregates in neurodegenerative diseases. Nature 501, 45–51.

Kad, N.M., Myers, S.L., Smith, D.P., Smith, D.A., Radford, S.E., Thomson, N.H., 2003. Hierarchical assembly of beta2-microglobulin amyloid in vitro revealed by atomic force microscopy. J. Mol. Biol. 330, 785–797.

Kalia, L.V., Kalia, S.K., McLean, P.J., Lozano, A.M., Lang, A.E., 2013. alpha-Synuclein-oligomers and clinical implications for Parkinson disease. Ann. Neurol. 73, 155–169.

Kara, E., Lewis, P.A., Ling, H., Proukakis, C., Houlden, H., Hardy, J., 2013. alpha-Synuclein mutations cluster around a putative protein loop. Neurosci. Lett. 546, 67–70.

Kayed, R., Head, E., Thompson, J.L., McIntire, T.M., Milton, S.C., Cotman, C.W., Glabe, C.G., 2003. Common structure of soluble amyloid oligomers implies common mechanism of pathogenesis. Science 300, 486–489.

Kayed, R., Sokolov, Y., Edmonds, B., McIntire, T.M., Milton, S.C., Hall, J.E., Glabe, C.G., 2004. Permeabilization of lipid bilayers is a common conformation-dependent activity of soluble amyloid oligomers in protein misfolding diseases. J. Biol. Chem. 279, 46363–46366.

Kayed, R., Head, E., Sarsoza, F., Saing, T., Cotman, C.W., Necula, M., Margol, L., Wu, J., Breydo, L., Thompson, J.L., et al., 2007. Fibril specific, conformation dependent antibodies recognize a generic epitope common to amyloid fibrils and fibrillar oligomers that is absent in prefibrillar oligomers. Mol. Neurodegener. 2, 18.

Kaylor, J., Bodner, N., Edridge, S., Yamin, G., Hong, D.P., Fink, A.L., 2005. Characterization of oligomeric intermediates in alpha-synuclein fibrillation: FRET studies of Y125W/Y133F/Y136F alpha-synuclein. J. Mol. Biol. 353, 357–372.

Kim, H.Y., Cho, M.K., Riedel, D., Fernandez, C.O., Zweckstetter, M., 2008. Dissociation of amyloid fibrils of alpha-synuclein in supercooled water. Angew. Chem. Int. Ed. Engl. 47, 5046–5048.

Kim, H.Y., Cho, M.K., Kumar, A., Maier, E., Siebenhaar, C., Becker, S., Fernandez, C.O., Lashuel, H.A., Benz, R., Lange, A., et al., 2009. Structural properties of pore-forming oligomers of alpha-synuclein. J. Am. Chem. Soc. 131, 17482–17489.
Knowles, T.P., Fitzpatrick, A.W., Meehan, S., Mott, H.R., Vendruscolo, M., Dobson, C.M., Welland, M.E., 2007. Role of intermolecular forces in defining material properties of protein nanofibrils. Science 318, 1900–1903.
Knowles, T.P., Waudby, C.A., Devlin, G.L., Cohen, S.I., Aguzzi, A., Vendruscolo, M., Terentjev, E.M., Welland, M.E., Dobson, C.M., 2009. An analytical solution to the kinetics of breakable filament assembly. Science 326, 1533–1537.
Knowles, T.P., De Simone, A., Fitzpatrick, A.W., Baldwin, A., Meehan, S., Rajah, L., Vendruscolo, M., Welland, M.E., Dobson, C.M., Terentjev, E.M., 2012. Twisting transition between crystalline and fibrillar phases of aggregated peptides. Phys. Rev. Lett. 109, 158101.
Knowles, T.P., Vendruscolo, M., Dobson, C.M., 2014. The amyloid state and its association with protein misfolding diseases. Nat. Rev. Mol. Cell Biol. 15, 384–396.
Kostka, M., Hogen, T., Danzer, K.M., Levin, J., Habeck, M., Wirth, A., Wagner, R., Glabe, C.G., Finger, S., Heinzelmann, U., et al., 2008. Single particle characterization of iron-induced pore-forming alpha-synuclein oligomers. J. Biol. Chem. 283, 10992–11003.
Kumar, S., Udgaonkar, J.B., 2009. Conformational conversion may precede or follow aggregate elongation on alternative pathways of amyloid protofibril formation. J. Mol. Biol. 385, 1266–1276.
Lacor, P.N., Buniel, M.C., Chang, L., Fernandez, S.J., Gong, Y., Viola, K.L., Lambert, M.P., Velasco, P.T., Bigio, E.H., Finch, C.E., et al., 2004. Synaptic targeting by Alzheimer's-related amyloid beta oligomers. J. Neurosci. 24, 10191–10200.
Lasagna-Reeves, C.A., Castillo-Carranza, D.L., Sengupta, U., Sarmiento, J., Troncoso, J., Jackson, G.R., Kayed, R., 2012. Identification of oligomers at early stages of tau aggregation in Alzheimer's disease. FASEB J. 26, 1946–1959.
Lashuel, H.A., Hartley, D., Petre, B.M., Walz, T., Lansbury Jr., P.T., 2002a. Neurodegenerative disease: amyloid pores from pathogenic mutations. Nature 418, 291.
Lashuel, H.A., Petre, B.M., Wall, J., Simon, M., Nowak, R.J., Walz, T., Lansbury Jr., P.T., 2002b. Alpha-synuclein, especially the Parkinson's disease-associated mutants, forms pore-like annular and tubular protofibrils. J. Mol. Biol. 322, 1089–1102.
Lashuel, H.A., Overk, C.R., Oueslati, A., Masliah, E., 2013. The many faces of alpha-synuclein: from structure and toxicity to therapeutic target. Nat. Rev. Neurosci. 14, 38–48.
Lee, J.C., Langen, R., Hummel, P.A., Gray, H.B., Winkler, J.R., 2004. Alpha-synuclein structures from fluorescence energy-transfer kinetics: implications for the role of the protein in Parkinson's disease. Proc. Natl. Acad. Sci. U.S.A. 101, 16466–16471.
Lee, H.J., Patel, S., Lee, S.J., 2005. Intravesicular localization and exocytosis of alpha-synuclein and its aggregates. J. Neurosci. 25, 6016–6024.
Lee, P.H., Lee, G., Park, H.J., Bang, O.Y., Joo, I.S., Huh, K., 2006. The plasma alpha-synuclein levels in patients with Parkinson's disease and multiple system atrophy. J. Neural Transm. (Vienna) 113, 1435–1439.
Lee, H.J., Suk, J.E., Bae, E.J., Lee, J.H., Paik, S.R., Lee, S.J., 2008. Assembly-dependent endocytosis and clearance of extracellular alpha-synuclein. Int. J. Biochem. Cell Biol. 40, 1835–1849.
Lee, D., Park, C.W., Paik, S.R., Choi, K.Y., 2009. The modification of alpha-synuclein by dicarbonyl compounds inhibits its fibril-forming process. Biochim. Biophys. Acta 1794, 421–430.
Lee, J., Culyba, E.K., Powers, E.T., Kelly, J.W., 2011. Amyloid-beta forms fibrils by nucleated conformational conversion of oligomers. Nat. Chem. Biol. 7, 602–609.

Lendel, C., Bertoncini, C.W., Cremades, N., Waudby, C.A., Vendruscolo, M., Dobson, C.M., Schenk, D., Christodoulou, J., Toth, G., 2009. On the mechanism of nonspecific inhibitors of protein aggregation: dissecting the interactions of alpha-synuclein with Congo red and lacmoid. Biochemistry 48, 8322–8334.

Lesage, S., Anheim, M., Letournel, F., Bousset, L., Honore, A., Rozas, N., Pieri, L., Madiona, K., Durr, A., Melki, R., et al., 2013. G51D alpha-synuclein mutation causes a novel Parkinsonian-pyramidal syndrome. Ann. Neurol. 73, 459–471.

Li, J., Zhu, M., Manning-Bog, A.B., Di Monte, D.A., Fink, A.L., 2004. Dopamine and L-dopa disaggregate amyloid fibrils: implications for Parkinson's and Alzheimer's disease. FASEB J. 18, 962–964.

Li, H.T., Lin, D.H., Luo, X.Y., Zhang, F., Ji, L.N., Du, H.N., Song, G.Q., Hu, J., Zhou, J.W., Hu, H.Y., 2005. Inhibition of alpha-synuclein fibrillization by dopamine analogs via reaction with the amino groups of alpha-synuclein. Implication for dopaminergic neurodegeneration. FEBS J. 272, 3661–3672.

Liu, Y., Carver, J.A., Calabrese, A.N., Pukala, T.L., 2014. Gallic acid interacts with alpha-synuclein to prevent the structural collapse necessary for its aggregation. Biochim. Biophys. Acta 1844, 1481–1485.

Lorenzen, N., Nielsen, S.B., Buell, A.K., Kaspersen, J.D., Arosio, P., Vad, B.S., Paslawski, W., Christiansen, G., Valnickova-Hansen, Z., Andreasen, M., et al., 2014a. The role of stable alpha-synuclein oligomers in the molecular events underlying amyloid formation. J. Am. Chem. Soc. 136, 3859–3868.

Lorenzen, N., Nielsen, S.B., Yoshimura, Y., Vad, B.S., Andersen, C.B., Betzer, C., Kaspersen, J.D., Christiansen, G., Pedersen, J.S., Jensen, P.H., et al., 2014b. How epigallocatechin gallate can inhibit alpha-synuclein oligomer toxicity in vitro. J. Biol. Chem. 289, 21299–21310.

Lowe, R., Pountney, D.L., Jensen, P.H., Gai, W.P., Voelcker, N.H., 2004. Calcium(II) selectively induces alpha-synuclein annular oligomers via interaction with the C-terminal domain. Protein Sci. 13, 3245–3252.

Lu, J.H., Ardah, M.T., Durairajan, S.S., Liu, L.F., Xie, L.X., Fong, W.F., Hasan, M.Y., Huang, J.D., El-Agnaf, O.M., Li, M., 2011. Baicalein inhibits formation of alpha-synuclein oligomers within living cells and prevents Abeta peptide fibrillation and oligomerisation. Chembiochem 12, 615–624.

Luk, K.C., Kehm, V., Carroll, J., Zhang, B., O'Brien, P., Trojanowski, J.Q., Lee, V.M., 2012. Pathological alpha-synuclein transmission initiates Parkinson-like neurodegeneration in nontransgenic mice. Science 338, 949–953.

Mannini, B., Mulvihill, E., Sgromo, C., Cascella, R., Khodarahmi, R., Ramazzotti, M., Dobson, C.M., Cecchi, C., Chiti, F., 2014. Toxicity of protein oligomers is rationalized by a function combining size and surface hydrophobicity. ACS Chem. Biol. 9, 2309–2317.

Maraganore, D.M., de Andrade, M., Elbaz, A., Farrer, M.J., Ioannidis, J.P., Kruger, R., Rocca, W.A., Schneider, N.K., Lesnick, T.G., Lincoln, S.J., et al., 2006. Collaborative analysis of alpha-synuclein gene promoter variability and Parkinson disease. JAMA 296, 661–670.

McCann, H., Stevens, C.H., Cartwright, H., Halliday, G.M., 2014. alpha-Synucleinopathy phenotypes. Parkinsonism Relat. Disord. 20 (Suppl. 1), S62–S67.

Miake, H., Mizusawa, H., Iwatsubo, T., Hasegawa, M., 2002. Biochemical characterization of the core structure of alpha-synuclein filaments. J. Biol. Chem. 277, 19213–19219.

Munch, G., Luth, H.J., Wong, A., Arendt, T., Hirsch, E., Ravid, R., Riederer, P., 2000. Crosslinking of alpha-synuclein by advanced glycation endproducts—an early pathophysiological step in Lewy body formation? J. Chem. Neuroanat. 20, 253–257.

Nakamura, K., 2013. alpha-Synuclein and mitochondria: partners in crime? Neurotherapeutics 10, 391–399.

Nasstrom, T., Wahlberg, T., Karlsson, M., Nikolajeff, F., Lannfelt, L., Ingelsson, M., Bergstrom, J., 2009. The lipid peroxidation metabolite 4-oxo-2-nonenal cross-links alpha-synuclein causing rapid formation of stable oligomers. Biochem. Biophys. Res. Commun. 378, 872–876.

Nasstrom, T., Fagerqvist, T., Barbu, M., Karlsson, M., Nikolajeff, F., Kasrayan, A., Ekberg, M., Lannfelt, L., Ingelsson, M., Bergstrom, J., 2011. The lipid peroxidation products 4-oxo-2-nonenal and 4-hydroxy-2-nonenal promote the formation of alpha-synuclein oligomers with distinct biochemical, morphological, and functional properties. Free Radic. Biol. Med. 50, 428–437.

Nekooki-Machida, Y., Kurosawa, M., Nukina, N., Ito, K., Oda, T., Tanaka, M., 2009. Distinct conformations of in vitro and in vivo amyloids of huntingtin-exon1 show different cytotoxicity. Proc. Natl. Acad. Sci. U.S.A. 106, 9679–9684.

Nelson, R., Sawaya, M.R., Balbirnie, M., Madsen, A.O., Riekel, C., Grothe, R., Eisenberg, D., 2005. Structure of the cross-beta spine of amyloid-like fibrils. Nature 435, 773–778.

Nemani, V.M., Lu, W., Berge, V., Nakamura, K., Onoa, B., Lee, M.K., Chaudhry, F.A., Nicoll, R.A., Edwards, R.H., 2010. Increased expression of alpha-synuclein reduces neurotransmitter release by inhibiting synaptic vesicle reclustering after endocytosis. Neuron 65, 66–79.

Nielsen, M.S., Vorum, H., Lindersson, E., Jensen, P.H., 2001. Ca^{2+} binding to alpha-synuclein regulates ligand binding and oligomerization. J. Biol. Chem. 276, 22680–22684.

Noguchi, A., Matsumura, S., Dezawa, M., Tada, M., Yanazawa, M., Ito, A., Akioka, M., Kikuchi, S., Sato, M., Ideno, S., et al., 2009. Isolation and characterization of patient-derived, toxic, high mass amyloid beta-protein (Abeta) assembly from Alzheimer disease brains. J. Biol. Chem. 284, 32895–32905.

Norris, E.H., Giasson, B.I., Hodara, R., Xu, S., Trojanowski, J.Q., Ischiropoulos, H., Lee, V.M., 2005. Reversible inhibition of alpha-synuclein fibrillization by dopaminochrome-mediated conformational alterations. J. Biol. Chem. 280, 21212–21219.

Nors Perdersen, M., Fodera, V., Horvath, I., van Maarschalkerweerd, A., Norgaard Toft, K., Weise, C., Almqvist, F., Wolf-Watz, M., Wittung-Stafshede, P., Vestergaard, B., 2015. Direct correlation between ligand-induced alpha-synuclein oligomers and amyloid-like fibril growth. Sci. Rep. 5, 10422.

Olofsson, A., Ippel, J.H., Wijmenga, S.S., Lundgren, E., Ohman, A., 2004. Probing solvent accessibility of transthyretin amyloid by solution NMR spectroscopy. J. Biol. Chem. 279, 5699–5707.

Ono, K., Yamada, M., 2006. Antioxidant compounds have potent anti-fibrillogenic and fibril-destabilizing effects for alpha-synuclein fibrils in vitro. J. Neurochem. 97, 105–115.

Orte, A., Birkett, N.R., Clarke, R.W., Devlin, G.L., Dobson, C.M., Klenerman, D., 2008. Direct characterization of amyloidogenic oligomers by single-molecule fluorescence. Proc. Natl. Acad. Sci. U.S.A. 105, 14424–14429.

Ouberai, M.M., Wang, J., Swann, M.J., Galvagnion, C., Guilliams, T., Dobson, C.M., Welland, M.E., 2013. alpha-Synuclein senses lipid packing defects and induces lateral expansion of lipids leading to membrane remodeling. J. Biol. Chem. 288, 20883–20895.

Padmaraju, V., Bhaskar, J.J., Prasada Rao, U.J., Salimath, P.V., Rao, K.S., 2011. Role of advanced glycation on aggregation and DNA binding properties of alpha-synuclein. J. Alzheimers Dis. 24 (Suppl. 2), 211–221.

Pagano, K., Bemporad, F., Fogolari, F., Esposito, G., Viglino, P., Chiti, F., Corazza, A., 2010. Structural and dynamics characteristics of acylphosphatase from Sulfolobus solfataricus in the monomeric state and in the initial native-like aggregates. J. Biol. Chem. 285, 14689–14700.

Paik, S.R., Shin, H.J., Lee, J.H., 2000. Metal-catalyzed oxidation of alpha-synuclein in the presence of copper(II) and hydrogen peroxide. Arch. Biochem. Biophys. 378, 269–277.

Paleologou, K.E., Kragh, C.L., Mann, D.M., Salem, S.A., Al-Shami, R., Allsop, D., Hassan, A.H., Jensen, P.H., El-Agnaf, O.M., 2009. Detection of elevated levels of soluble alpha-synuclein oligomers in post-mortem brain extracts from patients with dementia with Lewy bodies. Brain 132, 1093–1101.

Papp, M.I., Kahn, J.E., Lantos, P.L., 1989. Glial cytoplasmic inclusions in the CNS of patients with multiple system atrophy (striatonigral degeneration, olivopontocerebellar atrophy and Shy-Drager syndrome). J. Neurol. Sci. 94, 79–100.

Paslawski, W., Andreasen, M., Nielsen, S.B., Lorenzen, N., Thomsen, K., Kaspersen, J.D., Pedersen, J.S., Otzen, D.E., 2014a. High stability and cooperative unfolding of alpha-synuclein oligomers. Biochemistry 53, 6252–6263.

Paslawski, W., Mysling, S., Thomsen, K., Jorgensen, T.J., Otzen, D.E., 2014b. Co-existence of two different alpha-synuclein oligomers with different core structures determined by hydrogen/deuterium exchange mass spectrometry. Angew. Chem. Int. Ed. Engl. 53, 7560–7563.

Patterson, K.R., Ward, S.M., Combs, B., Voss, K., Kanaan, N.M., Morfini, G., Brady, S.T., Gamblin, T.C., Binder, L.I., 2011. Heat shock protein 70 prevents both tau aggregation and the inhibitory effects of preexisting tau aggregates on fast axonal transport. Biochemistry 50, 10300–10310.

Payton, J.E., Perrin, R.J., Woods, W.S., George, J.M., 2004. Structural determinants of PLD2 inhibition by alpha-synuclein. J. Mol. Biol. 337, 1001–1009.

Peelaerts, W., Bousset, L., Van der Perren, A., Moskalyuk, A., Pulizzi, R., Giugliano, M., Van den Haute, C., Melki, R., Baekelandt, V., 2015. alpha-Synuclein strains cause distinct synucleinopathies after local and systemic administration. Nature 522, 340–344.

Petty, S.A., Decatur, S.M., 2005. Intersheet rearrangement of polypeptides during nucleation of {beta}-sheet aggregates. Proc. Natl. Acad. Sci. U.S.A. 102, 14272–14277.

Pham, C.L., Cappai, R., 2013. The interplay between lipids and dopamine on alpha-synuclein oligomerization and membrane binding. Biosci. Rep. 33, 807–814.

Pham, C.L., Leong, S.L., Ali, F.E., Kenche, V.B., Hill, A.F., Gras, S.L., Barnham, K.J., Cappai, R., 2009. Dopamine and the dopamine oxidation product 5,6-dihydroxylindole promote distinct on-pathway and off-pathway aggregation of alpha-synuclein in a pH-dependent manner. J. Mol. Biol. 387, 771–785.

Pitschke, M., Prior, R., Haupt, M., Riesner, D., 1998. Detection of single amyloid beta-protein aggregates in the cerebrospinal fluid of Alzheimer's patients by fluorescence correlation spectroscopy. Nat. Med. 4, 832–834.

Plakoutsi, G., Bemporad, F., Calamai, M., Taddei, N., Dobson, C.M., Chiti, F., 2005. Evidence for a mechanism of amyloid formation involving molecular reorganisation within native-like precursor aggregates. J. Mol. Biol. 351, 910–922.

Plotegher, N., Gratton, E., Bubacco, L., 2014. Number and Brightness analysis of alpha-synuclein oligomerization and the associated mitochondrial morphology alterations in live cells. Biochim. Biophys. Acta 1840, 2014–2024.

Polymeropoulos, M.H., Lavedan, C., Leroy, E., Ide, S.E., Dehejia, A., Dutra, A., Pike, B., Root, H., Rubenstein, J., Boyer, R., et al., 1997. Mutation in the alpha-synuclein gene identified in families with Parkinson's disease. Science 276, 2045–2047.

Pountney, D.L., Lowe, R., Quilty, M., Vickers, J.C., Voelcker, N.H., Gai, W.P., 2004. Annular alpha-synuclein species from purified multiple system atrophy inclusions. J. Neurochem. 90, 502–512.

Prusiner, S.B., Woerman, A.L., Mordes, D.A., Watts, J.C., Rampersaud, R., Berry, D.B., Patel, S., Oehler, A., Lowe, J.K., Kravitz, S.N., et al., 2015. Evidence for alpha-synuclein prions causing multiple system atrophy in humans with parkinsonism. Proc. Natl. Acad. Sci. U.S.A. 112, E5308–E5317.

Qi, W., Zhang, A., Patel, D., Lee, S., Harrington, J.L., Zhao, L., Schaefer, D., Good, T.A., Fernandez, E.J., 2008. Simultaneous monitoring of peptide aggregate distributions, structure, and kinetics using amide hydrogen exchange: application to Abeta(1-40) fibrillogenesis. Biotechnol. Bioeng. 100, 1214–1227.

Qin, Z., Hu, D., Han, S., Reaney, S.H., Di Monte, D.A., Fink, A.L., 2007. Effect of 4-hydroxy-2-nonenal modification on alpha-synuclein aggregation. J. Biol. Chem. 282, 5862–5870.

Quist, A., Doudevski, I., Lin, H., Azimova, R., Ng, D., Frangione, B., Kagan, B., Ghiso, J., Lal, R., 2005. Amyloid ion channels: a common structural link for protein-misfolding disease. Proc. Natl. Acad. Sci. U.S.A. 102, 10427–10432.

Rao, J.N., Dua, V., Ulmer, T.S., 2008. Characterization of alpha-synuclein interactions with selected aggregation-inhibiting small molecules. Biochemistry 47, 4651–4656.

Roberts, R.F., Wade-Martins, R., Alegre-Abarrategui, J., 2015. Direct visualization of alpha-synuclein oligomers reveals previously undetected pathology in Parkinson's disease brain. Brain 138, 1642–1657.

Rodriguez, J.A., Ivanova, M.I., Sawaya, M.R., Cascio, D., Reyes, F.E., Shi, D., Sangwan, S., Guenther, E.L., Johnson, L.M., Zhang, M., et al., 2015. Structure of the toxic core of alpha-synuclein from invisible crystals. Nature 525, 486–490.

Ross, C.A., Poirier, M.A., 2005. Opinion: what is the role of protein aggregation in neurodegeneration? Nat. Rev. Mol. Cell Biol. 6, 891–898.

Sandal, M., Valle, F., Tessari, I., Mammi, S., Bergantino, E., Musiani, F., Brucale, M., Bubacco, L., Samori, B., 2008. Conformational equilibria in monomeric alpha-synuclein at the single-molecule level. PLoS Biol. 6, e6.

Sawaya, M.R., Sambashivan, S., Nelson, R., Ivanova, M.I., Sievers, S.A., Apostol, M.I., Thompson, M.J., Balbirnie, M., Wiltzius, J.J., McFarlane, H.T., et al., 2007. Atomic structures of amyloid cross-beta spines reveal varied steric zippers. Nature 447, 453–457.

Seidel, K., Schols, L., Nuber, S., Petrasch-Parwez, E., Gierga, K., Wszolek, Z., Dickson, D., Gai, W.P., Bornemann, A., Riess, O., et al., 2010. First appraisal of brain pathology owing to A30P mutant alpha-synuclein. Ann. Neurol. 67, 684–689.

Serio, T.R., Cashikar, A.G., Kowal, A.S., Sawicki, G.J., Moslehi, J.J., Serpell, L., Arnsdorf, M.F., Lindquist, S.L., 2000. Nucleated conformational conversion and the replication of conformational information by a prion determinant. Science 289, 1317–1321.

Serpell, L.C., Smith, J.M., 2000. Direct visualisation of the beta-sheet structure of synthetic Alzheimer's amyloid. J. Mol. Biol. 299, 225–231.

Serpell, L.C., Sunde, M., Fraser, P.E., Luther, P.K., Morris, E.P., Sangren, O., Lundgren, E., Blake, C.C., 1995. Examination of the structure of the transthyretin amyloid fibril by image reconstruction from electron micrographs. J. Mol. Biol. 254, 113–118.

Serpell, L.C., Berriman, J., Jakes, R., Goedert, M., Crowther, R.A., 2000a. Fiber diffraction of synthetic alpha-synuclein filaments shows amyloid-like cross-beta conformation. Proc. Natl. Acad. Sci. U.S.A. 97, 4897–4902.

Serpell, L.C., Sunde, M., Benson, M.D., Tennent, G.A., Pepys, M.B., Fraser, P.E., 2000b. The protofilament substructure of amyloid fibrils. J. Mol. Biol. 300, 1033–1039.

Shankar, G.M., Li, S., Mehta, T.H., Garcia-Munoz, A., Shepardson, N.E., Smith, I., Brett, F.M., Farrell, M.A., Rowan, M.J., Lemere, C.A., et al., 2008. Amyloid-beta protein dimers isolated directly from Alzheimer's brains impair synaptic plasticity and memory. Nat. Med. 14, 837–842.

Singh, P.K., Kotia, V., Ghosh, D., Mohite, G.M., Kumar, A., Maji, S.K., 2013. Curcumin modulates alpha-synuclein aggregation and toxicity. ACS Chem. Neurosci. 4, 393–407.

Singleton, A.B., Farrer, M., Johnson, J., Singleton, A., Hague, S., Kachergus, J., Hulihan, M., Peuralinna, T., Dutra, A., Nussbaum, R., et al., 2003. alpha-Synuclein locus triplication causes Parkinson's disease. Science 302, 841.

Souza, J.M., Giasson, B.I., Chen, Q., Lee, V.M., Ischiropoulos, H., 2000. Dityrosine cross-linking promotes formation of stable alpha -synuclein polymers. Implication of nitrative and oxidative stress in the pathogenesis of neurodegenerative synucleinopathies. J. Biol. Chem. 275, 18344–18349.

Spillantini, M.G., Schmidt, M.L., Lee, V.M., Trojanowski, J.Q., Jakes, R., Goedert, M., 1997. Alpha-synuclein in Lewy bodies. Nature 388, 839–840.

Spillantini, M.G., Crowther, R.A., Jakes, R., Cairns, N.J., Lantos, P.L., Goedert, M., 1998. Filamentous alpha-synuclein inclusions link multiple system atrophy with Parkinson's disease and dementia with Lewy bodies. Neurosci. Lett. 251, 205–208.

Stefanovic, A.N., Lindhoud, S., Semerdzhiev, S.A., Claessens, M.M., Subramaniam, V., 2015. Oligomers of Parkinson's disease-related alpha-synuclein mutants have similar structures but distinctive membrane permeabilization properties. Biochemistry 54, 3142–3150.

Stockl, M.T., Zijlstra, N., Subramaniam, V., 2013. alpha-Synuclein oligomers: an amyloid pore? Insights into mechanisms of alpha-synuclein oligomer-lipid interactions. Mol. Neurobiol. 47, 613–621.

Sunde, M., Serpell, L.C., Bartlam, M., Fraser, P.E., Pepys, M.B., Blake, C.C., 1997. Common core structure of amyloid fibrils by synchrotron X-ray diffraction. J. Mol. Biol. 273, 729–739.

Thakur, A.K., Jayaraman, M., Mishra, R., Thakur, M., Chellgren, V.M., Byeon, I.J., Anjum, D.H., Kodali, R., Creamer, T.P., Conway, J.F., et al., 2009. Polyglutamine disruption of the huntingtin exon 1N terminus triggers a complex aggregation mechanism. Nat. Struct. Mol. Biol. 16, 380–389.

Theillet, F.X., Binolfi, A., Bekei, B., Martorana, A., Rose, H.M., Stuiver, M., Verzini, S., Lorenz, D., van Rossum, M., Goldfarb, D., et al., 2016. Structural disorder of monomeric alpha-synuclein persists in mammalian cells. Nature 530, 45–50.

Tokuda, T., Salem, S.A., Allsop, D., Mizuno, T., Nakagawa, M., Qureshi, M.M., Locascio, J.J., Schlossmacher, M.G., El-Agnaf, O.M., 2006. Decreased alpha-synuclein in cerebrospinal fluid of aged individuals and subjects with Parkinson's disease. Biochem. Biophys. Res. Commun. 349, 162–166.

Townsend, M., Cleary, J.P., Mehta, T., Hofmeister, J., Lesne, S., O'Hare, E., Walsh, D.M., Selkoe, D.J., 2006. Orally available compound prevents deficits in memory caused by the Alzheimer amyloid-beta oligomers. Ann. Neurol. 60, 668–676.

Trexler, A.J., Rhoades, E., 2012. N-terminal acetylation is critical for forming alpha-helical oligomer of alpha-synuclein. Protein Sci. 21, 601–605.

Tsigelny, I.F., Bar-On, P., Sharikov, Y., Crews, L., Hashimoto, M., Miller, M.A., Keller, S.H., Platoshyn, O., Yuan, J.X., Masliah, E., 2007. Dynamics of alpha-synuclein aggregation and inhibition of pore-like oligomer development by beta-synuclein. FEBS J. 274, 1862–1877.

Tsigelny, I.F., Crews, L., Desplats, P., Shaked, G.M., Sharikov, Y., Mizuno, H., Spencer, B., Rockenstein, E., Trejo, M., Platoshyn, O., et al., 2008. Mechanisms of hybrid oligomer formation in the pathogenesis of combined Alzheimer's and Parkinson's diseases. PLoS One 3, e3135.

Tsigelny, I.F., Sharikov, Y., Wrasidlo, W., Gonzalez, T., Desplats, P.A., Crews, L., Spencer, B., Masliah, E., 2012. Role of alpha-synuclein penetration into the membrane in the mechanisms of oligomer pore formation. FEBS J. 279, 1000–1013.

Tuttle, M.D., Comellas, G., Nieuwkoop, A.J., Covell, D.J., Berthold, D.A., Kloepper, K.D., Courtney, J.M., Kim, J.K., Barclay, A.M., Kendall, A., et al., 2016. Solid-state NMR structure of a pathogenic fibril of full-length human alpha-synuclein. Nat. Struct. Mol. Biol. 23, 409–415.

Ueda, K., Fukushima, H., Masliah, E., Xia, Y., Iwai, A., Yoshimoto, M., Otero, D.A., Kondo, J., Ihara, Y., Saitoh, T., 1993. Molecular cloning of cDNA encoding an unrecognized component of amyloid in Alzheimer disease. Proc. Natl. Acad. Sci. U.S.A. 90, 11282–11286.

Uversky, V.N., Yamin, G., Munishkina, L.A., Karymov, M.A., Millett, I.S., Doniach, S., Lyubchenko, Y.L., Fink, A.L., 2005. Effects of nitration on the structure and aggregation of alpha-synuclein. Brain Res. Mol. Brain Res. 134, 84–102.

van Rooijen, B.D., van Leijenhorst-Groener, K.A., Claessens, M.M., Subramaniam, V., 2009. Tryptophan fluorescence reveals structural features of alpha-synuclein oligomers. J. Mol. Biol. 394, 826–833.

Vilar, M., Chou, H.T., Luhrs, T., Maji, S.K., Riek-Loher, D., Verel, R., Manning, G., Stahlberg, H., Riek, R., 2008. The fold of alpha-synuclein fibrils. Proc. Natl. Acad. Sci. U.S.A. 105, 8637–8642.

Visanji, N.P., Wislet-Gendebien, S., Oschipok, L.W., Zhang, G., Aubert, I., Fraser, P.E., Tandon, A., 2011. Effect of Ser-129 phosphorylation on interaction of alpha-synuclein with synaptic and cellular membranes. J. Biol. Chem. 286, 35863–35873.

Volles, M.J., Lee, S.J., Rochet, J.C., Shtilerman, M.D., Ding, T.T., Kessler, J.C., Lansbury Jr., P.T., 2001. Vesicle permeabilization by protofibrillar alpha-synuclein: implications for the pathogenesis and treatment of Parkinson's disease. Biochemistry 40, 7812–7819.

Walsh, P., Vanderlee, G., Yau, J., Campeau, J., Sim, V.L., Yip, C.M., Sharpe, S., 2014. The mechanism of membrane disruption by cytotoxic amyloid oligomers formed by prion protein(106-126) is dependent on bilayer composition. J. Biol. Chem. 289, 10419–10430.

Wang, X., Moualla, D., Wright, J.A., Brown, D.R., 2010. Copper binding regulates intracellular alpha-synuclein localisation, aggregation and toxicity. J. Neurochem. 113, 704–714.

Wang, W., Perovic, I., Chittuluru, J., Kaganovich, A., Nguyen, L.T., Liao, J., Auclair, J.R., Johnson, D., Landeru, A., Simorellis, A.K., et al., 2011. A soluble alpha-synuclein construct forms a dynamic tetramer. Proc. Natl. Acad. Sci. U.S.A. 108, 17797–17802.

Waudby, C.A., Camilloni, C., Fitzpatrick, A.W., Cabrita, L.D., Dobson, C.M., Vendruscolo, M., Christodoulou, J., 2013. In-cell NMR characterization of the secondary structure populations of a disordered conformation of alpha-synuclein within E. coli cells. PLoS One 8, e72286.

Wei, L., Jiang, P., Xu, W., Li, H., Zhang, H., Yan, L., Chan-Park, M.B., Liu, X.W., Tang, K., Mu, Y., et al., 2011. The molecular basis of distinct aggregation pathways of islet amyloid polypeptide. J. Biol. Chem. 286, 6291–6300.

Winner, B., Jappelli, R., Maji, S.K., Desplats, P.A., Boyer, L., Aigner, S., Hetzer, C., Loher, T., Vilar, M., Campioni, S., et al., 2011. In vivo demonstration that alpha -synuclein oligomers are toxic. Proc. Natl. Acad. Sci. U.S.A. 108, 4194–4199.

Woerman, A.L., Stohr, J., Aoyagi, A., Rampersaud, R., Krejciova, Z., Watts, J.C., Ohyama, T., Patel, S., Widjaja, K., Oehler, A., et al., 2015. Propagation of prions causing synucleinopathies in cultured cells. Proc. Natl. Acad. Sci. U.S.A. 112, E4949–E4958.

Wright, J.A., Wang, X., Brown, D.R., 2009. Unique copper-induced oligomers mediate alpha-synuclein toxicity. FASEB J. 23, 2384–2393.

Xiang, W., Schlachetzki, J.C., Helling, S., Bussmann, J.C., Berlinghof, M., Schaffer, T.E., Marcus, K., Winkler, J., Klucken, J., Becker, C.M., 2013. Oxidative stress-induced post-translational modifications of alpha-synuclein: specific modification of alpha-synuclein by 4-hydroxy-2-nonenal increases dopaminergic toxicity. Mol. Cell. Neurosci. 54, 71–83.

Xilouri, M., Brekk, O.R., Stefanis, L., 2013. alpha-Synuclein and protein degradation systems: a reciprocal relationship. Mol. Neurobiol. 47, 537–551.

Yoritaka, A., Hattori, N., Uchida, K., Tanaka, M., Stadtman, E.R., Mizuno, Y., 1996. Immunohistochemical detection of 4-hydroxynonenal protein adducts in Parkinson disease. Proc. Natl. Acad. Sci. U.S.A. 93, 2696–2701.

Yoshiike, Y., Kayed, R., Milton, S.C., Takashima, A., Glabe, C.G., 2007. Pore-forming proteins share structural and functional homology with amyloid oligomers. Neuromolecular Med. 9, 270–275.

Zampagni, M., Cascella, R., Casamenti, F., Grossi, C., Evangelisti, E., Wright, D., Becatti, M., Liguri, G., Mannini, B., Campioni, S., et al., 2011. A comparison of the biochemical modifications caused by toxic and non-toxic protein oligomers in cells. J. Cell. Mol. Med. 15, 2106–2116.

Zhang, R., Hu, X., Khant, H., Ludtke, S.J., Chiu, W., Schmid, M.F., Frieden, C., Lee, J.M., 2009. Interprotofilament interactions between Alzheimer's Abeta1-42 peptides in amyloid fibrils revealed by cryoEM. Proc. Natl. Acad. Sci. U.S.A. 106, 4653–4658.

Zhou, W., Long, C., Reaney, S.H., Di Monte, D.A., Fink, A.L., Uversky, V.N., 2010. Methionine oxidation stabilizes non-toxic oligomers of alpha-synuclein through strengthening the auto-inhibitory intra-molecular long-range interactions. Biochim. Biophys. Acta 1802, 322–330.

Zhu, M., Rajamani, S., Kaylor, J., Han, S., Zhou, F., Fink, A.L., 2004. The flavonoid baicalein inhibits fibrillation of alpha-synuclein and disaggregates existing fibrils. J. Biol. Chem. 279, 26846–26857.

Zibaee, S., Makin, O.S., Goedert, M., Serpell, L.C., 2007. A simple algorithm locates beta-strands in the amyloid fibril core of alpha-synuclein, Abeta, and tau using the amino acid sequence alone. Protein Sci. 16, 906–918.

Zijlstra, N., Blum, C., Segers-Nolten, I.M., Claessens, M.M., Subramaniam, V., 2012. Molecular composition of sub-stoichiometrically labeled alpha-synuclein oligomers determined by single-molecule photobleaching. Angew. Chem. Int. Ed. Engl. 51, 8821–8824.

CHAPTER FOUR

Effects of Intrinsic and Extrinsic Factors on Aggregation of Physiologically Important Intrinsically Disordered Proteins

L. Breydo*[,1], J.M. Redington*, V.N. Uversky*[,†,1]
*Byrd Alzheimer's Research Institute, Morsani College of Medicine, University of South Florida, Tampa, FL, United States
[†]Laboratory of Structural Dynamics, Stability and Folding of Proteins, Institute of Cytology, Russian Academy of Sciences, St. Petersburg, Russia
[1]Corresponding authors: e-mail address: lbreydo@health.usf.edu; vuversky@health.usf.edu

Contents

Abstract

Misfolding and aggregation of proteins and peptides play an important role in a number of diseases as well as in many physiological processes. Many of the proteins that misfold and aggregate in vivo are intrinsically disordered. Protein aggregation is a complex multistep process, and aggregates can significantly differ in morphology, structure, stability, cytotoxicity, and self-propagation ability. The aggregation process is influenced by both intrinsic (e.g., mutations and expression levels) and extrinsic (e.g., polypeptide chain truncation, macromolecular crowding, posttranslational modifications, as well as interaction with metal ions, other small molecules, lipid membranes, and chaperons) factors. This review examines the effect of a variety of these factors on aggregation of physiologically important intrinsically disordered proteins.

International Review of Cell and Molecular Biology, Volume 329
ISSN 1937-6448
http://dx.doi.org/10.1016/bs.ircmb.2016.08.011

ABBREVIATIONS

11pE-Aβ pyroglutamate-modified Aβ at position 11
3pE-Aβ pyroglutamate-modified Aβ at position 3
AD Alzheimer's disease
ALS amyotrophic lateral sclerosis
APP amyloid precursor protein
Aβ amyloid beta
C9orf72 chromosome 9 open reading frame 72
CPEB cytoplasmic polyadenylation element-binding protein
CREB cAMP response element-binding protein
CSF cerebrospinal fluid
EGCG epigallocatechin gallate
FTD frontotemporal dementia
FUS fused in sarcoma
Htt huntingtin
IDP intrinsically disordered protein
PD Parkinson's disease
Pmel17 premelanosome protein 17
ROS reactive oxygen species
SOD superoxide dismutase
TBP TATA-binding protein
TDP-43 TAR DNA-binding protein 43
TIA-1 T-cell-restricted intracellular antigen-1
TTR transthyretin

1. AGGREGATION OF INTRINSICALLY DISORDERED PROTEINS

A number of human diseases and physiological processes have been linked to protein aggregation. While some proteins involved in these processes (e.g., transthyretin (TTR), insulin, β-2-microglobulin, superoxide dismutase (SOD)) are natively folded, vast majority of them are intrinsically disordered. Illustrative examples of amyloidogenic intrinsically disordered proteins (IDPs) important to human diseases are listed in Table 1. The best-known examples include amyloid β (Aβ) peptides, tau (Alzheimer's disease (AD)), α-synuclein (Parkinson's disease (PD) and other synucleinopathies), TAR DNA-binding protein 43 (TDP-43), fused in sarcoma (FUS) protein (amyotrophic lateral sclerosis (ALS)), and huntingtin (Huntington's disease). A number of human functional amyloids (cytoplasmic polyadenylation element-binding protein (CPEB), T-cell-restricted intracellular antigen-1 (TIA-1), premelanosome protein 17 (Pmel17), secretory peptide hormones)

Table 1 Physiologically Important Amyloidogenic IDPs

Protein (Number of Residues)	Disease or Physiological Process	Disorder
Aβ (42)	Alzheimer's disease Dutch hereditary cerebral hemorrhage with amyloidosis Congophilic angiopathy	NMR and far-UV CD analyses revealed that the monomeric peptide is highly unfolded
Tau (758)	Tauopathies Alzheimer's disease Corticobasal degeneration Pick's disease Progressive supranuclear palsy	Tau protein was shown to be in a random coil-like conformation according to far-UV CD, FTIR, X-ray scattering, and biochemical assays
α-Synuclein (140)	Synucleinopathies Parkinson's disease Lewy body variant of Alzheimer's disease Diffuse Lewy body disease Dementia with Lewy bodies Multiple system atrophy Neurodegeneration with brain iron accumulation type I	Highly unfolded structure of entire protein is confirmed by NMR, FTIR, SAXS, far-UV CD, gel filtration, dynamic light scattering, FRET, limited proteolysis, and aberrant mobility in SDS-PAGE
Huntingtin (3144; polyQ tract: 16–37 glutamines in norm; >38 glutamines in pathology)	Huntington's disease	The far-UV CD spectra of poly(Gln) peptides with repeat lengths of 5, 15, 28, and 44 residues were shown to be nearly identical and were consistent with a high degree of random coil structure
DRPLA protein (1185; polyQ tract: 7–23 glutamines in norm; 49–75 glutamines in pathology)	Hereditary dentatorubral-pallidoluysian atrophy	Aberrant electrophoretic mobility. Apparent molecular mass estimated by SDS–PAGE is ~1.6-fold higher than the predicted molecular mass

Continued

Table 1 Physiologically Important Amyloidogenic IDPs—cont'd

Protein (Number of Residues)	Disease or Physiological Process	Disorder
Androgen receptor (919; polyQ tract: 15–31 glutamines in norm; 40–62 glutamines in pathology)	Kennedy's disease or X-linked spinal and bulbar muscular atrophy	Far-UV CD, gel filtration, limited proteolysis, ANS binding, and urea-induced unfolding studies revealed that the AF1 transactivation domain is in the molten globule state
Ataxin-1 (816; polyQ tract: 6–39 glutamines in norm; 41–81 glutamines in pathology)	Spinocerebellar ataxia 1 Neuronal intranuclear inclusion disease	
Ataxin-2 (1312; polyQ tract: 22–31 glutamines in norm; >32 glutamines in pathology)	Spinocerebellar ataxia 2	Ataxin-2 contains two globular domains, Lsm and LsmAD, in an acidic region (amino acid 254–475). The rest of ataxin-2 outside of the Lsm and LsmAD domains is predicted to be intrinsically disordered
Ataxin-3 (376; polyQ tract: 12–40 glutamines in norm; 55–84 glutamines in pathology)	Spinocerebellar ataxia 3	Far-UV CD and NMR spectroscopies suggest that ataxin-3 is only partially folded. The far-UV CD signal of the full-length protein is dominated by the Josephin motif (N-terminal domain 1–198), with the C-terminal portion of the protein making a smaller contribution, consistent with its largely unstructured conformation
P/Q-type calcium channel α1A subunit (2505; polyQ tract: 4–16 glutamines in norm; 21–28 glutamines in pathology)	Spinocerebellar ataxia 6	Aberrant electrophoretic mobility

Table 1 Physiologically Important Amyloidogenic IDPs—cont'd

Protein (Number of Residues)	Disease or Physiological Process	Disorder
Ataxin-7 (892; polyQ tract: 4–35 glutamines in norm; >36 glutamines in pathology)	Spinocerebellar ataxia 7	Aberrant electrophoretic mobility. Apparent molecular mass estimated by SDS-PAGE is 1.15-fold higher than that calculated from amino acid sequence
TATA-box-binding protein (339; polyQ tract: 25–42 glutamines in norm; >42 glutamines in pathology)	Spinocerebellar ataxia 17	Aberrant electrophoretic mobility. Apparent molecular mass estimated by SDS–PAGE is 1.3-fold higher than that calculated from amino acid sequence
ABri (34)	Familial British dementia	Far-UV CD and NMR spectroscopy revealed that ABri is in the random coil-like conformation at slightly acidic pH
ADan (34)	Familial Danish dementia	Far-UV CD revealed that ADan showed mostly random coil structure
CPEB (566)	Memory formation	
Pmel17 fragments (variable)	Melanosome formation	
Somatostatin 14 (14) and other peptide hormones	Growth-inhibiting hormone	Random coil by CD

Modified from Breydo, L., Uversky, V.N., 2011. Role of metal ions in aggregation of intrinsically disordered proteins in neurodegenerative diseases. Metallomics, 3, 1163–1180.

are also intrinsically disordered (Anoop et al., 2014; Gilks et al., 2004; Gottschald et al., 2010; Maji et al., 2009; McGlinchey et al., 2009; Si et al., 2003, 2010; Watt et al., 2009). In this chapter, we will focus on the aggregation of proteins and peptides where part of the sequence that forms an amyloid structure is intrinsically disordered. Some of these proteins (e.g., huntingtin (Htt)) may contain folded regions as well. For the sake of brevity, we will not cover folded proteins (e.g., prion protein) that contain intrinsically disordered regions that may be involved in protein aggregation (Prigent and Rezaei, 2011) but do not form a part of the amyloid structure.

Protein aggregation tends to start from partially folded, premolten globule-like conformations that expose the amyloidogenic regions of the protein to solvent (Uversky, 2008). Unlike natively folded or ordered proteins that need to undergo substantial unfolding to reach these conformations, IDPs amyloidogenic conformations are easily accessible and typically require partial folding. Depending on the protein and conditions, either conversion of the protein into the amyloidogenic conformation or formation of initial oligomeric aggregate could be the slow step in protein aggregation (Ferrone, 2015; Ramachandran et al., 2014; Vitalis and Caflisch, 2010; Vitalis et al., 2009). Over time, oligomers of amyloidogenic IDPs gradually increase in size and often gain secondary structure (usually a β-sheet-rich one). At some point in this process, the structures capable of serving as self-propagating nuclei may be generated. While many amyloid oligomers are kinetically stable, under appropriate conditions they may convert to fibrils. Kinetic relationship between oligomers and fibrils can be quite complex. Despite often containing β-sheet-rich structures, most stable oligomers are not directly on the pathway to fibrils.

Amyloid fibrils are extended, repetitive, β-sheet-rich structures, typically with the morphology of an extended twisted rope. Their high thermodynamic stability comes from association of multiple polypeptide molecules in a cross-β-structure. Fibrillar structures are organized on at least three levels: β-strands are stacked into β-sheets, β-sheets interact with each other to form protofilaments, and protofilaments stack on or twist around one another to form a fibril (Nelson et al., 2005; Sawaya et al., 2007). Differences in arrangement of β-sheets both on local and global level can result in populations of fibrils with distinct morphology. Fibrils can self-propagate by fragmentation followed by incorporation of monomers, and, in some cases, by secondary nucleation where fibrils present a catalytic surface for the formation of new aggregates (Arosio et al., 2015; Cohen et al., 2013). Propagation of fibrils usually preserves their conformation. However, conformational changes in the fibrils sometimes occur during propagation giving rise to a different structure either via a spontaneous structural change in the replicating fibril or due to a lack of fidelity during the replication process (Breydo, 2013; Ghaemmaghami et al., 2013; Giles et al., 2010; Kochneva-Pervukhova et al., 2001; Makarava and Baskakov, 2012; Makarava et al., 2011; Weissmann et al., 2011). Thus, the pathway of IDP aggregation is highly complex and a variety of factors can influence both kinetics of

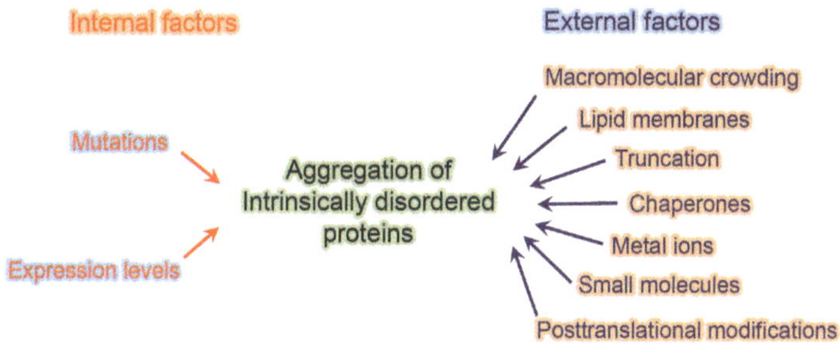

Fig. 1 Some of the internal and external factors affecting misfolding, aggregation, and fibrillation of intrinsically disordered proteins considered in this review.

aggregation and the structure of the aggregates. Effects of these factors will be examined later. Fig. 1 represents the factors affecting misfolding, aggregation, and fibrillation of IDPs that are considered in this review.

2. FACTORS INFLUENCING AGGREGATION OF IDPs

2.1 Intrinsic Factors

2.1.1 Mutations

Like other physical properties of proteins, propensity for aggregation is sequence dependent (Chiti et al., 2002, 2003; Dobson, 2002). Therefore, point mutations, deletions, and expansions can significantly influence the aggregation propensity of a protein or a peptide. For proteins whose aggregation is associated with various diseases, mutations promoting aggregation are often associated with increased disease pathology while mutations decreasing aggregation efficiency can be protective.

Aggregation of the intrinsically disordered Aβ peptides and tau protein plays an important role in AD. Tau protein in vivo is bound to microtubules, and mutations that either disrupt microtubule binding (e.g., K257T, K280Δ) or promote tau aggregation (e.g., L266V, N296Δ, P301L, Q336R, V337M, N410H) tend to be pathogenic (Hutton et al., 1998; Kouri et al., 2014; Momeni et al., 2009; Pastor et al., 2001; Pickering-Brown et al., 2000, 2004). In a number of mutants both functions are altered, and sometimes the mechanism of aggregation is affected as well with oligomers preferentially generated instead of fibrils. Aβ peptides (39–43 amino acids long) are generated by proteolytic cleavage of the amyloid precursor protein (APP). Mutations in both APP (e.g., K724N and V717L) (Murrell et al., 2000; Theuns et al., 2006) and in the respective protease (Ataka et al., 2004; Cai et al., 2015; Steiner et al., 2001) can alter the cleavage site, and thus changing the length of the resulting peptide. Since longer Aβ peptides (e.g., Aβ42) are significantly more amyloidogenic, mutations resulting in higher levels of Aβ42 are usually pathogenic (Cai et al., 2015; Guerreiro et al., 2010; Kumar-Singh et al., 2000). Other pathogenic Aβ mutations either increase aggregation propensity (e.g., E22Q, D23N) (Baumketner et al., 2008; Krone et al., 2008; Qiang et al., 2012) or cause a switch in the aggregation mechanism resulting in preferential formation of more toxic oligomeric aggregates (e.g., H6R, D7N, K16N, E22Δ) (Kaden et al., 2012; Tomiyama et al., 2008; Wakutani et al., 2004). These mutations are also associated with early onset AD (Bertram and Tanzi, 2012).

Three mutations in the α-synuclein gene (A53T, A30P, and E46K) were found in patients with familial PD (Kruger et al., 1998; Polymeropoulos et al., 1997; Zarranz et al., 2004). In fact, 85% of the patients who expressed the A53T mutant gene had clinical features of PD (Polymeropoulos et al., 1997). Analysis using a combination of biophysical techniques showed that A30P, E46K, and A53T mutations did not significantly alter the overall structure of α-synuclein monomer (Conway et al., 1998, 2000a,b,c; El-Agnaf et al., 1998; Giasson et al., 1999; Greenbaum et al., 2005; Li et al., 2001, 2002) but did cause some subtle structural changes (Bussell and Eliezer, 2001; Fredenburg et al., 2007; Rospigliosi et al., 2009). All three PD-related mutations were shown to accelerate α-synuclein aggregation (Choi et al., 2004; Conway et al., 1998, 2000b,c; Giasson et al., 1999; Greenbaum et al., 2005; Li et al., 2001, 2002; Narhi et al., 1999). A30P mutation promoted α-synuclein oligomer formation while A53T and E46K mutations promoted fibrillation. Recently, three novel pathogenic α-synuclein mutations (H50Q, G51D, and A53E) causing complex Parkinsonian phenotypes have been discovered (Petrucci et al., 2016). However, the effects of these mutations on the aggregation/fibrillation behavior of α-synuclein have not yet been studied.

Mutations in TDP-43 and FUS, closely related RNA-binding IDPs, were found in ALS and frontotemporal dementia (FTD) patients (Dewey et al., 2012). A number of TDP-43 mutations associated with increased incidence of ALS (e.g., N267S, G348C, and A328T) resulted in increased aggregation of this protein in vivo (Pesiridis et al., 2009) and formation of amyloid oligomers in vitro (Lim et al., 2016). Most disease causing mutations in FUS are clustered in the nuclear localization signal and result in the displacement of the protein from the nucleus leading to its aggregation

(Dormann et al., 2010). Moreover, many of these mutations increase FUS affinity for other proteins and decrease its affinity for RNA perhaps further promoting its aggregation (Sun et al., 2015).

A number of the aggregation-prone IDPs (Htt, several ataxins, TATA-binding protein (TBP), chromosome 9 open reading frame 72 (C9orf72) protein, etc.) are polyglutamine proteins (Edbauer and Haass, 2015; Kang and Hong, 2009; Li et al., 2015; Saunders and Bottomley, 2009; Trepte et al., 2014). While shorter polyQ stretches (10–20 residues) are common in healthy individuals, longer stretches of glutamines (30 + residues) are associated with both higher aggregation propensity and disease pathology (Table 1) (Edbauer and Haass, 2015; Matos et al., 2011; Rohrer et al., 2015; Saunders and Bottomley, 2009; Trepte et al., 2014).

2.1.2 Protein Expression Levels

Since aggregation is highly dependent on the concentration of the respective protein or peptide, higher levels of protein expression can also result in a significantly increased quantity of protein aggregates. One of the best-studied examples of this is early onset AD in Down syndrome patients (Hyman, 1992; Wiseman et al., 2015). Down syndrome (trisomy 21) patients have an extra copy of the APP gene resulting in increased expression of this protein and thus increased production of Aβ peptides. They consistently develop AD-like pathology and dementia in their 30s or 40s. While other factors may also be involved, there is a clear indication that overexpression of APP can lead to dementia. In addition, presenilin mutations that lead to the increased production of Aβ peptides are associated with AD pathology as well (Cai et al., 2015). Increased disease pathology and higher incidence of protein aggregates are also observed in animal models overexpressing α-synuclein (Aldrin-Kirk et al., 2014; Magen and Chesselet, 2010; Marmolino et al., 2015; Ulusoy and Di Monte, 2013). Overexpression of tau in animal models leads to cell dysfunction, amyotrophy, and/or axonopathy; although neurofibrillary tangles were not observed (Brion et al., 2010; de Silva and Farrer, 2002).

Overall, the vast majority of mutations in amyloidogenic IDPs linked to higher probability of the amyloid-related disease are also associated with either increased protein level in vivo, increased propensity of the protein to aggregate, or with the formation of more pathogenic aggregates.

2.1.3 Truncation

N-terminal and C-terminal truncations are common protein modifications, and they often have significant effects on the aggregation propensity and aggregation mechanism of the target protein. Amyloid β is probably the most obvious example of this phenomenon. Aβ peptides are produced by proteolysis of the nonamyloidogenic APP by β- and γ-secretases, and their aggregation propensity is highly dependent on the positions of both N-terminal and C-terminal cleavage sites. The C-terminal cleavage site is variable, and longer Aβ peptides (e.g., Aβ42) are consistently more amyloidogenic and associated with earlier incidence of AD. In fact, N-terminal truncations make up the majority of Aβ species in AD patients although not in AD mouse models (Schieb et al., 2011). N-terminal truncation of Aβ 42 at residue 3 produces a peptide that is more amyloidogenic and cytotoxic than the original (Wirths et al., 2009). Aβ42 peptides truncated at the residues 4 and 5 are present in AD patients as well (Kummer and Heneka, 2014; Lewis et al., 2006). In vivo, mice overexpressing Aβ4–42 suffer from CA1 pyramidal neuronal loss and memory dysfunction (Bouter et al., 2013; Kummer and Heneka, 2014). In addition, shorter Aβ fragments (e.g., Aβ 25–35) are produced by in vivo proteolysis (Kubo et al., 2002). These fragments form oligomeric aggregates (Do et al., 2015; Millucci et al., 2010) that were shown to be highly cytotoxic although not very effective at seeding aggregation of the full-length Aβ peptides (O'Nuallain et al., 2004). Proteolytic cleavage of Aβ by the α-secretase between the residues 16 and 17 produces the P3 peptide that is highly aggregation-prone (Vandersteen et al., 2012) but do not produce soluble oligomers (Dulin et al., 2008; Kummer and Heneka, 2014; Thinakaran and Koo, 2008).

IAPP is also produced by proteolysis of the precursor polypeptide (Westermark et al., 2011). Both C-terminal and N-terminal truncated fragments of this peptide have been observed (Miyazato et al., 1994; Wang et al., 1993), but their effect on IAPP aggregation or aggregate toxicity is unknown. Formation of a functional Pmel17 amyloid is also dependent on the proteolytic cleavage of the precursor Pmel17 protein (Kummer et al., 2009; Leonhardt et al., 2011).

Other amyloidogenic IDPs are not produced by the precursor cleavage, but their truncated versions are still quite common. For example, C-terminally truncated α-synuclein fragments (usually truncated between the residues 115 and 135) are often associated with Lewy bodies (Anderson et al., 2006; Li et al., 2005; Liu et al., 2005). These species are more amyloidogenic and more cytotoxic than the full-length protein and can initiate α-synuclein aggregation both in vivo and in vitro (Bertoncini et al., 2005; Li et al., 2005; Liu et al., 2005; Murray et al., 2003; Serpell et al., 2000; Wu et al., 2008). Experiments in transgenic animals expressing truncated α-synuclein showed neuronal loss and lower dopamine levels, especially if wild-type α-synuclein and its truncated versions were coexpressed (Periquet et al., 2007; Tofaris et al., 2006; Ulusoy et al., 2010).

Stepwise proteolysis of tau protein (initially at the N-terminal site followed by the C-terminal site) produces a variety of fragments that are more aggregation-prone than the full-length protein and capable of seeding its aggregation both in vitro and in cell models (Wang et al., 2009; Zhao et al., 2015b). Moreover, tau fragments were shown to coaggregate with amyloid β (Do et al., 2014). More extensive proteolysis of tau in vivo produces highly C-terminally truncated fragments incapable of aggregation (Kanmert et al., 2015).

Htt was shown to be cleaved by proteases resulting in N-terminal fragments of variable length containing the polyQ domain (Lunkes et al., 2002). In addition, N-terminal fragments of Htt are produced by aberrant splicing (Sathasivam et al., 2013). These N-terminal fragments (100 and 167 amino acids long) are aggregation-prone, form inclusions in the nucleus and cytoplasm, and their aggregates have been shown to be more cytotoxic than the full-length Htt (Barbaro et al., 2015; Landles et al., 2010). Proteolytic cleavage of ataxin 3 generated C-terminal fragments were more aggregation-prone than the full-length protein (Berke et al., 2004; Hubener et al., 2013; Simoes et al., 2014), and increased proteolytic cleavage was associated with an aggravated neurological phenotype in transgenic mice with an increased number of nuclear aggregates and accelerated neurodegeneration in the cerebellum (Goti et al., 2004; Hubener et al., 2013).

Overall, truncation of amyloidogenic IDPs usually produces the fragments that are more aggregation-prone than parent proteins and their aggregates tend to be more cytotoxic. The only exception to that rule is when the protein is cleaved in the middle of the amyloidogenic region (e.g., α-secretase cleavage of APP). In some cases (e.g., Aβ, IAPP, and Pmel17), only proteolytic fragments rather than a full-length protein are aggregation-prone in physiological conditions.

2.1.4 Posttranslational Modifications

Effect of posttranslational modifications (PTMs) on the protein secondary structure and susceptibility to conformational changes can be dramatic (Aebersold and Goodlett, 2001; Clark et al., 2005). Protein aggregation is a highly cooperative process, and even a small subpopulation of modified protein could have a substantial impact on kinetics and product distribution.

2.1.4.1 PTMs of Aβ Peptides

Increased oxidative stress plays an important role in the pathogenesis of AD (Zhao and Zhao, 2013). Increased levels of free radicals result in extensive protein oxidation, including methionine oxidation, tyrosine nitration and cross-linking, and other modifications. In the AD brain, most Aβ peptides are present in the Met35-oxidized form (Butterfield and Sultana, 2011). Several studies demonstrated that oxidation of Met35 impeded the formation of Aβ protofibrils and fibrils from monomers (Hou et al., 2002; Johansson et al., 2007), and promoted formation of less cytotoxic oligomers from Aβ42 (Bitan et al., 2003). Aβ nitrated at Tyr10 was able to initiate plaque formation in APP/PS1 mice and was detected in the core of amyloid plaques (Kummer et al., 2011). However, another study showed that this modification results in less cytotoxic Aβ oligomers (Zhao et al., 2015a). Tyrosine cross-links could also be detected in the core of amyloid plaques (Kummer et al., 2011) and may stabilize Aβ dimers (Al-Hilaly et al., 2013). Phosphorylation of Aβ at Ser8 promoted formation of oligomeric Aβ aggregates that could seed fibrillation of unmodified Aβ. This modified peptide showed increased toxicity in drosophila models as compared with nonphosphorylated Aβ (Kumar et al., 2011) and was detected in a subset of Aβ plaques and vascular deposits in late-stage AD (Ashby et al., 2015). Pyroglutamate-modified Aβ at position 3 (3pE-Aβ) is a highly neurotoxic Aβ isoform and is enriched in the brains of individuals with AD compared to healthy aged controls (Jawhar et al., 2011). Both 3pE-Aβ and 11pE-Aβ show increased propensity to aggregate in vitro, predominantly to oligomers (Kummer and Heneka, 2014; Nussbaum et al., 2012). 3pE-Aβ42 cooligomerizes with excess Aβ42 to form metastable oligomers that are structurally distinct and far more cytotoxic to cultured neurons than those made from Aβ42 alone. They can seed new cytotoxic oligomers through multiple serial dilutions into Aβ42 monomers. Enhanced 3pE-Aβ42 formation in mice triggered neuron loss and gliosis at 3 months (Nussbaum et al., 2012).

Isomerization of aspartates at positions 1, 7, and 23 of Aβ to isoaspartate increased the tendency of these peptides to aggregate in vitro (Fossati et al., 2013; Shimizu et al., 2002). Biochemical analyses revealed that Aβ purified from senile plaques and vascular amyloid are isomerized at Asp-1, Asp-7, and Asp-23 (Shimizu et al., 2000).

Overall, a variety of PTMs have been detected in Aβ peptides. Many of these PTMs alter the mechanism of their aggregation resulting in altered aggregate toxicity and their ability to self-propagate.

2.1.4.2 PTMs of Tau Protein

A variety of PTMs have been identified in tau protein (Fontaine et al., 2015; Mair et al., 2016; Morris et al., 2015; Song et al., 2015). The pathological tau was associated with multiple PTMs including phosphorylation, acetylation, and nitration. Tau phosphorylation is a dynamic, highly regulated process, requiring the precise interplay of a multitude of kinases and phosphatases. An increase in tau phosphorylation reduces its affinity for microtubules. Hyperphosphorylated tau is more abundant in the samples isolated from AD patients compared to controls (Mair et al., 2016; Morris et al., 2015). It specifically accumulated in neurofibrillary tangles and other tau deposits. Phosphorylation at most sites, especially in the C-terminal region and with the exception of some sites near the N-terminus, promoted tau aggregation in vitro (Haase et al., 2004; Liu et al., 2007). Increased aggregation propensity of highly phosphorylated tau is likely due to its facile conversion to a partially folded conformation, similar to that stabilized by anionic surfactants (Alonso et al., 1996; Eidenmuller et al., 2000; Hagestedt et al., 1989; Uversky et al., 1998).

Tau acetylation, especially at Lys174, was shown to occur at early stages of tauopathy (Min et al., 2010, 2015). The acetyl-mimicking mutant K174Q slowed tau turnover and induced cognitive deficits in vivo (Min et al., 2015). Tau acetylated at Lys280 was also shown to be neurotoxic in drosophila (Gorsky et al., 2016). Multiacetylation of tau at lysine residues by CREB-binding protein enhanced tau aggregation and reduced its ability to promote tubulin assembly (Cohen et al., 2011), whereas acetylation by p300 inhibited tau aggregation (Cook et al., 2014). It appears that tau acetylation promotes its pathological aggregation in most cases.

Increased tyrosine nitration at residues 18 and 29 was observed in glial and neuronal tau deposits in AD patients but it is not clear if this modification is contributing to pathology (Reyes et al., 2008; Reynolds et al., 2006). Tau has been found to be N-glycosylated in AD but not in controls, and the effect of this modification on tau aggregation is unknown (Wang et al., 1996). In contrast to N-glycosylation, levels of O-glycosylation of tau are decreased in AD brain compared to controls (Liu et al., 2009; Schedin-Weiss et al., 2014). In vitro and in vivo studies indicated that

O-GlcNAc modification of tau inhibited tau aggregation (Borghgraef et al., 2013; Yuzwa et al., 2012).

Overall, phosphorylation and acetylation of tau promoted its pathological aggregation while O-glycosylation interfered with it.

2.1.4.3 PTMs of α-Synuclein

Roles of PTMs in the aggregation of α-synuclein have been covered in a number of reviews (Beyer, 2006; Breydo et al., 2012). It was found to enhance aggregation of α-synuclein while reduction of Ser-129 phosphorylation in transgenic mice lead to a decrease in α-synuclein aggregation and improved motor performance (Azeredo da Silveira et al., 2009; Chung et al., 2001; Lee et al., 2011b; Smith et al., 2005). It has been estimated that 90% of α-synuclein in Lewy bodies is phosphorylated at Ser-129 (Fujiwara et al., 2002). Phosphorylation at Ser-87, on the other hand, inhibited aggregation in vitro (Paleologou et al., 2010). Phosphorylation of Tyr125 was shown to attenuate the conversion of α-synuclein to toxic oligomers (Chen et al., 2009; Nakamura et al., 2002). α-Synuclein nitrated at tyrosine residues formed oligomers instead of fibrils (Kovacs et al., 2014), but its presence accelerated fibril formation from unmodified protein (Hodara et al., 2004). Aggregates of nitrated α-synuclein were highly toxic to dopaminergic neurons and caused motor dysfunction in rats (Yu et al., 2010). Treatment of α-synuclein with oxidizing or nitrating agents can also result in oxidative cross-linking of tyrosines (Souza et al., 2000) and other residues (Norris et al., 2003). Tyrosine cross-linking has been shown to promote oligomerization of the protein and inhibit its transition to fibrils (Conway et al., 2001; Norris et al., 2003; Souza et al., 2000). Methionine-oxidized α-synuclein was found to be more highly unfolded than the nonoxidized protein (Glaser et al., 2005; Uversky et al., 2002b; Yamin et al., 2003), less prone to form oligomers and fibrils, and even able to inhibit the fibrillation of nonmodified α-synuclein (Uversky et al., 2002b). The inhibition of α-synuclein fibrillation by methionine oxidation was shown to be proportional to the number of oxidized methionines. It has been proposed that methionine oxidation disrupts end-to-end association of α-synuclein required for fibril formation, and thus directs its aggregation toward less structured, nontoxic oligomers (Rekas et al., 2010; Zhou et al., 2010). SUMOylation of α-synuclein was shown to promote its aggregation and decrease its toxicity in COS-7 cells (Oh et al., 2011). Ubiquitination of α-synuclein at Lys 6 interfered with its aggregation (Hejjaoui et al., 2011), but ubiquitination at other positions promoted α-synuclein aggregation (Breydo et al., 2012; Rott et al., 2008).

PTMs of α-synuclein usually modified the mechanism of its aggregation with variable effects on the aggregate toxicity.

2.1.4.4 PTMs of PolyQ Proteins

Effect of PTMs on polyQ proteins is highly variable. For ataxin-1, phosphorylation promoted aggregation (Duvick et al., 2010). SUMOylation prevented the proteasomal degradation of Htt and reduced the formation of insoluble aggregates although neurodegeneration was increased (possibly due to the formation of soluble oligomers) (Steffan et al., 2004). SUMOylation of ataxin-1, on the other hand, enhanced its aggregation (Ryu et al., 2010). Lack of palmitoylation of mutant Htt leads to its mislocalization and aggregation (Yanai et al., 2006). Phosphorylation of Htt at S13 and S16 reduced aggregate formation, while phosphorylation at T3 accelerated the aggregation process (Aiken et al., 2009; Ehrnhoefer et al., 2011; Gu et al., 2009).

2.2 External Factors

2.2.1 Binding of Metal Ions

Aβ peptides bind copper and zinc ions with nanomolar affinity at His 6 and His 13 (Tougu et al., 2008, 2011). Cu^{2+} binding was shown to direct Aβ aggregation toward formation of amorphous oligomers as well as transient dimers and trimers (Atwood et al., 1998; Bolognin et al., 2011; Chen et al., 2011; Sarell et al., 2010; Tougu et al., 2011). In addition, Aβ-bound copper ions participate in redox cycling, resulting in the generation of reactive oxygen species (ROS) (Nadal et al., 2008; Tabner et al., 2005). Zinc binding leads to rapid aggregation of Aβ into mixtures of oligomers (Bolognin et al., 2011; Bush et al., 1994; Chen et al., 2011; Noy et al., 2008). Aβ oligomers formed in the presence of Zn^{2+} were less stable than those formed in the absence of metal ions and reacted with ThT and the oligomer-specific A11 antibody (Bolognin et al., 2011; Chen et al., 2011). There appears to be a consensus that substoichiometric concentrations of Cu^{2+} and Zn^{2+} ions accelerate aggregation of Aβ into fibrils, but the excess of these ions promotes the formation of disordered oligomers instead (Chen et al., 2011; Cristovao et al., 2016; Sarell et al., 2010; Tougu et al., 2009). Iron and aluminum ions also bind in the N-terminal part of Aβ (Ali-Torres et al., 2011; Miura et al., 2001). Both ions were shown to promote the formation of cytotoxic β-sheet-rich oligomers and delay fibril formation (Bolognin et al., 2011; Chen et al., 2011; Drago et al., 2008; Liu et al., 2011; Ricchelli et al., 2005). In AD patients, it has been shown that Cu^{2+}, Zn^{2+}, and Fe^{2+} can be found in senile plaques colocalizing with Aβ (Cristovao et al., 2016).

Metal ions also promoted aggregation of tau to amyloid oligomers. Addition of aluminum salts promoted the formation of amorphous tau aggregates in vitro and in cell culture (Mizoroki et al., 2007; Shin et al., 1994). Fe^{3+} (but not Fe^{2+}) ions also bound to hyperphosphorylated tau and promoted its aggregation into oligomeric aggregates (Yamamoto et al., 2002). Calcium and magnesium ions also promoted aggregation of hyperphosphorylated tau into large oligomers (Yang and Ksiezak-Reding, 1999). Zinc was found to bind tau at micromolar concentrations in vitro via its cysteine and histidine residues and significantly accelerate its aggregation to fibrils. High concentrations of zinc promoted the formation of tau oligomers (Mo et al., 2009). Exposure to copper was shown to promote tau phosphorylation in vivo (Kitazawa et al., 2009; Voss et al., 2014) and induce its oligomerization in vitro (Soragni et al., 2008).

ABri is a 34 residue peptide that is the major component of amyloid deposits in familial British dementia (Plant et al., 1990). Metal ions have been shown to promote fibrillation of ABri at neutral pH (Khan et al., 2004). Incubation of ABri peptide in the presence of Al^{3+} or Fe^{3+} ions at neutral pH for an extended period of time resulted in a significant increase in the quantity of amyloid fibrils compared to metal-free controls (Khan et al., 2004).

Metal ions bind to the C-terminal region of α-synuclein with micromolar affinity. PTMs of α-synuclein (e.g., phosphorylation at Ser 129) increase its affinity for divalent metal ions (Carboni and Lingor, 2015; Lu et al., 2011). Metal binding induces α-synuclein to adopt a partially folded conformation accelerating its aggregation (Breydo and Uversky, 2011; Uversky et al., 2001). For example, incubation of α-synuclein with Al^{3+} leads to the formation of partially folded structures that is converted to amyloid oligomers over time (Paik et al., 1997; Uversky et al., 2001). Addition of Ca^{2+} directed the aggregation of α-synuclein to a mixture of annular and spherical oligomers (Lowe et al., 2004; Nath et al., 2011). Ferric ions were found to promote α-synuclein oligomer formation both in vitro and in living cells (Hillmer et al., 2009). Presence of Fe^{3+} also altered the morphology of α-synuclein fibrils making them shorter and thicker (Bharathi et al., 2007).

Addition of Mn^{2+} did not influence α-synuclein aggregation in vitro (Uversky et al., 2001), but induced its oligomerization in vivo, perhaps via indirect mechanisms such as ROS production (Xu et al., 2013). Zinc ions were shown to be effective promoters of α-synuclein fibril formation in vitro at the expense of oligomers (Hokenson et al., 2004; Kim et al., 2000; Uversky et al., 2001). Promotion of fibrillation of α-synuclein by zinc at the expense of oligomer formation may be protective since oligomers are more neurotoxic. On the other hand, Mg^{2+} inhibited α-synuclein aggregation (Golts et al., 2002) and protected dopaminergic neurons in the substantia nigra from MPP^{+}-mediated degeneration in transgenic rats (Hashimoto et al., 2008). Furthermore, Mg^{2+} counteracted the aggregation-promoting effect of other metal ions when they were present together (Andre et al., 2005). Cu^{+} and Cu^{2+} ions specifically bind to the N-terminus with nanomolar affinity with Cu^{2+} also binding to His50 and Cu^{+} binding to residues 119–121 (Bharathi and Rao, 2007; Binolfi et al., 2011; Davies et al., 2011; Dudzik et al., 2011; Lu et al., 2011; Rasia et al., 2005). Cu^{2+} was shown to be an effective accelerator of α-synuclein aggregation to oligomers even at physiologically relevant concentrations without altering the morphology of the resulting fibrils (Brown, 2009; Natalello et al., 2011; Rasia et al., 2005; Uversky et al., 2001). Neurotoxicity of α-synuclein oligomers was increased by the presence of Cu^{2+} as well (Wright et al., 2009). Cu depletion in vivo resulted in redistribution of α-Syn toward the membrane and reduces aggregate formation (Wang et al., 2010b).

Role of metal ions in the aggregation of polyQ proteins has not been extensively investigated. Copper was found to bind to the first 170 residues of Htt that include polyQ repeats with micromolar affinity and promote aggregation of this protein into oligomeric aggregates (Fox et al., 2007; Xiao et al., 2013). Divalent metal ions increased the exposure of hydrophobic surfaces in ataxin-3 and the amount of β-structure in this protein (Ricchelli et al., 2007; Stawoska et al., 2009). However, only Zn^{2+} and Al^{3+} promoted aggregation of ataxin-3 with 26 or 36 glutamines into amyloid fibrils by shortening the lag phase of fibril formation (Ricchelli et al., 2007).

Overall, metal ions tend to stabilize specific conformations of IDPs in the process of aggregation often altering both kinetics and mechanism of aggregation. In many cases, they direct aggregation away from amyloid fibrils and toward oligomeric aggregates.

2.2.2 Interaction With Small Molecules

2.2.2.1 Noncovalent Binding

Most small molecules capable of noncovalent binding to IDP aggregates do so in a relatively nonspecific manner by interaction with certain structural features (e.g., amyloid core) common for all protein aggregates (Andrich and Bieschke, 2015; Porat et al., 2006). Aromatic stacking plays an important role in amyloid assembly process for many proteins (Gazit, 2002; Porat et al., 2006; Tartaglia et al., 2004). Water-soluble aromatic molecules can often bind to hydrophobic interior of protein aggregates interfering with the aggregation process. For example, Aβ fibrillation was inhibited by catecholamine derivatives (Lashuel et al., 2002), resveratrol (Ladiwala et al., 2010; Marambaud et al., 2005), tannic acid (Ono et al., 2004a), curcumin (Ono et al., 2004b; Yang et al., 2005), epigallocatechin gallate (EGCG) (Andrich and Bieschke, 2015), and other aromatic compounds. Many of these compounds inhibited fibrillation of other amyloidogenic IDPs (e.g., α-synuclein, amylin, and tau) although not necessarily with the same efficiency. A more detailed look at these inhibitors show that they often do not prevent aggregation altogether but rather alter its mechanism. A number of small molecules specifically inhibited formation of either fibrils or oligomeric aggregates but rarely both (Necula et al., 2007a,b). For example, EGCG directed Aβ aggregation toward oligomers where the residues 22–39 were folded into a β-sheet while the rest of the peptide was unstructured (Lopez del Amo et al., 2012). Curcumin also promoted formation of "off-pathway" soluble Aβ oligomers and prefibrillar aggregates that were nontoxic (Thapa et al., 2015). Methylene blue, on the other hand, inhibited Aβ oligomerization by promoting fibril formation (Necula et al., 2007a). A variety of aromatic molecules such as EGCG (Bieschke et al., 2010; Ehrnhoefer et al., 2008; Lorenzen et al., 2014), baicalein (Hong et al., 2008), dopamine (Cappai et al., 2005; Lee et al., 2011a; Rekas et al., 2010), other polyphenols (Meng et al., 2010), thioflavins (Grelle et al., 2011), curcumin (Ono et al., 2008; Pandey et al., 2008; Wang et al., 2010a), an anti-PD drug selegiline (Braga et al., 2011), β-hairpins (Sivanesam et al., 2015), and many other compounds (Bodner et al., 2006; Masuda et al., 2006) induced formation of mostly disordered α-synuclein oligomers, although in some cases (Hong et al., 2008) residual β-structure was found. Many of these compounds preferentially bound to the C-terminus of α-synuclein (Breydo and Uversky, 2015; Ehrnhoefer et al., 2008; Herrera et al., 2008; Sivanesam et al., 2015). EGCG also promoted the formation of primarily disordered oligomers from a variety of proteins, such as tau (Wobst et al., 2015), ataxin-3 (Bonanomi et al., 2014), and many others (Andrich and Bieschke, 2015) indicating a common mechanism of action. Highly charged small molecules, on the other hand, tend to promote fibril formation. It applies to both negatively charged molecules (e.g., heparan sulfate) (Holmes and Diamond, 2014; Zhang and Li, 2010) and to positively charged molecules (e.g., polyamines) (Luo et al., 2013).

In addition to generic amyloid-binding agents, development of selective ligands targeting specific protein aggregates is also a feasible approach due to structural differences between these aggregates (Irwin et al., 2015; Kumar et al., 2015; Nath et al., 2015). For example, small molecules (either designed or selected by high-throughput screening) can target Aβ, α-synuclein, and other proteins stabilizing their structure and preventing their aggregation (Cheruvara et al., 2015; de Almeida et al., 2015; McKoy et al., 2014; Saunders et al., 2016; Wang et al., 2014; Zheng et al., 2015). In some cases, this targeting is based on binding to specific residues (e.g., lysines for molecular tweezers) (Attar et al., 2014; Lopes et al., 2015; Zheng et al., 2015). Alternatively, small molecules can be designed to target the heterogeneous conformational ensemble of an unfolded monomeric protein (Nath et al., 2015). By altering the aggregate structure or preventing their formation altogether both generic- and sequence-specific aggregation modulators were able to decrease aggregate cytotoxicity for a variety of proteins and peptides (Ardah et al., 2015; Cheruvara et al., 2015; Guzior et al., 2015; Scherzer-Attali et al., 2010; Wang et al., 2014).

Small molecules can alter the mechanism of IDP aggregation by interacting with these proteins with various degrees of affinity and specificity. A variety of small molecules are being developed in order to direct IDP aggregation toward stable, nontoxic oligomers, or to prevent it altogether.

2.2.2.2 Covalent Modification

Small molecules are also capable of covalent modification of proteins, usually involving lysine or cysteine residues. For amyloidogenic IDPs, these modifications usually interfere with the formation of fibrils directing aggregation toward amyloid oligomers instead. It is likely that this happens because bulky adducts interfere with β-sheet stacking necessary for fibril formation while looser oligomer structure might be more accommodating to these modifications. For example, dopamine is highly susceptible to oxidation and its oxidation products can form adducts with α-synuclein (Bisaglia et al., 2010; Rekas et al., 2010). These adducts drive aggregation of α-synuclein into primarily unstructured, SDS-resistant oligomers (Bisaglia et al., 2010;

Lee et al., 2011a; Rekas et al., 2010). Lipid-derived aldehydes such as HNE reacted with α-synuclein to form lysine adducts and promoted formation of either β-sheet-rich (Nasstrom et al., 2009, 2011; Qin et al., 2007; Siegel et al., 2007) or primarily disordered (Bae et al., 2013) oligomers, depending on the reaction conditions. Interestingly, the oligomers formed after α-synuclein modification with HNE were more stable and less toxic than those formed in the presence of 4-oxo-2-nonenal (Nasstrom et al., 2011). Schiff-base formation between Aβ and the aldehyde-bearing cholesterol oxidation products promoted the formation of cytotoxic oligomers (Usui et al., 2009). Specific covalent aggregation inhibitors have also been designed, and in some cases, they have been shown to promote formation of disordered, less toxic amyloid oligomers (Arai et al., 2014).

Overall, covalent modification of IDPs with small molecules interferes with their aggregation usually resulting in oligomeric aggregates. In some cases, aggregation may be inhibited altogether.

2.2.3 Macromolecular Crowding

The cellular environment is crowded with macromolecules, such as proteins, nucleic acids, and carbohydrates, which occupy up to 30% of the available volume. In crowded environments, the structures of both folded and natively unfolded proteins become more compact and the effective protein concentrations increase (Minton, 1997; Zhou et al., 2008). Aggregation of IDPs consists of multiple steps: (1) the structural collapse of the monomer producing aggregation-prone partially folded species, (2) the formation of oligomers inducing the nucleation process, and (3) the growth and elongation of fibrils. The excluded volume effect is expected to favor the first two steps due to structural compaction of the protein in the crowded environment, but other effects of polymers (e.g., direct interaction with the protein or changes in protein solvation) could have an effect as well (Fig. 2). Simulations showed that the net influence of macromolecular crowding on the aggregation process is the result of two competing effects: aggregate stabilization and solution viscosity increase (Magno et al., 2010). Analysis of the effect of macromolecular crowding on aggregation of 150 proteins showed that the effects were relatively mild and depended on the nature of both protein and crowding agent (Niwa et al., 2015). Aggregation of various IDPs (Aβ, α-synuclein, β-synuclein, and tau) is accelerated by typical coil-like crowding agents, such as dextran and polyethylene glycol, or a neutral, highly branched polysaccharide Ficoll, or inert proteins (Engel et al., 2008; Lee et al., 2012; Minton, 2000; Morar et al., 2001;

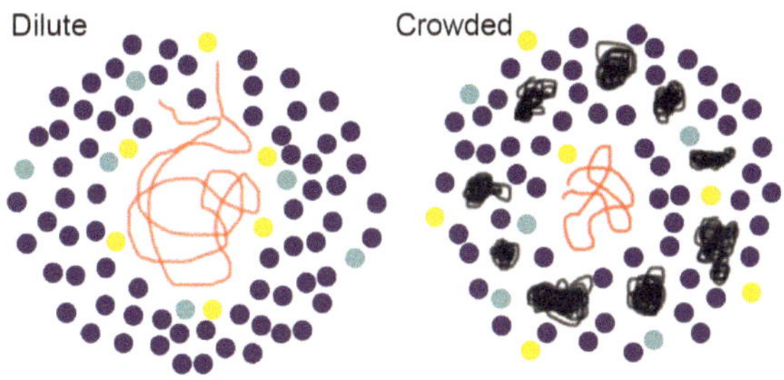

Fig. 2 Effect of macromolecular crowding on conformation of IDPs. The presence of other large molecules can force the protein to compact into a partially folded species that is prone to aggregation. *Yellow spheres* are negatively charged, and *light blue spheres* are positively charged.

Mukherjee et al., 2009; Munishkina et al., 2004, 2008a,b, 2009; Shtilerman et al., 2002; Uversky et al., 2002a; Yamin et al., 2005; Zhou et al., 2009). For example, formation of α-synuclein protofibrils (Shtilerman et al., 2002) and fibrillation of tau (especially hyperphosphorylated tau) (Wu et al., 2015; Zhou et al., 2009), although not fibrillation of IAPP (Gao and Winter, 2015; Seeliger et al., 2013) were accelerated in the presence of crowding agents. Oligomeric aggregates formed by IAPP in the presence of crowding agents were shown to be more cytotoxic than fibrils (Weise et al., 2010). Fibril formation from α-synuclein was inhibited by rod-like polymers (Breydo et al., 2014) as well as moderately hydrophobic polymers with aggregation in both cases directed toward oligomers instead (Breydo et al., 2015; Ferreira et al., 2015).

2.2.4 Lipid Membranes

Lipid membranes play an important role in protein aggregation in vivo. Moreover, membrane lipids are often incorporated into amyloid aggregates (Gellermann et al., 2005). Effect of membranes on IDP aggregation also strongly depends on the membrane charge and composition. For example, negatively charged membranes significantly accelerated aggregation of IAPP and Aβ to amyloid fibrils while neutral ones had almost no effect (Gao and Winter, 2015; Jha et al., 2009; McLaurin and Chakrabartty, 1997). The presence of gangliosides and cholesterol in the lipid membrane accelerated IAPP and Aβ aggregation into fibrils (Choo-Smith et al., 1997; Sasahara et al., 2014; Wakabayashi and Matsuzaki, 2009). IDPs (e.g., IAPP and α-synuclein) can convert to α-helical conformations upon membrane binding (Dikiy and Eliezer, 2012; Duan et al., 2012; Jao et al., 2008; Nanga et al., 2008, 2009; Ulmer et al., 2005) and aggregate further into metastable α-helical oligomers (Gao and Winter, 2015; Varkey et al., 2013). Thereafter, rapid conversion to a β-sheet conformation takes place, followed by formation of ordered fibrillar structures. This process is often accompanied by dissolution of the membrane along with incorporation of membrane lipids into the protein aggregates (Gao and Winter, 2015; Mahul-Mellier et al., 2015). Preformed amyloid oligomers are also often capable of membrane binding. Association with the membrane can accelerate conversion of these oligomers either to amyloid pores (Kayed et al., 2009) or to more compact and cytotoxic oligomeric aggregates (Sasahara et al., 2014).

2.2.5 Chaperones

Protein chaperones act to promote folding, block aggregation, disaggregate proteins, and facilitate protein degradation (Duncan et al., 2015; Finka et al., 2015). In many cases, they do so by redirecting aggregation toward globular, amorphous oligomers. Alternatively, ATP-dependent chaperones sequester proteins in their binding sites and allow them to fold (Hayer-Hartl et al., 2016; Stull et al., 2016) or stabilize them in monomeric form (Chaari et al., 2016) (see Fig. 3). Moreover, chaperones can disassemble α-synuclein fibrils to generate monomeric protein (Gao et al., 2015). For example, a protein chaperone serum amyloid P component redirected aggregation of IAPP from fibrils to primarily disordered oligomers (Gao and Winter, 2015). In addition, they are often able to suppress the cytotoxic effects of protein

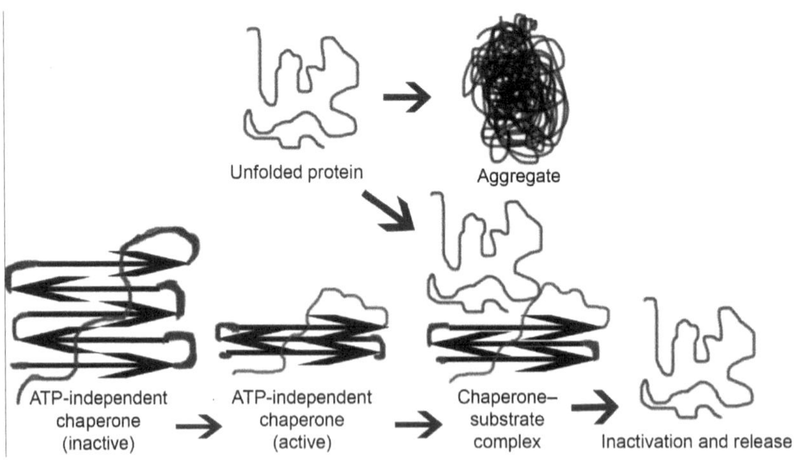

Fig. 3 Effect of ATP-independent chaperones on IDP aggregation. A chaperone acts to prevent the aggregation of an IDP by binding to it.

aggregates after these are formed (Binger et al., 2013; Cascella et al., 2013; Lindberg et al., 2015; Ojha et al., 2011). Incubation of Aβ oligomers with Hsp27 suppressed their cytotoxicity by converting them into larger aggregates (Ojha et al., 2011). The same happened with Aβ and IAPP amyloid oligomers upon incubation with α-B-crystallin, Hsp70, and other chaperones (Mannini et al., 2014). Interestingly, α-B-crystallin also suppressed the toxicity of protein fibrils by converting them to large fibril clusters (Binger et al., 2013). Hsp90 and its co-chaperones promote the conversion of toxic tau oligomers in the brain into larger, nontoxic aggregates (Blair et al., 2013; Lindberg et al., 2015). Brichos domain, a part of many APPs, can inhibit fibril-catalyzed production of amyloid oligomers and limit Aβ fibril toxicity. It achieves this inhibition by binding to the surfaces of fibrils and blocking their catalysis of oligomer formation (Cohen et al., 2015). One of HSP40 chaperones, HSJ1a, was shown to prevent TDP-43 aggregation in vivo (Chen et al., 2016).

Chaperones thus play a crucial role in detoxification of protein aggregates in vivo.

3. CONCLUSIONS

Because of their lack of stable 3D structure, IDPs are extremely sensitive to their environment, much more so than ordered proteins. Also, this lack of fixed structure is often associated with extreme binding promiscuity of IDPs. As a result, aggregation of IDPs is easily influenced by both internal factors, such as mutations and expression levels, and by diverse external features, such as macromolecular crowding and interaction with metal ions, other small molecules, lipid membranes, and chaperons. Overall, functional and pathological behaviors of IDPs at physiological conditions are influenced by a variety of factors. Their relative importance and the possibility for interplay should be the subject of further investigation.

ACKNOWLEDGMENTS

This work was supported by a grant from the Russian Science Foundation RSCF no. 14-24-00131.

REFERENCES

Ali-Torres, J., Rodriguez-Santiago, L., Sodupe, M., et al., 2011. Structures and stabilities of $Fe(2+/3+)$ complexes relevant to Alzheimer's disease: an ab initio study. J. Phys. Chem. A 115 (45), 12523–12530.

Aebersold, R., Goodlett, D.R., 2001. Mass spectrometry in proteomics. Chem. Rev. 101, 269–295.

Aiken, C.T., Steffan, J.S., Guerrero, C.M., et al., 2009. Phosphorylation of threonine 3: implications for Huntingtin aggregation and neurotoxicity. J. Biol. Chem. 284, 29427–29436.

Aldrin-Kirk, P., Davidsson, M., Holmqvist, S., et al., 2014. Novel AAV-based rat model of forebrain synucleinopathy shows extensive pathologies and progressive loss of cholinergic interneurons. PLoS One 9, e100869.

Al-Hilaly, Y.K., Williams, T.L., Stewart-Parker, M., et al., 2013. A central role for dityrosine crosslinking of Amyloid-beta in Alzheimer's disease. Acta Neuropathol. Commun. 1, 83.

Alonso, A.C., Grundke-Iqbal, I., Iqbal, K., 1996. Alzheimer's disease hyperphosphorylated tau sequesters normal tau into tangles of filaments and disassembles microtubules. Nat. Med. 2, 783–787.

Anderson, J.P., Walker, D.E., Goldstein, J.M., et al., 2006. Phosphorylation of Ser-129 is the dominant pathological modification of alpha-synuclein in familial and sporadic Lewy body disease. J. Biol. Chem. 281, 29739–29752.

Andre, C., Truong, T.T., Robert, J.F., et al., 2005. Effect of metals on herbicides-alpha-synuclein association: a possible factor in neurodegenerative disease studied by capillary electrophoresis. Electrophoresis 26, 3256–3264.

Andrich, K., Bieschke, J., 2015. The effect of (−)-Epigallo-catechin-(3)-gallate on amyloidogenic proteins suggests a common mechanism. Adv. Exp. Med. Biol. 863, 139–161.

Anoop, A., Ranganathan, S., Das Dhaked, B., et al., 2014. Elucidating the role of disulfide bond on amyloid formation and fibril reversibility of somatostatin-14: relevance to its storage and secretion. J. Biol. Chem. 289, 16884–16903.

Arai, T., Sasaki, D., Araya, T., et al., 2014. A cyclic KLVFF-derived peptide aggregation inhibitor induces the formation of less-toxic off-pathway amyloid-beta oligomers. Chembiochem 15, 2577–2583.

Ardah, M.T., Paleologou, K.E., Lv, G., et al., 2015. Ginsenoside Rb1 inhibits fibrillation and toxicity of alpha-synuclein and disaggregates preformed fibrils. Neurobiol. Dis. 74, 89–101.

Arosio, P., Knowles, T.P., Linse, S., 2015. On the lag phase in amyloid fibril formation. Phys. Chem. Chem. Phys. 17, 7606–7618.

Ashby, E.L., Miners, J.S., Kumar, S., et al., 2015. Investigation of Abeta phosphorylated at serine 8 (pAbeta) in Alzheimer's disease, dementia with Lewy bodies and vascular dementia. Neuropathol. Appl. Neurobiol. 41, 428–444.

Ataka, S., Tomiyama, T., Takuma, H., et al., 2004. A novel presenilin-1 mutation (Leu85Pro) in early-onset Alzheimer disease with spastic paraparesis. Arch. Neurol. 61, 1773–1776.

Attar, A., Chan, W.T., Klarner, F.G., et al., 2014. Safety and pharmacological characterization of the molecular tweezer CLR01—a broad-spectrum inhibitor of amyloid proteins' toxicity. BMC Pharmacol. Toxicol. 15, 23.

Atwood, C.S., Moir, R.D., Huang, X., et al., 1998. Dramatic aggregation of Alzheimer abeta by Cu(II) is induced by conditions representing physiological acidosis. J. Biol. Chem. 273, 12817–12826.

Azeredo Da Silveira, S., Schneider, B.L., Cifuentes-Diaz, C., et al., 2009. Phosphorylation does not prompt, nor prevent, the formation of alpha-synuclein toxic species in a rat model of Parkinson's disease. Hum. Mol. Genet. 18, 872–887.

Bae, E.J., Ho, D.H., Park, E., et al., 2013. Lipid peroxidation product 4-hydroxy-2-nonenal promotes seeding-capable oligomer formation and cell-to-cell transfer of alpha-synuclein. Antioxid. Redox Signal. 18, 770–783.

Barbaro, B.A., Lukacsovich, T., Agrawal, N., et al., 2015. Comparative study of naturally occurring huntingtin fragments in Drosophila points to exon 1 as the most pathogenic species in Huntington's disease. Hum. Mol. Genet. 24, 913–925.

Baumketner, A., Krone, M.G., Shea, J.E., 2008. Role of the familial Dutch mutation E22Q in the folding and aggregation of the 15-28 fragment of the Alzheimer amyloid-beta protein. Proc. Natl. Acad. Sci. U.S.A. 105, 6027–6032.

Berke, S.J., Schmied, F.A., Brunt, E.R., et al., 2004. Caspase-mediated proteolysis of the polyglutamine disease protein ataxin-3. J. Neurochem. 89, 908–918.

Bertoncini, C.W., Jung, Y.S., Fernandez, C.O., et al., 2005. Release of long-range tertiary interactions potentiates aggregation of natively unstructured alpha-synuclein. Proc. Natl. Acad. Sci. U.S.A. 102, 1430–1435.

Bertram, L., Tanzi, R.E., 2012. The genetics of Alzheimer's disease. Prog. Mol. Biol. Transl. Sci. 107, 79–100.

Beyer, K., 2006. Alpha-synuclein structure, posttranslational modification and alternative splicing as aggregation enhancers. Acta Neuropathol. (Berl) 112, 237–251.

Bharathi, Rao, K.S., 2007. Thermodynamics imprinting reveals differential binding of metals to alpha-synuclein: relevance to Parkinson's disease. Biochem. Biophys. Res. Commun. 359, 115–120.

Bharathi, Indi, S.S., Rao, K.S., 2007. Copper- and iron-induced differential fibril formation in alpha-synuclein: TEM study. Neurosci. Lett. 424, 78–82.

Bieschke, J., Russ, J., Friedrich, R.P., et al., 2010. EGCG remodels mature alpha-synuclein and amyloid-beta fibrils and reduces cellular toxicity. Proc. Natl. Acad. Sci. U.S.A. 107, 7710–7715.

Binger, K.J., Ecroyd, H., Yang, S., et al., 2013. Avoiding the oligomeric state: alphaB-crystallin inhibits fragmentation and induces dissociation of apolipoprotein C-II amyloid fibrils. FASEB J. 27, 1214–1222.

Binolfi, A., Valiente-Gabioud, A.A., Duran, R., et al., 2011. Exploring the structural details of Cu(I) binding to alpha-synuclein by NMR spectroscopy. J. Am. Chem. Soc. 133, 194–196.

Bisaglia, M., Tosatto, L., Munari, F., et al., 2010. Dopamine quinones interact with alpha-synuclein to form unstructured adducts. Biochem. Biophys. Res. Commun. 394, 424–428.

Bitan, G., Tarus, B., Vollers, S.S., et al., 2003. A molecular switch in amyloid assembly: Met35 and amyloid beta-protein oligomerization. J. Am. Chem. Soc. 125, 15359–15365.

Blair, L.J., Nordhues, B.A., Hill, S.E., et al., 2013. Accelerated neurodegeneration through chaperone-mediated oligomerization of tau. J. Clin. Invest. 123, 4158–4169.

Bodner, R.A., Outeiro, T.F., Altmann, S., et al., 2006. Pharmacological promotion of inclusion formation: a therapeutic approach for Huntington's and Parkinson's diseases. Proc. Natl. Acad. Sci. U.S.A. 103, 4246–4251.

Bolognin, S., Messori, L., Drago, D., et al., 2011. Aluminum, copper, iron and zinc differentially alter amyloid-Abeta(1-42) aggregation and toxicity. Int. J. Biochem. Cell Biol. 43, 877–885.

Bonanomi, M., Natalello, A., Visentin, C., et al., 2014. Epigallocatechin-3-gallate and tetracycline differently affect ataxin-3 fibrillogenesis and reduce toxicity in spinocerebellar ataxia type 3 model. Hum. Mol. Genet. 23, 6542–6552.

Borghgraef, P., Menuet, C., Theunis, C., et al., 2013. Increasing brain protein O-GlcNAcylation mitigates breathing defects and mortality of Tau.P301L mice. PLoS One 8, e84442.

Bouter, Y., Dietrich, K., Wittnam, J.L., et al., 2013. N-truncated amyloid beta (Abeta) 4-42 forms stable aggregates and induces acute and long-lasting behavioral deficits. Acta Neuropathol. 126, 189–205.

Braga, C.A., Follmer, C., Palhano, F.L., et al., 2011. The anti-Parkinsonian drug selegiline delays the nucleation phase of alpha-synuclein aggregation leading to the formation of nontoxic species. J. Mol. Biol. 405, 254–273.
Breydo, L., 2013. Strain phenomenon in protein aggregation: interplay between sequence and conformation. Intrinsically Disord. Proteins 1, 39.
Breydo, L., Uversky, V.N., 2011. Role of metal ions in aggregation of intrinsically disordered proteins in neurodegenerative diseases. Metallomics 3, 1163–1180.
Breydo, L., Uversky, V.N., 2015. Structural, morphological, and functional diversity of amyloid oligomers. FEBS Lett. 589, 2640–2648.
Breydo, L., Wu, J.W., Uversky, V.N., 2012. Alpha-synuclein misfolding and Parkinson's disease. Biochim. Biophys. Acta 1822, 261–285.
Breydo, L., Reddy, K.D., Piai, A., et al., 2014. The crowd you're in with: effects of different types of crowding agents on protein aggregation. Biochim. Biophys. Acta 1844, 346–357.
Breydo, L., Sales, A.E., Frege, T., et al., 2015. Effects of polymer hydrophobicity on protein structure and aggregation kinetics in crowded milieu. Biochemistry 54, 2957–2966.
Brion, J.P., Ando, K., Heraud, C., et al., 2010. Modulation of tau pathology in tau transgenic models. Biochem. Soc. Trans. 38, 996–1000.
Brown, D.R., 2009. Metal binding to alpha-synuclein peptides and its contribution to toxicity. Biochem. Biophys. Res. Commun. 380, 377–381.
Bush, A.I., Pettingell, W.H., Multhaup, G., et al., 1994. Rapid induction of Alzheimer A beta amyloid formation by zinc [see comments]. Science 265, 1464–1467.
Bussell Jr., R., Eliezer, D., 2001. Residual structure and dynamics in Parkinson's disease-associated mutants of alpha-synuclein. J. Biol. Chem. 276, 45996–46003.
Butterfield, D.A., Sultana, R., 2011. Methionine-35 of abeta(1-42): importance for oxidative stress in Alzheimer disease. J. Amino Acids 2011, 198430.
Cai, Y., An, S.S., Kim, S., 2015. Mutations in presenilin 2 and its implications in Alzheimer's disease and other dementia-associated disorders. Clin. Interv. Aging 10, 1163–1172.
Cappai, R., Leck, S.L., Tew, D.J., et al., 2005. Dopamine promotes alpha-synuclein aggregation into SDS-resistant soluble oligomers via a distinct folding pathway. FASEB J. 19, 1377–1379.
Carboni, E., Lingor, P., 2015. Insights on the interaction of alpha-synuclein and metals in the pathophysiology of Parkinson's disease. Metallomics 7, 395–404.
Cascella, R., Conti, S., Mannini, B., et al., 2013. Transthyretin suppresses the toxicity of oligomers formed by misfolded proteins in vitro. Biochim. Biophys. Acta 1832, 2302–2314.
Chaari, A., Eliezer, D., Ladjimi, M., 2016. The C-terminal alpha-helices of mammalian Hsc70 play a critical role in the stabilization of alpha-synuclein binding and inhibition of aggregation. Int. J. Biol. Macromol. 83, 433–441.
Chen, L., Periquet, M., Wang, X., et al., 2009. Tyrosine and serine phosphorylation of alpha-synuclein have opposing effects on neurotoxicity and soluble oligomer formation. J. Clin. Invest. 119, 3257–3265.
Chen, W.T., Liao, Y.H., Yu, H.M., et al., 2011. Distinct effects of Zn2+, Cu2+, Fe3+, and Al3+ on amyloid-beta stability, oligomerization, and aggregation: amyloid-beta destabilization promotes annular protofibril formation. J. Biol. Chem. 286, 9646–9656.
Chen, H.J., Mitchell, J.C., Novoselov, S., et al., 2016. The heat shock response plays an important role in TDP-43 clearance: evidence for dysfunction in amyotrophic lateral sclerosis. Brain 139, 1417–1432.
Cheruvara, H., Allen-Baume, V.L., Kad, N.M., et al., 2015. Intracellular screening of a peptide library to derive a potent peptide inhibitor of alpha-synuclein aggregation. J. Biol. Chem. 290, 7426–7435.

Chiti, F., Taddei, N., Baroni, F., et al., 2002. Kinetic partitioning of protein folding and aggregation. Nat. Struct. Biol. 9, 137–143.
Chiti, F., Stefani, M., Taddei, N., et al., 2003. Rationalization of the effects of mutations on peptide and protein aggregation rates. Nature 424, 805–808.
Choi, W., Zibaee, S., Jakes, R., et al., 2004. Mutation E46K increases phospholipid binding and assembly into filaments of human alpha-synuclein. FEBS Lett. 576, 363–368.
Choo-Smith, L.P., Garzon-Rodriguez, W., Glabe, C.G., et al., 1997. Acceleration of amyloid fibril formation by specific binding of Abeta-(1-40) peptide to ganglioside-containing membrane vesicles. J. Biol. Chem. 272, 22987–22990.
Chung, K.K., Zhang, Y., Lim, K.L., et al., 2001. Parkin ubiquitinates the alpha-synuclein-interacting protein, synphilin-1: implications for Lewy-body formation in Parkinson disease. Nat. Med. 7, 1144–1150.
Clark, R.S., Bayir, H., Jenkins, L.W., 2005. Posttranslational protein modifications. Crit. Care Med. 33, S407–S409.
Cohen, T.J., Guo, J.L., Hurtado, D.E., et al., 2011. The acetylation of tau inhibits its function and promotes pathological tau aggregation. Nat. Commun. 2, 252.
Cohen, S.I., Linse, S., Luheshi, L.M., et al., 2013. Proliferation of amyloid-beta42 aggregates occurs through a secondary nucleation mechanism. Proc. Natl. Acad. Sci. U.S.A. 110, 9758–9763.
Cohen, S.I., Arosio, P., Presto, J., et al., 2015. A molecular chaperone breaks the catalytic cycle that generates toxic Abeta oligomers. Nat. Struct. Mol. Biol. 22, 207–213.
Conway, K.A., Harper, J.D., Lansbury, P.T., 1998. Accelerated in vitro fibril formation by a mutant alpha-synuclein linked to early-onset Parkinson disease. Nat. Med. 4, 1318–1320.
Conway, K.A., Harper, J.D., Lansbury Jr., P.T., 2000a. Fibrils formed in vitro from alpha-synuclein and two mutant forms linked to Parkinson's disease are typical amyloid. Biochemistry 39, 2552–2563.
Conway, K.A., Lee, S.J., Rochet, J.C., et al., 2000b. Accelerated oligomerization by Parkinson's disease linked alpha-synuclein mutants. Ann. N.Y. Acad. Sci. 920, 42–45.
Conway, K.A., Lee, S.J., Rochet, J.C., et al., 2000c. Acceleration of oligomerization, not fibrillization, is a shared property of both alpha-synuclein mutations linked to early-onset Parkinson's disease: implications for pathogenesis and therapy. Proc. Natl. Acad. Sci. U.S.A. 97, 571–576.
Conway, K.A., Rochet, J.C., Bieganski, R.M., et al., 2001. Kinetic stabilization of the alpha-synuclein protofibril by a dopamine-alpha-synuclein adduct. Science 294, 1346–1349.
Cook, C., Carlomagno, Y., Gendron, T.F., et al., 2014. Acetylation of the KXGS motifs in tau is a critical determinant in modulation of tau aggregation and clearance. Hum. Mol. Genet. 23, 104–116.
Cristovao, J.S., Santos, R., Gomes, C.M., 2016. Metals and neuronal metal binding proteins implicated in Alzheimer's disease. Oxid. Med. Cell. Longev. 2016, 9812178.
Davies, P., Wang, X., Sarell, C.J., et al., 2011. The synucleins are a family of redox-active copper binding proteins. Biochemistry 50, 37–47.
De Almeida, N.E., Do, T.D., Tro, M., et al., 2015. Opposing effects of cucurbit[7]uril and 1,2,3,4,6-penta-O-galloyl-beta-d-glucopyranose on amyloid beta assembly. ACS Chem. Neurosci. 7, 218–226.
De Silva, R., Farrer, M., 2002. Tau neurotoxicity without the lesions: a fly challenges a tangled web. Trends Neurosci. 25, 327–329.
Dewey, C.M., Cenik, B., Sephton, C.F., et al., 2012. TDP-43 aggregation in neurodegeneration: are stress granules the key? Brain Res. 1462, 16–25.
Dikiy, I., Eliezer, D., 2012. Folding and misfolding of alpha-synuclein on membranes. Biochim. Biophys. Acta 1818, 1013–1018.

Do, T.D., Economou, N.J., Chamas, A., et al., 2014. Interactions between amyloid-beta and Tau fragments promote aberrant aggregates: implications for amyloid toxicity. J. Phys. Chem. B 118, 11220–11230.

Do, T.D., Lapointe, N.E., Nelson, R., et al., 2015. Amyloid beta-protein C-terminal fragments: formation of cylindrins and beta-barrels. J. Am. Chem. Soc. 138, 549–557.

Dobson, C.M., 2002. Getting out of shape. Nature 418, 729–730.

Dormann, D., Rodde, R., Edbauer, D., et al., 2010. ALS-associated fused in sarcoma (FUS) mutations disrupt Transportin-mediated nuclear import. EMBO J. 29, 2841–2857.

Drago, D., Bettella, M., Bolognin, S., et al., 2008. Potential pathogenic role of beta-amyloid(1-42)-aluminum complex in Alzheimer's disease. Int. J. Biochem. Cell Biol. 40, 731–746.

Duan, M., Fan, J., Huo, S., 2012. Conformations of islet amyloid polypeptide monomers in a membrane environment: implications for fibril formation. PLoS One 7, e47150.

Dudzik, C.G., Walter, E.D., Millhauser, G.L., 2011. Coordination features and affinity of the Cu(2)+ site in the alpha-synuclein protein of Parkinson's disease. Biochemistry 50, 1771–1777.

Dulin, F., Leveille, F., Ortega, J.B., et al., 2008. P3 peptide, a truncated form of A beta devoid of synaptotoxic effect, does not assemble into soluble oligomers. FEBS Lett. 582, 1865–1870.

Duncan, E.J., Cheetham, M.E., Chapple, J.P., et al., 2015. The role of HSP70 and its co-chaperones in protein misfolding, aggregation and disease. Subcell. Biochem. 78, 243–273.

Duvick, L., Barnes, J., Ebner, B., et al., 2010. SCA1-like disease in mice expressing wild-type ataxin-1 with a serine to aspartic acid replacement at residue 776. Neuron 67, 929–935.

Edbauer, D., Haass, C., 2015. An amyloid-like cascade hypothesis for C9orf72 ALS/FTD. Curr. Opin. Neurobiol. 36, 99–106.

Ehrnhoefer, D.E., Bieschke, J., Boeddrich, A., et al., 2008. EGCG redirects amyloidogenic polypeptides into unstructured, off-pathway oligomers. Nat. Struct. Mol. Biol. 15, 558–566.

Ehrnhoefer, D.E., Sutton, L., Hayden, M.R., 2011. Small changes, big impact: posttranslational modifications and function of huntingtin in Huntington disease. Neuroscientist 17, 475–492.

Eidenmuller, J., Fath, T., Hellwig, A., et al., 2000. Structural and functional implications of tau hyperphosphorylation: information from phosphorylation-mimicking mutated tau proteins. Biochemistry 39, 13166–13175.

El-Agnaf, O.M., Jakes, R., Curran, M.D., et al., 1998. Effects of the mutations Ala30 to Pro and Ala53 to Thr on the physical and morphological properties of alpha-synuclein protein implicated in Parkinson's disease. FEBS Lett. 440, 67–70.

Engel, R., Westphal, A.H., Huberts, D.H., et al., 2008. Macromolecular crowding compacts unfolded apoflavodoxin and causes severe aggregation of the off-pathway intermediate during apoflavodoxin folding. J. Biol. Chem. 283, 27383–27394.

Ferreira, L.A., Madeira, P.P., Breydo, L., et al., 2015. Role of solvent properties of aqueous media in macromolecular crowding effects. J. Biomol. Struct. Dyn, 1–12.

Ferrone, F.A., 2015. Assembly of Abeta proceeds via monomeric nuclei. J. Mol. Biol. 427, 287–290.

Finka, A., Sharma, S.K., Goloubinoff, P., 2015. Multi-layered molecular mechanisms of polypeptide holding, unfolding and disaggregation by HSP70/HSP110 chaperones. Front. Mol. Biosci. 2, 29.

Fontaine, S.N., Sabbagh, J.J., Baker, J., et al., 2015. Cellular factors modulating the mechanism of tau protein aggregation. Cell. Mol. Life Sci. 72, 1863–1879.

Fossati, S., Todd, K., Sotolongo, K., et al., 2013. Differential contribution of isoaspartate post-translational modifications to the fibrillization and toxic properties of amyloid beta and the Asn23 Iowa mutation. Biochem. J. 456, 347–360.

Fox, J.H., Kama, J.A., Lieberman, G., et al., 2007. Mechanisms of copper ion mediated Huntington's disease progression. PLoS One 2, e334.
Fredenburg, R.A., Rospigliosi, C., Meray, R.K., et al., 2007. The impact of the E46K mutation on the properties of alpha-synuclein in its monomeric and oligomeric states. Biochemistry 46, 7107–7118.
Fujiwara, H., Hasegawa, M., Dohmae, N., et al., 2002. alpha-Synuclein is phosphorylated in synucleinopathy lesions. Nat. Cell Biol. 4, 160–164.
Gao, M., Winter, R., 2015. The effects of lipid membranes, crowding and osmolytes on the aggregation, and fibrillation propensity of human IAPP. J. Diabetes Res. 2015, 849017.
Gao, X., Carroni, M., Nussbaum-Krammer, C., et al., 2015. Human Hsp70 disaggregase reverses Parkinson's-linked alpha-Synuclein amyloid fibrils. Mol. Cell 59, 781–793.
Gazit, E., 2002. A possible role for pi-stacking in the self-assembly of amyloid fibrils. FASEB J. 16, 77–83.
Gellermann, G.P., Appel, T.R., Tannert, A., et al., 2005. Raft lipids as common components of human extracellular amyloid fibrils. Proc. Natl. Acad. Sci. U.S.A. 102, 6297–6302.
Ghaemmaghami, S., Colby, D.W., Nguyen, H.O., et al., 2013. Convergent replication of mouse synthetic prion strains. Am. J. Pathol. 182, 866–874.
Giasson, B.I., Uryu, K., Trojanowski, J.Q., et al., 1999. Mutant and wild type human alpha-synucleins assemble into elongated filaments with distinct morphologies in vitro. J. Biol. Chem. 274, 7619–7622.
Giles, K., Glidden, D.V., Patel, S., et al., 2010. Human prion strain selection in transgenic mice. Ann. Neurol. 68, 151–161.
Gilks, N., Kedersha, N., Ayodele, M., et al., 2004. Stress granule assembly is mediated by prion-like aggregation of TIA-1. Mol. Biol. Cell 15, 5383–5398.
Glaser, C.B., Yamin, G., Uversky, V.N., et al., 2005. Methionine oxidation, alpha-synuclein and Parkinson's disease. Biochim. Biophys. Acta 1703, 157–169.
Golts, N., Snyder, H., Frasier, M., et al., 2002. Magnesium inhibits spontaneous and iron-induced aggregation of alpha-synuclein. J. Biol. Chem. 277, 16116–16123.
Gorsky, M.K., Burnouf, S., Dols, J., et al., 2016. Acetylation mimic of lysine 280 exacerbates human Tau neurotoxicity in vivo. Sci. Rep. 6, 22685.
Goti, D., Katzen, S.M., Mez, J., et al., 2004. A mutant ataxin-3 putative-cleavage fragment in brains of Machado-Joseph disease patients and transgenic mice is cytotoxic above a critical concentration. J. Neurosci. 24, 10266–10279.
Gottschald, O.R., Malec, V., Krasteva, G., et al., 2010. TIAR and TIA-1 mRNA-binding proteins co-aggregate under conditions of rapid oxygen decline and extreme hypoxia and suppress the HIF-1alpha pathway. J. Mol. Cell Biol. 2, 345–356.
Greenbaum, E.A., Graves, C.L., Mishizen-Eberz, A.J., et al., 2005. The E46K mutation in alpha-synuclein increases amyloid fibril formation. J. Biol. Chem. 280, 7800–7807.
Grelle, G., Otto, A., Lorenz, M., et al., 2011. Black tea theaflavins inhibit formation of toxic amyloid-beta and alpha-synuclein fibrils. Biochemistry 50, 10624–10636.
Gu, X., Greiner, E.R., Mishra, R., et al., 2009. Serines 13 and 16 are critical determinants of full-length human mutant huntingtin induced disease pathogenesis in HD mice. Neuron 64, 828–840.
Guerreiro, R.J., Baquero, M., Blesa, R., et al., 2010. Genetic screening of Alzheimer's disease genes in Iberian and African samples yields novel mutations in presenilins and APP. Neurobiol. Aging 31, 725–731.
Guzior, N., Bajda, M., Skrok, M., et al., 2015. Development of multifunctional, heterodimeric isoindoline-1,3-dione derivatives as cholinesterase and beta-amyloid aggregation inhibitors with neuroprotective properties. Eur. J. Med. Chem. 92, 738–749.
Haase, C., Stieler, J.T., Arendt, T., et al., 2004. Pseudophosphorylation of tau protein alters its ability for self-aggregation. J. Neurochem. 88, 1509–1520.

Hagestedt, T., Lichtenberg, B., Wille, H., et al., 1989. Tau protein becomes long and stiff upon phosphorylation: correlation between paracrystalline structure and degree of phosphorylation. J. Cell Biol. 109, 1643–1651.
Hashimoto, T., Nishi, K., Nagasao, J., et al., 2008. Magnesium exerts both preventive and ameliorating effects in an in vitro rat Parkinson disease model involving 1-methyl-4-phenylpyridinium (MPP+) toxicity in dopaminergic neurons. Brain Res. 1197, 143–151.
Hayer-Hartl, M., Bracher, A., Hartl, F.U., 2016. The GroEL-GroES chaperonin machine: a nano-cage for protein folding. Trends Biochem. Sci. 41, 62–76.
Hejjaoui, M., Haj-Yahya, M., Kumar, K.S., et al., 2011. Towards elucidation of the role of ubiquitination in the pathogenesis of Parkinson's disease with semisynthetic ubiquitinated alpha-synuclein. Angew. Chem. Int. Ed. Engl. 50, 405–409.
Herrera, F.E., Chesi, A., Paleologou, K.E., et al., 2008. Inhibition of alpha-synuclein fibrillization by dopamine is mediated by interactions with five C-terminal residues and with E83 in the NAC region. PLoS One 3, e3394.
Hillmer, A.S., Putcha, P., Levin, J., et al., 2009. Converse modulation of toxic alpha-synuclein oligomers in living cells by N′-benzylidene-benzohydrazide derivates and ferric iron. Biochem. Biophys. Res. Commun 391, 461–466.
Hodara, R., Norris, E.H., Giasson, B.I., et al., 2004. Functional consequences of alpha-synuclein tyrosine nitration: diminished binding to lipid vesicles and increased fibril formation. J. Biol. Chem. 279, 47746–47753.
Hokenson, M.J., Uversky, V.N., Goers, J., et al., 2004. Role of individual methionines in the fibrillation of methionine-oxidized alpha-synuclein. Biochemistry 43, 4621–4633.
Holmes, B.B., Diamond, M.I., 2014. Prion-like properties of Tau protein: the importance of extracellular Tau as a therapeutic target. J. Biol. Chem. 289, 19855–19861.
Hong, D.P., Fink, A.L., Uversky, V.N., 2008. Structural characteristics of alpha-synuclein oligomers stabilized by the flavonoid baicalein. J. Mol. Biol. 383, 214–223.
Hou, L., Kang, I., Marchant, R.E., et al., 2002. Methionine 35 oxidation reduces fibril assembly of the amyloid abeta-(1-42) peptide of Alzheimer's disease. J. Biol. Chem. 277, 40173–40176.
Hubener, J., Weber, J.J., Richter, C., et al., 2013. Calpain-mediated ataxin-3 cleavage in the molecular pathogenesis of spinocerebellar ataxia type 3 (SCA3). Hum. Mol. Genet. 22, 508–518.
Hutton, M., Lendon, C.L., Rizzu, P., et al., 1998. Association of missense and 5′-splice-site mutations in tau with the inherited dementia FTDP-17. Nature 393, 702–705.
Hyman, B.T., 1992. Down syndrome and Alzheimer disease. Prog. Clin. Biol. Res. 379, 123–142.
Irwin, J.A., Edward Wong, H., Kwon, I., 2015. Determining binding sites of polycyclic aromatic small molecule-based amyloid-beta peptide aggregation modulators using sequence-specific antibodies. Anal. Biochem. 470, 61–70.
Jao, C.C., Hegde, B.G., Chen, J., et al., 2008. Structure of membrane-bound alpha-synuclein from site-directed spin labeling and computational refinement. Proc. Natl. Acad. Sci. U.S.A. 105, 19666–19671.
Jawhar, S., Wirths, O., Bayer, T.A., 2011. Pyroglutamate amyloid-beta (Abeta): a hatchet man in Alzheimer disease. J. Biol. Chem. 286, 38825–38832.
Jha, S., Sellin, D., Seidel, R., et al., 2009. Amyloidogenic propensities and conformational properties of ProIAPP and IAPP in the presence of lipid bilayer membranes. J. Mol. Biol. 389, 907–920.
Johansson, A.S., Bergquist, J., Volbracht, C., et al., 2007. Attenuated amyloid-beta aggregation and neurotoxicity owing to methionine oxidation. Neuroreport 18, 559–563.
Kaden, D., Harmeier, A., Weise, C., et al., 2012. Novel APP/Abeta mutation K16N produces highly toxic heteromeric Abeta oligomers. EMBO Mol. Med. 4, 647–659.

Kang, S., Hong, S., 2009. Molecular pathogenesis of spinocerebellar ataxia type 1 disease. Mol. Cells 27, 621–627.

Kanmert, D., Cantlon, A., Muratore, C.R., et al., 2015. C-terminally truncated forms of tau, but not full-length tau or its c-terminal fragments, are released from neurons independently of cell death. J. Neurosci. 35, 10851–10865.

Kayed, R., Pensalfini, A., Margol, L., et al., 2009. Annular protofibrils are a structurally and functionally distinct type of amyloid oligomer. J. Biol. Chem. 284, 4230–4237.

Khan, A., Ashcroft, A.E., Korchazhkina, O.V., et al., 2004. Metal-mediated formation of fibrillar ABri amyloid. J. Inorg. Biochem. 98, 2006–2010.

Kim, T.D., Paik, S.R., Yang, C.H., et al., 2000. Structural changes in alpha-synuclein affect its chaperone-like activity in vitro. Protein Sci. 9, 2489–2496.

Kitazawa, M., Cheng, D., Laferla, F.M., 2009. Chronic copper exposure exacerbates both amyloid and tau pathology and selectively dysregulates cdk5 in a mouse model of AD. J. Neurochem. 108, 1550–1560.

Kochneva-Pervukhova, N.V., Chechenova, M.B., Valouev, I.A., et al., 2001. [Psi(+)] prion generation in yeast: characterization of the 'strain' difference. Yeast 18, 489–497.

Kouri, N., Carlomagno, Y., Baker, M., et al., 2014. Novel mutation in MAPT exon 13 (p.N410H) causes corticobasal degeneration. Acta Neuropathol. 127, 271–282.

Kovacs, G.G., Breydo, L., Green, R., et al., 2014. Intracellular processing of disease-associated alpha-synuclein in the human brain suggests prion-like cell-to-cell spread. Neurobiol. Dis. 69, 76–92.

Krone, M.G., Baumketner, A., Bernstein, S.L., et al., 2008. Effects of familial Alzheimer's disease mutations on the folding nucleation of the amyloid beta-protein. J. Mol. Biol. 381, 221–228.

Kruger, R., Kuhn, W., Muller, T., et al., 1998. Ala30Pro mutation in the gene encoding alpha-synuclein in Parkinson's disease. Nat. Genet. 18, 106–108.

Kubo, T., Nishimura, S., Kumagae, Y., et al., 2002. In vivo conversion of racemized beta-amyloid ([D-Ser 26]A beta 1-40) to truncated and toxic fragments ([D-Ser 26]A beta 25-35/40) and fragment presence in the brains of Alzheimer's patients. J. Neurosci. Res. 70, 474–483.

Kumar, S., Rezaei-Ghaleh, N., Terwel, D., et al., 2011. Extracellular phosphorylation of the amyloid beta-peptide promotes formation of toxic aggregates during the pathogenesis of Alzheimer's disease. EMBO J. 30, 2255–2265.

Kumar, S., Schlamadinger, D.E., Brown, M.A., et al., 2015. Islet amyloid-induced cell death and bilayer integrity loss share a molecular origin targetable with oligopyridylamide-based alpha-helical mimetics. Chem. Biol. 22, 369–378.

Kumar-Singh, S., De Jonghe, C., Cruts, M., et al., 2000. Nonfibrillar diffuse amyloid deposition due to a gamma(42)-secretase site mutation points to an essential role for N-truncated A beta(42) in Alzheimer's disease. Hum. Mol. Genet. 9, 2589–2598.

Kummer, M.P., Heneka, M.T., 2014. Truncated and modified amyloid-beta species. Alzheimers Res. Ther. 6, 28.

Kummer, M.P., Maruyama, H., Huelsmann, C., et al., 2009. Formation of Pmel17 amyloid is regulated by juxtamembrane metalloproteinase cleavage, and the resulting C-terminal fragment is a substrate for gamma-secretase. J. Biol. Chem. 284, 2296–2306.

Kummer, M.P., Hermes, M., Delekarte, A., et al., 2011. Nitration of tyrosine 10 critically enhances amyloid beta aggregation and plaque formation. Neuron 71, 833–844.

Ladiwala, A.R.A., Lin, J.C., Bale, S.S., et al., 2010. Resveratrol selectively remodels soluble oligomers and fibrils of amyloid abeta into off-pathway conformers. J. Biol. Chem. 285, 24228–24237.

Landles, C., Sathasivam, K., Weiss, A., et al., 2010. Proteolysis of mutant huntingtin produces an exon 1 fragment that accumulates as an aggregated protein in neuronal nuclei in Huntington disease. J. Biol. Chem. 285, 8808–8823.

Lashuel, H.A., Hartley, D.M., Balakhaneh, D., et al., 2002. New class of inhibitors of amyloid-beta fibril formation. Implications for the mechanism of pathogenesis in Alzheimer's disease. J. Biol. Chem. 277, 42881–42890.
Lee, H.J., Baek, S.M., Ho, D.H., et al., 2011a. Dopamine promotes formation and secretion of non-fibrillar alpha-synuclein oligomers. Exp. Mol. Med. 43, 216–222.
Lee, K.W., Chen, W., Junn, E., et al., 2011b. Enhanced phosphatase activity attenuates {alpha}-Synucleinopathy in a mouse model. J. Neurosci. 31, 6963–6971.
Lee, C.F., Bird, S., Shaw, M., et al., 2012. Combined effects of agitation, macromolecular crowding, and interfaces on amyloidogenesis. J. Biol. Chem. 287, 38006–38019.
Leonhardt, R.M., Vigneron, N., Rahner, C., et al., 2011. Proprotein convertases process Pmel17 during secretion. J. Biol. Chem. 286, 9321–9337.
Lewis, H., Beher, D., Cookson, N., et al., 2006. Quantification of Alzheimer pathology in ageing and dementia: age-related accumulation of amyloid-beta(42) peptide in vascular dementia. Neuropathol. Appl. Neurobiol. 32, 103–118.
Li, J., Uversky, V.N., Fink, A.L., 2001. Effect of familial Parkinson's disease point mutations A30P and A53T on the structural properties, aggregation, and fibrillation of human alpha-synuclein. Biochemistry 40, 11604–11613.
Li, J., Uversky, V.N., Fink, A.L., 2002. Conformational behavior of human alpha-synuclein is modulated by familial Parkinson's disease point mutations A30P and A53T. Neurotoxicology 23, 553–567.
Li, W., West, N., Colla, E., et al., 2005. Aggregation promoting C-terminal truncation of alpha-synuclein is a normal cellular process and is enhanced by the familial Parkinson's disease-linked mutations. Proc. Natl. Acad. Sci. U.S.A. 102, 2162–2167.
Li, X., Liu, H., Fischhaber, P.L., et al., 2015. Toward therapeutic targets for SCA3: insight into the role of Machado-Joseph disease protein ataxin-3 in misfolded proteins clearance. Prog. Neurobiol. 132, 34–58.
Lim, L., Wei, Y., Lu, Y., et al., 2016. ALS-causing mutations significantly perturb the self-assembly and interaction with nucleic acid of the intrinsically disordered prion-like domain of TDP-43. PLoS Biol. 14, e1002338.
Lindberg, I., Shorter, J., Wiseman, R.L., et al., 2015. Chaperones in neurodegeneration. J. Neurosci. 35, 13853–13859.
Liu, C.W., Giasson, B.I., Lewis, K.A., et al., 2005. A precipitating role for truncated alpha-synuclein and the proteasome in alpha-synuclein aggregation: implications for pathogenesis of Parkinson disease. J. Biol. Chem. 280, 22670–22678.
Liu, F., Li, B., Tung, E.J., et al., 2007. Site-specific effects of tau phosphorylation on its microtubule assembly activity and self-aggregation. Eur. J. Neurosci. 26, 3429–3436.
Liu, F., Shi, J., Tanimukai, H., et al., 2009. Reduced O-GlcNAcylation links lower brain glucose metabolism and tau pathology in Alzheimer's disease. Brain 132, 1820–1832.
Liu, B., Moloney, A., Meehan, S., et al., 2011. Iron promotes the toxicity of amyloid beta peptide by impeding its ordered aggregation. J. Biol. Chem. 286, 4248–4256.
Lopes, D.H., Attar, A., Nair, G., et al., 2015. Molecular tweezers inhibit islet amyloid polypeptide assembly and toxicity by a new mechanism. ACS Chem. Biol. 10, 1555–1569.
Lopez Del Amo, J.M., Fink, U., Dasari, M., et al., 2012. Structural properties of EGCG-induced, nontoxic Alzheimer's disease Abeta oligomers. J. Mol. Biol. 421, 517–524.
Lorenzen, N., Nielsen, S.B., Yoshimura, Y., et al., 2014. How epigallocatechin gallate can inhibit alpha-synuclein oligomer toxicity in vitro. J. Biol. Chem. 289, 21299–21310.
Lowe, R., Pountney, D.L., Jensen, P.H., et al., 2004. Calcium(II) selectively induces alpha-synuclein annular oligomers via interaction with the C-terminal domain. Protein Sci. 13, 3245–3252.
Lu, Y., Prudent, M., Fauvet, B., et al., 2011. Phosphorylation of alpha-Synuclein at Y125 and S129 alters its metal binding properties: implications for understanding the role of alpha-Synuclein in the pathogenesis of Parkinson's disease and related disorders. ACS Chem. Neurosci. 2, 667–675.

Lunkes, A., Lindenberg, K.S., Ben-Haiem, L., et al., 2002. Proteases acting on mutant huntingtin generate cleaved products that differentially build up cytoplasmic and nuclear inclusions. Mol. Cell 10, 259–269.
Luo, J., Yu, C.H., Yu, H., et al., 2013. Cellular polyamines promote amyloid-beta (Abeta) peptide fibrillation and modulate the aggregation pathways. ACS Chem. Neurosci. 4, 454–462.
Magen, I., Chesselet, M.F., 2010. Genetic mouse models of Parkinson's disease the state of the art. Prog. Brain Res. 184, 53–87.
Magno, A., Caflisch, A., Pellarin, R., 2010. Crowding effects on amyloid aggregation kinetics. J. Phys. Chem. Lett. 1, 3027–3032.
Mahul-Mellier, A.L., Vercruysse, F., Maco, B., et al., 2015. Fibril growth and seeding capacity play key roles in alpha-synuclein-mediated apoptotic cell death. Cell Death Differ. 22, 2107–2122.
Mair, W., Muntel, J., Tepper, K., et al., 2016. FLEXITau: quantifying post-translational modifications of tau protein in vitro and in human disease. Anal. Chem. 88, 3704–3714.
Maji, S.K., Perrin, M.H., Sawaya, M.R., et al., 2009. Functional amyloids as natural storage of peptide hormones in pituitary secretory granules. Science 325, 328–332.
Makarava, N., Baskakov, I.V., 2012. Genesis of transmissible protein states via deformed templating. Prion 6, 252–255.
Makarava, N., Kovacs, G.G., Savtchenko, R., et al., 2011. Genesis of mammalian prions: from non-infectious amyloid fibrils to a transmissible prion disease. PLoS Pathog. 7, e1002419.
Mannini, B., Mulvihill, E., Sgromo, C., et al., 2014. Toxicity of protein oligomers is rationalized by a function combining size and surface hydrophobicity. ACS Chem. Biol. 9, 2309–2317.
Marambaud, P., Zhao, H., Davies, P., 2005. Resveratrol promotes clearance of Alzheimer's disease amyloid-beta peptides. J. Biol. Chem. 280, 37377–37382.
Marmolino, D., Foerch, P., Atienzar, F.A., et al., 2015. Alpha synuclein dimers and oligomers are increased in overexpressing conditions in vitro and in vivo. Mol. Cell. Neurosci. 71, 92–101.
Masuda, M., Suzuki, N., Taniguchi, S., et al., 2006. Small molecule inhibitors of alpha-synuclein filament assembly. Biochemistry 45, 6085–6094.
Matos, C.A., De Macedo-Ribeiro, S., Carvalho, A.L., 2011. Polyglutamine diseases: the special case of ataxin-3 and Machado-Joseph disease. Prog. Neurobiol. 95, 26–48.
Mcglinchey, R.P., Shewmaker, F., Mcphie, P., et al., 2009. The repeat domain of the melanosome fibril protein Pmel17 forms the amyloid core promoting melanin synthesis. Proc. Natl. Acad. Sci. U.S.A. 106, 13731–13736.
Mckoy, A.F., Chen, J., Schupbach, T., et al., 2014. Structure-activity relationships for a series of compounds that inhibit aggregation of the Alzheimer's peptide, Abeta42. Chem. Biol. Drug Des. 84, 505–512.
Mclaurin, J., Chakrabartty, A., 1997. Characterization of the interactions of Alzheimer beta-amyloid peptides with phospholipid membranes. Eur. J. Biochem. 245, 355–363.
Meng, X., Munishkina, L.A., Fink, A.L., et al., 2010. Effects of various flavonoids on the alpha-Synuclein fibrillation process. Parkinsons Dis. 2010, 650794.
Millucci, L., Ghezzi, L., Bernardini, G., et al., 2010. Conformations and biological activities of amyloid beta peptide 25-35. Curr. Protein Pept. Sci. 11, 54–67.
Min, S.W., Cho, S.H., Zhou, Y., et al., 2010. Acetylation of tau inhibits its degradation and contributes to tauopathy. Neuron 67, 953–966.
Min, S.W., Chen, X., Tracy, T.E., et al., 2015. Critical role of acetylation in tau-mediated neurodegeneration and cognitive deficits. Nat. Med. 21, 1154–1162.
Minton, A.P., 1997. Influence of excluded volume upon macromolecular structure and associations in 'crowded' media. Curr. Opin. Biotechnol. 8, 65–69.

Minton, A.P., 2000. Implications of macromolecular crowding for protein assembly. Curr. Opin. Struct. Biol. 10, 34–39.
Miura, T., Suzuki, K., Takeuchi, H., 2001. Binding of iron(III) to the single tyrosine residue of amyloid beta-peptide probed by Raman spectroscopy. J. Mol. Struct. 598, 79–84.
Miyazato, M., Nakazato, M., Shiomi, K., et al., 1994. Isolation and sequence determination of two N-terminal fragments of islet amyloid polypeptide in rat pancreas. Regul. Pept. 49, 203–210.
Mizoroki, T., Meshitsuka, S., Maeda, S., et al., 2007. Aluminum induces tau aggregation in vitro but not in vivo. J. Alzheimers Dis. 11, 419–427.
Mo, Z.Y., Zhu, Y.Z., Zhu, H.L., et al., 2009. Low micromolar zinc accelerates the fibrillization of human tau via bridging of Cys-291 and Cys-322. J. Biol. Chem. 284, 34648–34657.
Momeni, P., Pittman, A., Lashley, T., et al., 2009. Clinical and pathological features of an Alzheimer's disease patient with the MAPT Delta K280 mutation. Neurobiol. Aging 30, 388–393.
Morar, A.S., Olteanu, A., Young, G.B., et al., 2001. Solvent-induced collapse of alpha-synuclein and acid-denatured cytochrome c. Protein Sci. 10, 2195–2199.
Morris, M., Knudsen, G.M., Maeda, S., et al., 2015. Tau post-translational modifications in wild-type and human amyloid precursor protein transgenic mice. Nat. Neurosci. 18, 1183–1189.
Mukherjee, S., Chowdhury, P., Gai, F., 2009. Effect of dehydration on the aggregation kinetics of two amyloid peptides. J. Phys. Chem. B 113, 531–535.
Munishkina, L.A., Cooper, E.M., Uversky, V.N., et al., 2004. The effect of macromolecular crowding on protein aggregation and amyloid fibril formation. J. Mol. Recognit. 17, 456–464.
Munishkina, L.A., Ahmad, A., Fink, A.L., et al., 2008a. Guiding protein aggregation with macromolecular crowding. Biochemistry 47, 8993–9006.
Munishkina, L.A., Fink, A.L., Uversky, V.N., 2008b. Concerted action of metals and macromolecular crowding on the fibrillation of alpha-synuclein. Protein Pept. Lett. 15, 1079–1085.
Munishkina, L.A., Fink, A.L., Uversky, V.N., 2009. Accelerated fibrillation of alpha-synuclein induced by the combined action of macromolecular crowding and factors inducing partial folding. Curr. Alzheimer Res. 6, 252–260.
Murray, I.V., Giasson, B.I., Quinn, S.M., et al., 2003. Role of alpha-synuclein carboxy-terminus on fibril formation in vitro. Biochemistry 42, 8530–8540.
Murrell, J.R., Hake, A.M., Quaid, K.A., et al., 2000. Early-onset Alzheimer disease caused by a new mutation (V717L) in the amyloid precursor protein gene. Arch. Neurol. 57, 885–887.
Nadal, R.C., Rigby, S.E., Viles, J.H., 2008. Amyloid beta-Cu^{2+} complexes in both monomeric and fibrillar forms do not generate H2O2 catalytically but quench hydroxyl radicals. Biochemistry 47, 11653–11664.
Nakamura, T., Yamashita, H., Nagano, Y., et al., 2002. Activation of Pyk2/RAFTK induces tyrosine phosphorylation of alpha-synuclein via Src-family kinases. FEBS Lett. 521, 190–194.
Nanga, R.P., Brender, J.R., Xu, J., et al., 2008. Structures of rat and human islet amyloid polypeptide IAPP(1-19) in micelles by NMR spectroscopy. Biochemistry 47, 12689–12697.
Nanga, R.P., Brender, J.R., Xu, J., et al., 2009. Three-dimensional structure and orientation of rat islet amyloid polypeptide protein in a membrane environment by solution NMR spectroscopy. J. Am. Chem. Soc. 131, 8252–8261.
Narhi, L., Wood, S.J., Steavenson, S., et al., 1999. Both familial Parkinson's disease mutations accelerate alpha-synuclein aggregation. J. Biol. Chem. 274, 9843–9846.

Nasstrom, T., Wahlberg, T., Karlsson, M., et al., 2009. The lipid peroxidation metabolite 4-oxo-2-nonenal cross-links alpha-synuclein causing rapid formation of stable oligomers. Biochem. Biophys. Res. Commun. 378, 872–876.

Nasstrom, T., Fagerqvist, T., Barbu, M., et al., 2011. The lipid peroxidation products 4-oxo-2-nonenal and 4-hydroxy-2-nonenal promote the formation of alpha-synuclein oligomers with distinct biochemical, morphological, and functional properties. Free Radic. Biol. Med. 50, 428–437.

Natalello, A., Benetti, F., Doglia, S.M., et al., 2011. Compact conformations of alpha-synuclein induced by alcohols and copper. Proteins 79, 611–621.

Nath, S., Goodwin, J., Engelborghs, Y., et al., 2011. Raised calcium promotes alpha-synuclein aggregate formation. Mol. Cell. Neurosci. 46, 516–526.

Nath, A., Schlamadinger, D.E., Rhoades, E., et al., 2015. Structure-based small molecule modulation of a pre-amyloid state: pharmacological enhancement of IAPP membrane-binding and toxicity. Biochemistry 54, 3555–3564.

Necula, M., Breydo, L., Milton, S., et al., 2007a. Methylene blue inhibits amyloid abeta oligomerization by promoting fibrillization. Biochemistry 46, 8850–8860.

Necula, M., Kayed, R., Milton, S., et al., 2007b. Small molecule inhibitors of aggregation indicate that amyloid beta oligomerization and fibrillization pathways are independent and distinct. J. Biol. Chem. 282, 10311–10324.

Nelson, R., Sawaya, M.R., Balbirnie, M., et al., 2005. Structure of the cross-beta spine of amyloid-like fibrils. Nature 435, 773–778.

Niwa, T., Sugimoto, R., Watanabe, L., et al., 2015. Large-scale analysis of macromolecular crowding effects on protein aggregation using a reconstituted cell-free translation system. Front. Microbiol. 6, 1113.

Norris, E.H., Giasson, B.I., Ischiropoulos, H., et al., 2003. Effects of oxidative and nitrative challenges on alpha-synuclein fibrillogenesis involve distinct mechanisms of protein modifications. J. Biol. Chem. 278, 27230–27240.

Noy, D., Solomonov, I., Sinkevich, O., et al., 2008. Zinc-amyloid beta interactions on a millisecond time-scale stabilize non-fibrillar Alzheimer-related species. J. Am. Chem. Soc. 130, 1376–1383.

Nussbaum, J.M., Schilling, S., Cynis, H., et al., 2012. Prion-like behaviour and tau-dependent cytotoxicity of pyroglutamylated amyloid-beta. Nature 485, 651–655.

O'nuallain, B., Williams, A.D., Westermark, P., et al., 2004. Seeding specificity in amyloid growth induced by heterologous fibrils. J. Biol. Chem. 279, 17490–17499.

Oh, Y., Kim, Y.M., Mouradian, M.M., et al., 2011. Human Polycomb protein 2 promotes alpha-synuclein aggregate formation through covalent SUMOylation. Brain Res. 1381, 78–89.

Ojha, J., Masilamoni, G., Dunlap, D., et al., 2011. Sequestration of toxic oligomers by HspB1 as a cytoprotective mechanism. Mol. Cell. Biol. 31, 3146–3157.

Ono, K., Hasegawa, K., Naiki, H., et al., 2004a. Anti-amyloidogenic activity of tannic acid and its activity to destabilize Alzheimer's beta-amyloid fibrils in vitro. Biochim. Biophys. Acta 1690, 193–202.

Ono, K., Hasegawa, K., Naiki, H., et al., 2004b. Curcumin has potent anti-amyloidogenic effects for Alzheimer's beta-amyloid fibrils in vitro. J. Neurosci. Res. 75, 742–750.

Ono, K., Hirohata, M., Yamada, M., 2008. Alpha-synuclein assembly as a therapeutic target of Parkinson's disease and related disorders. Curr. Pharm. Des. 14, 3247–3266.

Paik, S.R., Lee, J.H., Kim, D.H., et al., 1997. Aluminum-induced structural alterations of the precursor of the non-A beta component of Alzheimer's disease amyloid. Arch. Biochem. Biophys. 344, 325–334.

Paleologou, K.E., Oueslati, A., Shakked, G., et al., 2010. Phosphorylation at S87 is enhanced in synucleinopathies, inhibits alpha-synuclein oligomerization, and influences synuclein-membrane interactions. J. Neurosci. 30, 3184–3198.

Pandey, N., Strider, J., Nolan, W.C., et al., 2008. Curcumin inhibits aggregation of alpha-synuclein. Acta Neuropathol. 115, 479–489.

Pastor, P., Pastor, E., Carnero, C., et al., 2001. Familial atypical progressive supranuclear palsy associated with homozygosity for the delN296 mutation in the tau gene. Ann. Neurol. 49, 263–267.

Periquet, M., Fulga, T., Myllykangas, L., et al., 2007. Aggregated alpha-synuclein mediates dopaminergic neurotoxicity in vivo. J. Neurosci. 27, 3338–3346.

Pesiridis, G.S., Lee, V.M., Trojanowski, J.Q., 2009. Mutations in TDP-43 link glycine-rich domain functions to amyotrophic lateral sclerosis. Hum. Mol. Genet. 18, R156–R162.

Petrucci, S., Ginevrino, M., Valente, E.M., 2016. Phenotypic spectrum of alpha-synuclein mutations: new insights from patients and cellular models. Parkinsonism Relat. Disord. 22 (Suppl. 1), S16–S20.

Pickering-Brown, S., Baker, M., Yen, S.H., et al., 2000. Pick's disease is associated with mutations in the tau gene. Ann. Neurol. 48, 859–867.

Pickering-Brown, S.M., Baker, M., Nonaka, T., et al., 2004. Frontotemporal dementia with Pick-type histology associated with Q336R mutation in the tau gene. Brain 127, 1415–1426.

Plant, G.T., Revesz, T., Barnard, R.O., et al., 1990. Familial cerebral amyloid angiopathy with nonneuritic amyloid plaque formation. Brain 113 (Pt. 3), 721–747.

Polymeropoulos, M.H., Lavedan, C., Leroy, E., et al., 1997. Mutation in the alpha-synuclein gene identified in families with Parkinson's disease. Science 276, 2045–2047.

Porat, Y., Abramowitz, A., Gazit, E., 2006. Inhibition of amyloid fibril formation by polyphenols: structural similarity and aromatic interactions as a common inhibition mechanism. Chem. Biol. Drug Des. 67, 27–37.

Prigent, S., Rezaei, H., 2011. PrP assemblies: spotting the responsible regions in prion propagation. Prion 5, 69–75.

Qiang, W., Yau, W.M., Luo, Y., et al., 2012. Antiparallel beta-sheet architecture in Iowa-mutant beta-amyloid fibrils. Proc. Natl. Acad. Sci. U.S.A. 109, 4443–4448.

Qin, Z., Hu, D., Han, S., et al., 2007. Effect of 4-hydroxy-2-nonenal modification on alpha-synuclein aggregation. J. Biol. Chem. 282, 5862–5870.

Ramachandran, G., Milan-Garces, E.A., Udgaonkar, J.B., et al., 2014. Resonance Raman spectroscopic measurements delineate the structural changes that occur during tau fibril formation. Biochemistry 53, 6550–6565.

Rasia, R.M., Bertoncini, C.W., Marsh, D., et al., 2005. Structural characterization of copper(II) binding to alpha-synuclein: insights into the bioinorganic chemistry of Parkinson's disease. Proc. Natl. Acad. Sci. U.S.A. 102, 4294–4299.

Rekas, A., Knott, R.B., Sokolova, A., et al., 2010. The structure of dopamine induced alpha-synuclein oligomers. Eur. Biophys. J. 39, 1407–1419.

Reyes, J.F., Reynolds, M.R., Horowitz, P.M., et al., 2008. A possible link between astrocyte activation and tau nitration in Alzheimer's disease. Neurobiol. Dis. 31, 198–208.

Reynolds, M.R., Reyes, J.F., Fu, Y., et al., 2006. Tau nitration occurs at tyrosine 29 in the fibrillar lesions of Alzheimer's disease and other tauopathies. J. Neurosci. 26, 10636–10645.

Ricchelli, F., Drago, D., Filippi, B., et al., 2005. Aluminum-triggered structural modifications and aggregation of beta-amyloids. Cell. Mol. Life Sci. 62, 1724–1733.

Ricchelli, F., Fusi, P., Tortora, P., et al., 2007. Destabilization of non-pathological variants of ataxin-3 by metal ions results in aggregation/fibrillogenesis. Int. J. Biochem. Cell Biol. 39, 966–977.

Rohrer, J.D., Isaacs, A.M., Mizielinska, S., et al., 2015. C9orf72 expansions in frontotemporal dementia and amyotrophic lateral sclerosis. Lancet Neurol. 14, 291–301.

Rospigliosi, C.C., Mcclendon, S., Schmid, A.W., et al., 2009. E46K Parkinson's-linked mutation enhances C-terminal-to-N-terminal contacts in alpha-synuclein. J. Mol. Biol. 388, 1022–1032.

Rott, R., Szargel, R., Haskin, J., et al., 2008. Monoubiquitylation of alpha-synuclein by seven in absentia homolog (SIAH) promotes its aggregation in dopaminergic cells. J. Biol. Chem. 283, 3316–3328.
Ryu, J., Cho, S., Park, B.C., et al., 2010. Oxidative stress-enhanced SUMOylation and aggregation of ataxin-1: implication of JNK pathway. Biochem. Biophys. Res. Commun. 393, 280–285.
Sarell, C.J., Wilkinson, S.R., Viles, J.H., 2010. Substoichiometric levels of Cu^{2+} ions accelerate the kinetics of fiber formation and promote cell toxicity of amyloid-{beta} from Alzheimer disease. J. Biol. Chem. 285, 41533–41540.
Sasahara, K., Morigaki, K., Shinya, K., 2014. Amyloid aggregation and deposition of human islet amyloid polypeptide at membrane interfaces. FEBS J. 281, 2597–2612.
Sathasivam, K., Neueder, A., Gipson, T.A., et al., 2013. Aberrant splicing of HTT generates the pathogenic exon 1 protein in Huntington disease. Proc. Natl. Acad. Sci. U.S.A. 110, 2366–2370.
Saunders, H.M., Bottomley, S.P., 2009. Multi-domain misfolding: understanding the aggregation pathway of polyglutamine proteins. Protein Eng. Des. Sel. 22, 447–451.
Saunders, J.C., Young, L.M., Mahood, R.A., et al., 2016. An in vivo platform for identifying inhibitors of protein aggregation. Nat. Chem. Biol. 12, 94–101.
Sawaya, M.R., Sambashivan, S., Nelson, R., et al., 2007. Atomic structures of amyloid cross-beta spines reveal varied steric zippers. Nature 447, 453–457.
Schedin-Weiss, S., Winblad, B., Tjernberg, L.O., 2014. The role of protein glycosylation in Alzheimer disease. FEBS J. 281, 46–62.
Scherzer-Attali, R., Pellarin, R., Convertino, M., et al., 2010. Complete phenotypic recovery of an Alzheimer's disease model by a quinone-tryptophan hybrid aggregation inhibitor. PLoS One 5, e11101.
Schieb, H., Kratzin, H., Jahn, O., et al., 2011. Beta-amyloid peptide variants in brains and cerebrospinal fluid from amyloid precursor protein (APP) transgenic mice: comparison with human Alzheimer amyloid. J. Biol. Chem. 286, 33747–33758.
Seeliger, J., Werkmuller, A., Winter, R., 2013. Macromolecular crowding as a suppressor of human IAPP fibril formation and cytotoxicity. PLoS One 8, e69652.
Serpell, L.C., Berriman, J., Jakes, R., et al., 2000. Fiber diffraction of synthetic alpha-synuclein filaments shows amyloid-like cross-beta conformation. Proc. Natl. Acad. Sci. U.S.A. 97, 4897–4902.
Shimizu, T., Watanabe, A., Ogawara, M., et al., 2000. Isoaspartate formation and neurodegeneration in Alzheimer's disease. Arch. Biochem. Biophys. 381, 225–234.
Shimizu, T., Fukuda, H., Murayama, S., et al., 2002. Isoaspartate formation at position 23 of amyloid beta peptide enhanced fibril formation and deposited onto senile plaques and vascular amyloids in Alzheimer's disease. J. Neurosci. Res. 70, 451–461.
Shin, R.W., Lee, V.M., Trojanowski, J.Q., 1994. Aluminum modifies the properties of Alzheimer's disease PHF tau proteins in vivo and in vitro. J. Neurosci. 14, 7221–7233.
Shtilerman, M.D., Ding, T.T., Lansbury Jr., P.T., 2002. Molecular crowding accelerates fibrillization of alpha-synuclein: could an increase in the cytoplasmic protein concentration induce Parkinson's disease? Biochemistry 41, 3855–3860.
Si, K., Lindquist, S., Kandel, E.R., 2003. A neuronal isoform of the aplysia CPEB has prion-like properties. Cell 115, 879–891.
Si, K., Choi, Y.B., White-Grindley, E., et al., 2010. Aplysia CPEB can form prion-like multimers in sensory neurons that contribute to long-term facilitation. Cell 140, 421–435.
Siegel, S.J., Bieschke, J., Powers, E.T., et al., 2007. The oxidative stress metabolite 4-hydroxynonenal promotes Alzheimer protofibril formation. Biochemistry 46, 1503–1510.

Simoes, A.T., Goncalves, N., Nobre, R.J., et al., 2014. Calpain inhibition reduces ataxin-3 cleavage alleviating neuropathology and motor impairments in mouse models of Machado-Joseph disease. Hum. Mol. Genet. 23, 4932–4944.

Sivanesam, K., Byrne, A., Bisaglia, M., et al., 2015. Binding interactions of agents that alter alpha-Synuclein aggregation. RSC Adv. 5, 11577–11590.

Smith, W.W., Margolis, R.L., Li, X., et al., 2005. Alpha-synuclein phosphorylation enhances eosinophilic cytoplasmic inclusion formation in SH-SY5Y cells. J. Neurosci. 25, 5544–5552.

Song, L., Lu, S.X., Ouyang, X., et al., 2015. Analysis of tau post-translational modifications in rTg4510 mice, a model of tau pathology. Mol. Neurodegener. 10, 14.

Soragni, A., Zambelli, B., Mukrasch, M.D., et al., 2008. Structural characterization of binding of Cu(II) to tau protein. Biochemistry 47, 10841–10851.

Souza, J.M., Giasson, B.I., Chen, Q., et al., 2000. Dityrosine cross-linking promotes formation of stable alpha-synuclein polymers. Implication of nitrative and oxidative stress in the pathogenesis of neurodegenerative synucleinopathies. J. Biol. Chem. 275, 18344–18349.

Stawoska, I., Weselucha-Birczynska, A., Regonesi, M.E., et al., 2009. Interaction of selected divalent metal ions with human ataxin-3 Q36. J. Biol. Inorg. Chem. 14, 1175–1185.

Steffan, J.S., Agrawal, N., Pallos, J., et al., 2004. SUMO modification of Huntingtin and Huntington's disease pathology. Science 304, 100–104.

Steiner, H., Revesz, T., Neumann, M., et al., 2001. A pathogenic presenilin-1 deletion causes aberrant Abeta 42 production in the absence of congophilic amyloid plaques. J. Biol. Chem. 276, 7233–7239.

Stull, F., Koldewey, P., Humes, J.R., et al., 2016. Substrate protein folds while it is bound to the ATP-independent chaperone Spy. Nat. Struct. Mol. Biol. 23, 53–58.

Sun, S., Ling, S.C., Qiu, J., et al., 2015. ALS-causative mutations in FUS/TLS confer gain and loss of function by altered association with SMN and U1-snRNP. Nat. Commun. 6, 6171.

Tabner, B.J., El-Agnaf, O.M., Turnbull, S., et al., 2005. Hydrogen peroxide is generated during the very early stages of aggregation of the amyloid peptides implicated in Alzheimer disease and familial British dementia. J. Biol. Chem. 280, 35789–35792.

Tartaglia, G.G., Cavalli, A., Pellarin, R., et al., 2004. The role of aromaticity, exposed surface, and dipole moment in determining protein aggregation rates. Protein Sci. 13, 1939–1941.

Thapa, A., Jett, S.D., Chi, E.Y., 2015. Curcumin attenuates amyloid-beta aggregate toxicity and modulates amyloid-beta aggregation pathway. ACS Chem. Neurosci. 7, 56–68.

Theuns, J., Marjaux, E., Vandenbulcke, M., et al., 2006. Alzheimer dementia caused by a novel mutation located in the APP C-terminal intracytosolic fragment. Hum. Mutat. 27, 888–896.

Thinakaran, G., Koo, E.H., 2008. Amyloid precursor protein trafficking, processing, and function. J. Biol. Chem. 283, 29615–29619.

Tofaris, G.K., Garcia Reitbock, P., Humby, T., et al., 2006. Pathological changes in dopaminergic nerve cells of the substantia nigra and olfactory bulb in mice transgenic for truncated human alpha-synuclein(1-120): implications for Lewy body disorders. J. Neurosci. 26, 3942–3950.

Tomiyama, T., Nagata, T., Shimada, H., et al., 2008. A new amyloid beta variant favoring oligomerization in Alzheimer's-type dementia. Ann. Neurol. 63, 377–387.

Tougu, V., Karafin, A., Palumaa, P., 2008. Binding of zinc(II) and copper(II) to the full-length Alzheimer's amyloid-beta peptide. J. Neurochem. 104, 1249–1259.

Tougu, V., Karafin, A., Zovo, K., et al., 2009. Zn(II)- and Cu(II)-induced non-fibrillar aggregates of amyloid-beta (1-42) peptide are transformed to amyloid fibrils, both spontaneously and under the influence of metal chelators. J. Neurochem. 110, 1784–1795.

Tougu, V., Tiiman, A., Palumaa, P., 2011. Interactions of Zn(II) and Cu(II) ions with Alzheimer's amyloid-beta peptide. Metal ion binding, contribution to fibrillization and toxicity. Metallomics 3, 250–261.
Trepte, P., Strempel, N., Wanker, E.E., 2014. Spontaneous self-assembly of pathogenic huntingtin exon 1 protein into amyloid structures. Essays Biochem. 56, 167–180.
Ulmer, T.S., Bax, A., Cole, N.B., et al., 2005. Structure and dynamics of micelle-bound human alpha-synuclein. J. Biol. Chem. 280, 9595–9603.
Ulusoy, A., Di Monte, D.A., 2013. alpha-Synuclein elevation in human neurodegenerative diseases: experimental, pathogenetic, and therapeutic implications. Mol. Neurobiol. 47, 484–494.
Ulusoy, A., Febbraro, F., Jensen, P.H., et al., 2010. Co-expression of C-terminal truncated alpha-synuclein enhances full-length alpha-synuclein-induced pathology. Eur. J. Neurosci. 32, 409–422.
Usui, K., Hulleman, J.D., Paulsson, J.F., et al., 2009. Site-specific modification of Alzheimer's peptides by cholesterol oxidation products enhances aggregation energetics and neurotoxicity. Proc. Natl. Acad. Sci. U.S.A. 106, 18563–18568.
Uversky, V.N., 2008. Amyloidogenesis of natively unfolded proteins. Curr. Alzheimer Res. 5, 260–287.
Uversky, V.N., Winter, S., Galzitskaya, O.V., et al., 1998. Hyperphosphorylation induces structural modification of tau-protein. FEBS Lett. 439, 21–25.
Uversky, V.N., Li, J., Fink, A.L., 2001. Metal-triggered structural transformations, aggregation, and fibrillation of human alpha-synuclein. A possible molecular NK between Parkinson's disease and heavy metal exposure. J. Biol. Chem. 276, 44284–44296.
Uversky, V.N., Cooper, E.M., Bower, K.S., et al., 2002a. Accelerated alpha-synuclein fibrillation in crowded milieu. FEBS Lett. 515, 99–103.
Uversky, V.N., Yamin, G., Souillac, P.O., et al., 2002b. Methionine oxidation inhibits fibrillation of human alpha-synuclein in vitro. FEBS Lett. 517, 239–244.
Vandersteen, A., Hubin, E., Sarroukh, R., et al., 2012. A comparative analysis of the aggregation behavior of amyloid-beta peptide variants. FEBS Lett. 586, 4088–4093.
Varkey, J., Mizuno, N., Hegde, B.G., et al., 2013. alpha-Synuclein oligomers with broken helical conformation form lipoprotein nanoparticles. J. Biol. Chem. 288, 17620–17630.
Vitalis, A., Caflisch, A., 2010. Micelle-like architecture of the monomer ensemble of Alzheimer's amyloid-beta peptide in aqueous solution and its implications for Abeta aggregation. J. Mol. Biol. 403, 148–165.
Vitalis, A., Lyle, N., Pappu, R.V., 2009. Thermodynamics of beta-sheet formation in polyglutamine. Biophys. J. 97, 303–311.
Voss, K., Harris, C., Ralle, M., et al., 2014. Modulation of tau phosphorylation by environmental copper. Transl. Neurodegener. 3, 24.
Wakabayashi, M., Matsuzaki, K., 2009. Ganglioside-induced amyloid formation by human islet amyloid polypeptide in lipid rafts. FEBS Lett. 583, 2854–2858.
Wakutani, Y., Watanabe, K., Adachi, Y., et al., 2004. Novel amyloid precursor protein gene missense mutation (D678N) in probable familial Alzheimer's disease. J. Neurol. Neurosurg. Psychiatry 75, 1039–1042.
Wang, Z.L., Bennet, W.M., Ghatei, M.A., et al., 1993. Influence of islet amyloid polypeptide and the 8-37 fragment of islet amyloid polypeptide on insulin release from perfused rat islets. Diabetes 42, 330–335.
Wang, J.Z., Grundke-Iqbal, I., Iqbal, K., 1996. Glycosylation of microtubule-associated protein tau: an abnormal posttranslational modification in Alzheimer's disease. Nat. Med. 2, 871–875.
Wang, Y., Martinez-Vicente, M., Kruger, U., et al., 2009. Tau fragmentation, aggregation and clearance: the dual role of lysosomal processing. Hum. Mol. Genet. 18, 4153–4170.

Wang, M.S., Boddapati, S., Emadi, S., et al., 2010a. Curcumin reduces alpha-synuclein induced cytotoxicity in Parkinson's disease cell model. BMC Neurosci. 11, 57.

Wang, X., Moualla, D., Wright, J.A., et al., 2010b. Copper binding regulates intracellular alpha-synuclein localisation, aggregation and toxicity. J. Neurochem. 113, 704–714.

Wang, Q., Liang, G., Zhang, M., et al., 2014. De novo design of self-assembled hexapeptides as beta-amyloid (Abeta) peptide inhibitors. ACS Chem. Neurosci. 5, 972–981.

Watt, B., Van Niel, G., Fowler, D.M., et al., 2009. N-terminal domains elicit formation of functional Pmel17 amyloid fibrils. J. Biol. Chem. 284, 35543–35555.

Weise, K., Radovan, D., Gohlke, A., et al., 2010. Interaction of hIAPP with model raft membranes and pancreatic β-cells: cytotoxicity of hIAPP oligomers. ChemBioChem 11, 1280–1290.

Weissmann, C., Li, J., Mahal, S.P., et al., 2011. Prions on the move. EMBO Rep. 12, 1109–1117.

Westermark, P., Andersson, A., Westermark, G.T., 2011. Islet amyloid polypeptide, islet amyloid, and diabetes mellitus. Physiol. Rev. 91, 795–826.

Wirths, O., Breyhan, H., Cynis, H., et al., 2009. Intraneuronal pyroglutamate-Abeta 3-42 triggers neurodegeneration and lethal neurological deficits in a transgenic mouse model. Acta Neuropathol. 118, 487–496.

Wiseman, F.K., Al-Janabi, T., Hardy, J., et al., 2015. A genetic cause of Alzheimer disease: mechanistic insights from Down syndrome. Nat. Rev. Neurosci. 16, 564–574.

Wobst, H.J., Sharma, A., Diamond, M.I., et al., 2015. The green tea polyphenol (−)-epigallocatechin gallate prevents the aggregation of tau protein into toxic oligomers at substoichiometric ratios. FEBS Lett. 589, 77–83.

Wright, J.A., Wang, X., Brown, D.R., 2009. Unique copper-induced oligomers mediate alpha-synuclein toxicity. FASEB J. 23, 2384–2393.

Wu, K.P., Kim, S., Fela, D.A., et al., 2008. Characterization of conformational and dynamic properties of natively unfolded human and mouse alpha-synuclein ensembles by NMR: implication for aggregation. J. Mol. Biol. 378, 1104–1115.

Wu, Y., Teng, N., Li, S., 2015. Effects of macromolecular crowding and osmolyte on human Tau fibrillation. Int. J. Biol. Macromol. 90, 27–36.

Xiao, G., Fan, Q., Wang, X., et al., 2013. Huntington disease arises from a combinatory toxicity of polyglutamine and copper binding. Proc. Natl. Acad. Sci. U.S.A. 110, 14995–15000.

Xu, B., Wu, S.W., Lu, C.W., et al., 2013. Oxidative stress involvement in manganese-induced alpha-synuclein oligomerization in organotypic brain slice cultures. Toxicology 305, 71–78.

Yamamoto, A., Shin, R.W., Hasegawa, K., et al., 2002. Iron (III) induces aggregation of hyperphosphorylated tau and its reduction to iron (II) reverses the aggregation: implications in the formation of neurofibrillary tangles of Alzheimer's disease. J. Neurochem. 82, 1137–1147.

Yamin, G., Glaser, C.B., Uversky, V.N., et al., 2003. Certain metals trigger fibrillation of methionine-oxidized alpha-synuclein. J. Biol. Chem. 278, 27630–27635.

Yamin, G., Munishkina, L.A., Karymov, M.A., et al., 2005. Forcing nonamyloidogenic beta-synuclein to fibrillate. Biochemistry 44, 9096–9107.

Yanai, A., Huang, K., Kang, R., et al., 2006. Palmitoylation of huntingtin by HIP14 is essential for its trafficking and function. Nat. Neurosci. 9, 824–831.

Yang, L.S., Ksiezak-Reding, H., 1999. Ca^{2+} and Mg^{2+} selectively induce aggregates of PHF-tau but not normal human tau. J. Neurosci. Res. 55, 36–43.

Yang, F., Lim, G.P., Begum, A.N., et al., 2005. Curcumin inhibits formation of amyloid beta oligomers and fibrils, binds plaques, and reduces amyloid in vivo. J. Biol. Chem. 280, 5892–5901.

Yu, Z., Xu, X., Xiang, Z., et al., 2010. Nitrated alpha-synuclein induces the loss of dopaminergic neurons in the substantia nigra of rats. PLoS One 5, e9956.
Yuzwa, S.A., Shan, X., Macauley, M.S., et al., 2012. Increasing O-GlcNAc slows neurodegeneration and stabilizes tau against aggregation. Nat. Chem. Biol. 8, 393–399.
Zarranz, J.J., Alegre, J., Gomez-Esteban, J.C., et al., 2004. The new mutation, E46K, of alpha-synuclein causes Parkinson and Lewy body dementia. Ann. Neurol. 55, 164–173.
Zhang, X., Li, J.P., 2010. Heparan sulfate proteoglycans in amyloidosis. Prog. Mol. Biol. Transl. Sci. 93, 309–334.
Zhao, Y., Zhao, B., 2013. Oxidative stress and the pathogenesis of Alzheimer's disease. Oxid. Med. Cell. Longev. 2013, 316523.
Zhao, J., Wang, P., Li, H., et al., 2015a. Nitration of Y10 in Abeta1-40: is it a compensatory reaction against oxidative/nitrative stress and Abeta aggregation? Chem. Res. Toxicol. 28, 401–407.
Zhao, Y., Tseng, I.C., Heyser, C.J., et al., 2015b. Appoptosin-mediated caspase cleavage of tau contributes to progressive supranuclear palsy pathogenesis. Neuron 87, 963–975.
Zheng, X., Liu, D., Klarner, F.G., et al., 2015. Amyloid beta-protein assembly: the effect of molecular tweezers CLR01 and CLR03. J. Phys. Chem. B 119, 4831–4841.
Zhou, H.X., Rivas, G., Minton, A.P., 2008. Macromolecular crowding and confinement: biochemical, biophysical, and potential physiological consequences. Annu. Rev. Biophys. 37, 375–397.
Zhou, Z., Fan, J.B., Zhu, H.L., et al., 2009. Crowded cell-like environment accelerates the nucleation step of amyloidogenic protein misfolding. J. Biol. Chem. 284, 30148–30158.
Zhou, W., Long, C., Reaney, S.H., et al., 2010. Methionine oxidation stabilizes non-toxic oligomers of alpha-synuclein through strengthening the auto-inhibitory intra-molecular long-range interactions. Biochim. Biophys. Acta 1802, 322–330.

CHAPTER FIVE

The Nucleation of Protein Aggregates - From Crystals to Amyloid Fibrils

Alexander K. Buell*[,1]
*Institute of Physical Biology, University of Düsseldorf, Universitätsstr.1, 40225, Düsseldorf, Germany
[1]Corresponding author: e-mail address: alexander.buell@uni-duesseldorf.de

Contents

Abstract

The condensation and aggregation of individual protein molecules into dense insoluble phases is of relevance in such diverse fields as materials science, medicine, structural biology and pharmacology. A common feature of these condensation phenomena is that they usually are nucleated processes, i.e. the first piece of the condensed phase is energetically costly to create and hence forms slowly compared to its subsequent growth. Here we give a compact overview of the differences and similarities of various protein nucleation phenomena, their theoretical description in the framework of colloid and polymer science and their experimental study. Particular emphasis is put on the nucleation of a specific type of filamentous protein aggregates, amyloid fibrils. The current experimentally derived knowledge on amyloid fibril nucleation is critically assessed, and we argue that it is less advanced than is generally believed. This is due to (I) the lack of emphasis that has been put on the distinction between homogeneous and heterogeneous nucleation in experimental studies (II) the use of oversimplifying and/or inappropriate theoretical frameworks for the analysis of kinetic data of amyloid fibril nucleation. A strategy is outlined and advocated of how our understanding of this important class of processes can be improved in the future.

International Review of Cell and Molecular Biology, Volume 329
ISSN 1937-6448
http://dx.doi.org/10.1016/bs.ircmb.2016.08.014

1. PROTEIN CONDENSATION

Life as we know it would be unthinkable without proteins. Many biological functions, with the notable exceptions of long-term information and energy storage, crucially depend on specific proteins. The modular build of proteins as hetero-polymers of the 20 or so chemically diverse natural amino acids enables this versatility; however, there is also an inherent risk associated with the potential to interact in many different ways with many different partners. This risk is the formation of a condensed phase, held together by favourable interactions between individual protein molecules. From a purely thermodynamic point of view, any system of molecules or particles that have an attractive interaction for each other will display a critical concentration above which the condensed phase becomes thermodynamically stable. Indeed, even for ideal hard spheres, i.e. particles with no attractive interactions, it has been predicted that condensed phases can form at very high volume fractions, driven entirely by entropy Hoover and Ree (1968). The question is, therefore, why proteins exist in most cases in a non-condensed state *in vivo*. The combined volume fractions of proteins and other molecules inside a living cell (up to 0.3 Ellis (2001)) are not far from those at which even hard spheres would crystallise (ca. 0.5 Pusey and van Megen (1986); Auer and Frenkel (2001)). Under such highly crowded conditions, it can be estimated that a binding energy of the order of 10 kJ/mol, less than half a net hydrogen bond, would be sufficient to make the condensed phase the thermodynamic minimum and it is therefore somewhat surprising that the content of a living cell is not just a single large precipitate. Despite the gross oversimplification of this picture, the question as to whether a living cell is at thermodynamic equilibrium with respect to the phase behaviour of its protein content, is a legitimate one. Of course a living cell is a non-equilibrium system at many different levels related to its chemical composition. However, certain cellular sub-systems can often be treated as if they were at equilibrium, depending on how fast the kinetics of the given set of processes is. We are therefore concerned with the question as to the kinetics of protein condensation[1], or more precisely, the kinetics of the initiation of condensation.

[1] In this work, we often employ the term condensation, rather than aggregation. This terminology, inspired by Gunton et al. (2007), is more general than aggregation, as it comprises also crystallisation, and is therefore preferrable in the most general sections of the present work.

But before going into the details of the nucleation of condensed phases of proteins, it is important to get an overview over the different types of condensed phases a protein can adopt. In this work, we will limit ourselves to homo-molecular condensed phases, even though the formation of hetero-molecular condensed protein phases are also of biological relevance. In particular the formation of liquid-like condensed phases in the form of 'membrane-less organelles', involving different types of proteins as well as nucleic acids, has been receiving increasing attention in recent years Brangwynne et al. (2015).

The formation and behaviour of condensed protein phases is best discussed within a physico-chemical framework. The 20th century saw not only an enormous development of protein science, but importantly also the birth of both modern polymer and colloid science; these three branches of science are closely linked (Fig. 1). Quantitative physical theories were developed that allow understanding of the behaviour of polymers and colloids as a function of various chemical and physical parameters, such as concentration, ionic strength, solvent type and temperature. It was realised that the structures and dynamics of polymers are governed by the competition between entropic contributions stemming from the large number of internal degrees of freedom of a polymer (rotation around all the bonds between monomers) and the energetic contributions from the interactions of the building blocks with each other and with the solvent Flory (1953). Colloidal behaviour, on the other hand, is governed by the interfacial energy of the dispersed phase in contact with the solvent Israelachvili (1992). The kinetic stability against condensation ('coagulation') of a colloidal suspension is determined by the balance of repulsive and attractive interactions Derjaguin and Landau (1941); Verwey and Overbeek (1948). Interestingly, despite the absence of internal degrees of freedom in most colloids (colloids are well approximated as structureless particles) entropic factors also play an important role in colloidal behaviour. Indeed, the repulsive interactions between colloid particles in condensed phases are often entropic in nature. This is in particular, and somewhat counter-intuitively, also true for the electrostatic contribution to the repulsion Gitlin et al. (2006). This can be understood by considering that the repulsion between two charged colloid particles is not caused by their surface charges, but by the diffuse clouds of counterions that surround each particle and that balance the net charge of the colloid. The repulsion can therefore be linked to an increase in osmotic pressure upon overlap of the counterion clouds, an entropic effect Israelachvili (1992).

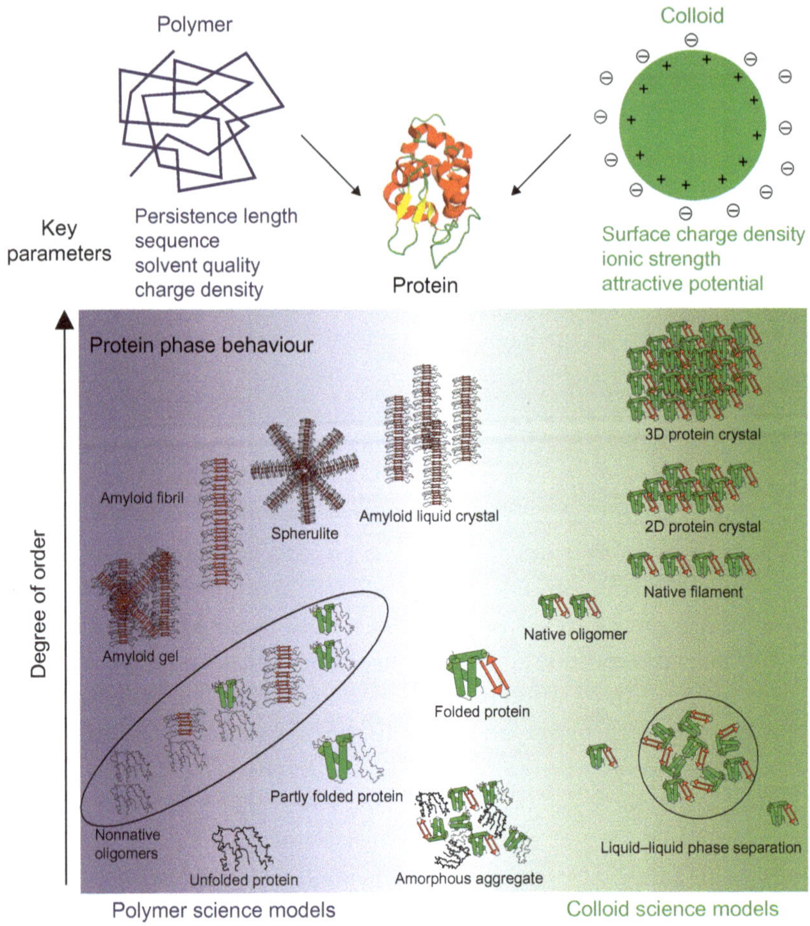

Fig. 1 Physical protein science uses tools and concepts from both polymer science and colloid science, as proteins (here illustrated with hen lysozyme, PDB 3WMK) possess features of both colloids and polymers. Distinct physical and chemical parameters determine the behaviour of colloids and polymers. The description of protein condensation into dense and/or supramolecular phases can be well-described with concepts from colloid science, if no internal degrees of freedom change during the condensation (green part of the figure). If, on the other hand, significant changes in internal degrees of freedom are involved in the condensation reaction, as is the case in the formation of amyloid fibrils, polymer science theories will need to be employed (blue part of the figure). The condensation phenomena can display several levels of hierarchy; for example after the protein has assembled into amyloid fibrils (a process involving changes in internal degrees of freedom), those fibrils can adopt liquid crystalline order (a process that can be understood using colloid science theories Jung and Mezzenga (2010)). One of the biggest outstanding challenges in protein science is the establishment of global phase diagrams that comprise the entirety of the accessible structures, and also discriminate between kinetic and thermodynamic effects.

The theoretical descriptions developed for colloids and polymers remain even to-date the most important conceptual tools for the understanding of phase transition and condensation phenomena of proteins. As is illustrated in Fig. 1, proteins share features with both polymers and colloids. Protein molecules are linear polymers of several dozen to several hundreds or even thousands of amino acid residues. In many cases, they are folded into a unique and well-defined three dimensional and often globular structure, justifying their description as colloid particles, albeit very complicated ones, with surface patches of varied polarity and charge Bianchi et al. (2011). On the other hand, it is increasingly established that some fraction of all proteins can also occur in unstructured, disordered states in solution under physiological conditions, so called intrinsically disordered proteins (IDPs) Dunker et al. (2001). Furthermore, structured proteins can be induced to unfold by the effect of high temperature, pH or chemical denaturants. Polymer physics theories Brant and Flory (1965) and their extension to polyelectrolytes Dobrynin and Rubinstein (2005) are appropriate physical approaches to describe these disordered states of the polypeptide chain. Therefore, depending on the extent by which internal degrees of freedom change upon protein condensation, colloidal or polymer theories, or a combination of the two need to be employed to describe these phenomena.

In Fig. 1 we present an overview over the different types of condensed protein phases. The very rich phase behaviour is striking, with both ordered and disordered condensed phases being accessible to polypeptide chains. Ordered phases include crystals (three-dimensional, as well as two-dimensional crystals of membrane proteins Hasler et al. (1998)), layers (e.g. surface coatings by hydrophobins Mackay et al. (2001)), and (pseudo) one-dimensional fibrillar phases, in which the monomeric protein molecules either largely retain their native fold (e.g. actin filaments Hanson and Lowy (1963)) or completely change their secondary structures (amyloid fibrils Jimenez et al. (2002)). At high concentrations of fibrillar structures, higher order assembly into gels and flocculation Buell et al. (2014b) or the formation of spherulite structures Domike and Donald (2007) or liquid crystals Jung and Mezzenga (2010) can be observed. Disordered phases can be solid-like, as in amorphous aggregates Yoshimura et al. (2012); Borgia et al. (2013), within which monomers can retain their native fold or adopt a different structure, or else they can be liquid-like, as in liquid-liquid phase separation phenomena, where a protein solution spontaneously phase separates into a dilute and a highly concentrated solution phase Thomson et al. (1987); Dumetz et al. (2008). Many proteins can form several, if not all of these different condensed phases and the question as to which external influences drive a given protein into a given phase has been a central problem in protein science in the last decades Mezzenga and Fischer (2013). Individual sub-parts of this phase behaviour form today entire fields of research, such as protein crystallography, which is concerned with the formation of protein crystals and their use for structure determination, or the study of protein misfolding and aggregation into amyloid fibrils, which are a hallmark of many severe disorders, such as Parkinson's disease. Indeed, this fragmentation of protein science into individual sub-fields has led to the fact that currently, no complete phase diagram (including all the accessible states) is available for any protein. Even for some of the best-studied proteins, such as lysozyme and α-B crystallin, where detailed phase diagrams have been established Gunton et al. (2007), amyloid fibrils have not yet been included, even though it has been established that these proteins can indeed form amyloid fibrils under certain conditions Booth et al. (1997); Meehan et al. (2004). Furthermore, an equilibrium phase diagram, even if it did include all possible soluble and condensed states of a given protein, would still fall short of providing the complete picture of relevant protein behaviour. This is because the kinetics of interconversion between the different phases are an important factor, and might in any given practical case determine the observed behaviour. This is the rationale why this article is concerned mainly with kinetics, rather than thermodynamics, and in particular with the initial steps in condensation reactions, the nucleation of the condensed phase. The aim of this article is to discuss the current state of knowledge on the nucleation of amyloid fibrils but at the same time put this process into the wider context of protein condensation phenomena in general. Understanding and controlling nucleation is of great relevance; whether the formation of a new condensed protein phase is desired or should be avoided, nucleation is the crucial first step that needs to be understood in order to be enhanced or inhibited in a systematic way.

2. NUCLEATION - THE EMERGENCE OF A NEW PHASE

One aspect that all protein condensation phenomena have in common, whether they involve changes in the internal degrees of freedom of the polypeptide chain or not, is that they are associated with the creation of an interface between the solution phase and the new condensed phase. The concentrations of all components of the solution (the proteins themselves, as well as water molecules and salt ions) exhibit a discontinuity at the interface and hence the creation of such an interface is associated with a free energy cost (Fig. 2A). However, under conditions under which the condensed phase forms spontaneously, it must be associated with a lower free energy compared to the solution phase. Unless the condensed phase is truly one-dimensional, the surface-to-volume ratio decreases as the phase

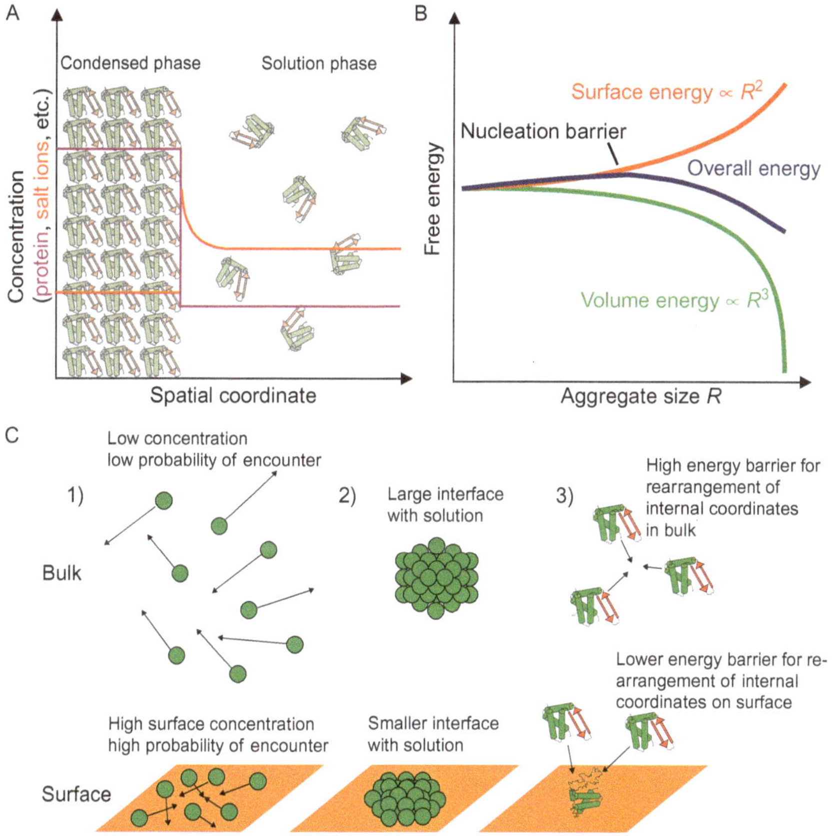

Fig. 2 The (classical) nucleation of a new protein phase. (A) There is an energetic cost associated with the creation of a new interface, between the solution phase and the condensed phase. At the interface, the molecular and ionic species in solution display a discontinuity in their concentrations, illustrated here for the proteins and salt ions. (B) The different functional dependencies of the surface- and volume-energies of the aggregate on its size leads to a barrier for nucleation. (C) The presence of a surface can accelerate nucleation in three different ways: 1) Through an increase in local concentration and an increase in the probability of encounter. 2) Through a decrease in interfacial energy, due to a favourable interfacial energy between the condensed phase and the surface. 3) Through a change in internal coordinates induced by surface binding, that facilitates the subsequent nucleation reaction.

grows, leading to the existence of a nucleation barrier, an energetic net cost for the creation of the first piece of the condensed phase (Kashchiev et al., 2013, Fig. 2B). This barrier is the reason why the formation of a condensed phase needs to be divided into at least two processes, nucleation and growth. In many cases, the nucleation barrier can be circumvented by adding preformed pieces of the new phase, which can then grow by the addition of monomeric protein, a process referred to as 'seeding'. However, other scenarios without nucleation are also possible; the interfacial free energy and the energy of the solution phase depend on the protein concentration, and therefore a regime exists at high concentrations where the nucleation barrier vanishes, i.e. where the system changes from a state of meta-stability to one of instability. In liquid-liquid phase separation, for example, this would correspond to the regime in which the dense liquid phase forms through spinodal decomposition (i.e. the simultaneous formation of many regions of the dense phase), rather than through nucleation of each individual dense solution domain. Other than by increasing the protein concentration to very high values, such a behaviour can also be induced by an abrupt change in solvent condition, a principle that has been exploited for example in the formation of rigid spherical peptide structures from aromatic dipeptides Levin et al. (2014). Furthermore, certain protocols for the preparation of oligomeric protein structures in the context of toxic amyloid species, that involved co-solvents might be based on this phenomenon Campioni et al. (2010). Nevertheless, in the large majority of known cases, a significant delay is observed in the formation of the new condensed protein phase and hence it is important to understand the nucleation of such phases.

Depending on what type of condensed protein phase is forming, the nucleation (as well as the growth) process can be associated with substantial rearrangements of the internal degrees of freedom of the protein. This is most prominently the case in the formation of amyloid fibrils, whose study has arguably been the most active field of research within the general context of protein condensation in the last decade. Amyloid fibrils are filamentous protein aggregates, where the individual polypeptide molecules are engaged in intermolecular β-sheet interactions Sunde et al. (1997). Those structures are a hallmark of dozens of severe disorders, some of which are age-related and hence strongly increase in incidence in many countries, such as Alzheimer's disease Knowles et al. (2014). However, an ever increasing number of proteins is described to be able to form amyloid fibrils, where this process is not linked to any known disease. Indeed, it has been proposed that the amyloid state might be the true thermodynamic minimum of most, if not all, polypeptide chains Baldwin et al. (2011), at least within the parameter space of the protein phase diagram that is relevant for biology, i.e. for moderate values of pH and temperature, in aqueous solution. Extremes of temperature Morel et al. (2010), pH Shammas et al. (2011) and changes in solvent composition Hirota-Nakaoka et al. (2003) are able to induce the dissociation of aggregates, indicating that under these conditions, the monomeric state of the protein represents the thermodynamically most favourable state. The generality of the amyloid phenomenon implies that amyloid fibrils are formed by proteins of a very large structural diversity. Therefore, while amyloid fibrils are characterised by β-sheet secondary structure, their formation is observed from precursor proteins of any secondary and tertiary structure, including in particular also precursors that lack any distinct structural elements, such as IDPs. For natively folded proteins, amyloid fibril formation is often observed under conditions that decrease the stability of the native fold Zurdo et al. (2001); Buell et al. (2011). However, this does not mean that amyloid fibrils can only form under conditions where the given protein is mostly unfolded Chiti and Dobson (2009). Nevertheless, these observations suggest that the rearrangements of internal degrees of freedom necessary for adopting the structure of the amyloid fibril can contribute to the free energy barrier associated with the nucleation and growth of amyloid fibrils Buell et al. (2012). Therefore, the *de novo* formation and growth of an amyloid fibril can only be treated approximately within a colloidal picture of classical nucleation. This is in contrast to other filament-forming proteins, such as actin, tubulin and sickle haemoglobin, which essentially maintain their native fold upon condensation into filaments (neglecting small differences due to nucleotide-binding in the cytoskeletal filaments). For amyloid fibrils a more appropriate conceptual description aims at combining external degrees of freedom, such as the distance between the centres of mass of two protein molecules and the angle that the molecular axes form, with a suitable treatment of internal degrees of freedom, such as the Φ and Ψ angles of the individual peptide bonds, with the latter part being inspired by polymer physics Buell et al. (2010). An energy landscape can then be constructed, where each multiparticle configuration of the protein system is attributed an energy as a function of all internal and external degrees of freedom. Of course such a detailed description leads to very large numbers of degrees of freedom, even for the shortest possible peptides that can still form amyloid fibrils Tjernberg et al. (2002), in particular if the solvent molecules are explicitly included. Such systems can only be treated computationally.

3. HETEROGENEOUS NUCLEATION AT INTERFACES

In order to understand nucleation processes in all relevant scenarios, it is important to not restrict the discussion to homogeneous nucleation in bulk solution, but rather to also consider the effects of surfaces. Depending on the type of condensed phase that is formed, all, or part, of the nucleation barrier is caused by the energy cost of creating a new interface in a previously homogeneous solution. It can therefore be expected that the introduction of additional interfaces into a metastable protein solution might be able to influence both the kinetic and the thermodynamic behaviour of this solution. Many proteins have an affinity for hydrophilic/hydrophobic interfaces, the most ubiquitous of which in *in vitro* studies is the air-water interface. Due to their high affinities for gas-liquid interfaces, proteins can for example stabilise foams Damodaran (2005). The presence of an interface to which the protein of interest has an affinity can influence the nucleation of a new, condensed protein phase essentially in three distinct ways (Fig. 2C):

1) Through an increase in local concentration on the surface, and, if the protein molecules have sufficient mobility once they are bound to the surface, by restricting the diffusion of the protein to two, rather than three dimensions. The character of reaction-diffusion dynamics depends very strongly on the dimensionality of the space in which it occurs, with lower dimensionalities considerably increasing the likelihood of encounters Krapivsky et al. (2010).
2) Through a decrease in the interfacial energy. If the new condensed phase forms on an interface, its surface area in contact with the solution phase is smaller than it would be if the new phase had formed in the bulk of the solution. If the interfacial energy at the contact of the new condensed phase with the added interface is lower than that with the solution (i.e. if the condensed phase 'wets' the surface), the free energy barrier of the nucleation is reduced.
3) The interaction of the protein with the interface might open new reactive trajectories for the nucleation process. The absorbed proteins might, for example, be oriented favourably for a nucleation event to occur. Furthermore, if changes in the internal degrees of freedom are required for the nucleation process, such as is the case for amyloid fibrils, the adsorption onto the surface might induce structural rearrangements that render conformations of the protein accessible that would be too high in energy in bulk solution. This effect can lower the effective energy barrier of the nucleation reaction, by changing the mechanism of nucleation.

It is therefore not surprising that it has been found for many protein condensation reactions that the nucleation process can be considerably enhanced by the presence of suitable surfaces. For example the rate of protein crystal nucleation has been found to be accelerated by impurities Parmar et al. (2007), by the surface of the reaction vessel Tsekova et al. (1999); Tosi et al. (2011) or by microporous media Chayen et al. (2006). The high efficiency of heterogeneous nucleation poses a challenge to the determination of accurate homogeneous nucleation rates. It is difficult to completely avoid heterogeneous nucleation; a promising approach in this context is the encapsulation of the protein solution into microdroplets. Of course such droplets have a very large surface-to-volume ratio, but their inner surface can be made protein repellent through appropriate surfactant chemistry Roach et al. (2005), and is virtually defect-free. Such approaches are increasingly used in studies of protein crystallisation Selimović et al. (2009).

Surface effects are also very important for amyloid fibril nucleation. Even if non-binding polymer surfaces are used in *in vitro* studies, experiments are usually conducted in the presence of an air-water interface, and it has been shown that a range of amyloid polypeptides have an affinity for the air-water interface, such as the Aβ peptide Schladitz et al. (1999); Jean et al. (2012), α-synuclein Wang et al. (2010), IAPP Jean et al. (2012); Trigg et al. (2013) and others. Furthermore, it has been shown that the air water interface can induce secondary structure changes in these proteins Schladitz et al. (1999); Wang et al. (2010) and also enhance their nucleation rate. This effect is most dramatically illustrated in the case of the intrinsically disordered protein α-synuclein which under the commonly utilised aggregation conditions (with mechanical agitation) nucleates into amyloid fibrils at the air-water interface Campioni et al. (2014). In the absence of mechanical agitation, aggregation is very slow Vácha et al. (2014); Buell et al. (2014b), presumably reflecting the increase in the air-water interface, as well as additional processes, such as fibril fragmentation that can be induced by the agitation. However, heterogeneous nucleation of amyloid fibrils is not restricted to the air-water interface. A large variety of other surfaces has been shown to induce nucleation, ranging from inorganic, hydrophilic surfaces Lin et al. (2014) to hydrophobic surfaces Nayak et al. (2008) and nanoparticles of various chemistry Linse et al. (2007); Cabaleiro-Lago et al. (2010); Vácha et al. (2014). Of particular relevance to *in vivo* behaviour is the finding that lipid membranes of various composition have the potential to accelerate the amyloid fibril formation of many of the most relevant amyloid forming proteins Zhu et al. (2003); Galvagnion et al. (2015); Terakawa et al. (2015); Grey et al. (2015). This phenomenon has so far been studied mostly qualitatively, but recently the enhancement of the primary nucleation of α-synuclein by a model lipid has been quantified and shown to be at least 10^3, compared to the presence of only an air-water interface, under quiescent conditions Galvagnion et al. (2015). Heterogeneous nucleation of amyloid fibrils on lipid membranes is distinct from heterogeneous nucleation on other surfaces, in that the nucleation process itself is able to remodel the lipid bilayer, and even to extract individual lipid molecules Hellstrand et al. (2013); Iyer et al. (2014). It is therefore more appropriate to speak of protein-lipid co-aggregation rather than surface catalysis.

Biology also exploits specific types of heterogeneous nucleation. In the case of functional protein filaments, be they non amyloid structures, such as actin filaments Firat-Karalar (2011) and microtubules Job et al. (2003), or amyloid fibrils, such as bacterial curli Wang et al. (2008), they are also not forming through a classical homogeneous nucleation process, but rather their nucleation is tightly regulated by specific 'nucleator' proteins. Besides the biologically relevant nucleation inducing factors, a number of other compounds can induce the nucleation of protein filaments. An important example is the case of α-synuclein, where surfactant molecules, such as SDS, are able to induce aggregation, probably triggered by the formation of mixed micelles, leading to the formation of fibrillar structures with distinct morphology Giehm et al. (2010). Fig. 3B) gives an overview of various scenarios of heterogeneous nucleation of amyloid fibrils. A distinct feature of heterogeneous nucleation compared to homogeneous nucleation is that the heterogeneous nucleation rate can be saturated, if all of the binding sites on the nucleation inducing surface are occupied Galvagnion et al. (2015); Meisl et al. (2014), i.e. an increase in monomer concentration leads no longer to an increase in the nucleation rate. Observed saturation can be used to deduce that a given nucleation process is heterogeneous.

The extraordinary kinetic stability of the unstructured protein α-synuclein against amyloid fibril formation, in the absence of a suitable interface or co-factors that can enhance the nucleation process, is remarkable, and a prime example that kinetic and thermodynamic stabilities are strongly decoupled in the amyloid phenomenon. If the nucleation step is circumvented by the addition of pre-formed seed fibrils, the soluble protein converts rapidly and almost quantitatively (until the critical concentration of the order of 1 μM is reached Baldwin et al. (2011), or until a dense gel is formed, that can slow down the kinetics dramatically Buell et al. (2014b)) into amyloid fibrils. A similarly high kinetic stability is observed in the tau protein, which is also an IDP, and which has a very slow intrinsic

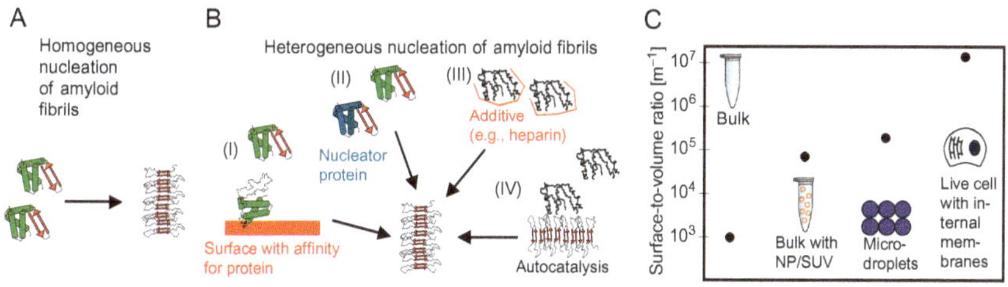

Fig. 3 The importance of heterogeneous nucleation in amyloid fibril formation. (A) Homogeneous nucleation describes the nucleation directly from bulk solution. (B) Heterogeneous nucleation can be induced by a variety of different external factors, such as (I) surfaces to which the proteins have the appropriate affinity (e.g. lipid bilayers Galvagnion et al. (2015) or polymer nanoparticles Vácha et al. (2014) (II) specific nucleator proteins (illustrated in blue) a scenario encountered for functional amyloid formation *in vivo* Wang et al. (2008) (III) co-factors (illustrated in red), such as heparin for the tau protein Ramachandran and Udgaonkar (2011) or SDS for α-synuclein Giehm et al. (2010) (IV) pre-existing amyloid fibrils, leading to autocatalytic secondary nucleation and exponential aggregate amplification. This is the case for a variety of disease-related amyloid systems Ruschak and Miranker (2007); Ramachandran and Udgaonkar (2012); Cohen et al. (2013); Buell et al. (2014b). (C) Surface-to-volume ratios in various practically encountered situations where amyloid fibril formation is observed. Classical bulk experiments can be made to resemble *in vivo* experiments better if relevant surfaces are added in a controlled manner, e.g. through the addition of lipid vesicles (SUV) Galvagnion et al. (2015) or nanoparticles (NP) Linse et al. (2007); Vácha et al. (2014). Microdroplets have even higher surface-to-volume ratios, but their inner surfaces can also be made completely inert Meier et al. (2009). A live cell with all internal surfaces, such as the ER, has two orders of magnitude higher surface-to-volume ratio, highlighting the importance of surfaces in protein aggregation *in vivo*.

aggregation rate in the absence of the polyelectrolyte heparin sulfate Ramachandran and Udgaonkar (2011). Despite the fact that the effect of heparin does not correspond to the classical scenario of heterogeneous nucleation on a surface, it nevertheless underlines the exceedingly slow rates of homogeneous nucleation, in particular of some unstructured proteins. These observations potentially reveal a feature (or 'design' principle) common to at least some IDPs, namely a very high kinetic barrier against nucleation, which is responsible for the metastability of these proteins even at very high concentrations Buell et al. (2014b); Campioni et al. (2014); Galvagnion et al. (2015). While the nature of the nucleation barriers of IDPs are not yet understood, they are likely to be qualitatively different from the barriers that folded proteins have to overcome, as it was shown that in those cases a significant part of the barrier is caused by the very existence of structure Buell et al. (2012).

Therefore, while it is likely that some amyloid-forming proteins do indeed feature homogeneous primary nucleation, most experimental studies to-date might not have been well-designed to detect it unambigiously, given the prominence of interfaces which are known to lead to heterogeneous nucleation. However, heterogeneous nucleation of amyloid fibrils is likely to be more similar to the processes *in vivo* than homogeneous nucleation in bulk solution, given the very high surface to volume ratio in biological cells and the large variety of different types of surfaces featured by living systems. Fig. 3C compares the various surface-to-volume ratios of systems in which amyloid fibril formation is observed, from a test tube to a living cell.

Given the very high surface-to-volume ratio in cells, the arguments discussed above suggest that purely homogeneous nucleation of protein filaments might be the exception, rather than the rule in biology. Furthermore, interesting recent findings suggest that even scenarios intermediate between homogeneous and heterogeneous nucleation are possible, such as the nucleation of filaments from dense liquid phases that form through liquid-liquid phase separation Patel et al. (2015).

4. SECONDARY NUCLEATION AS A SPECIAL CASE OF HETEROGENEOUS NUCLEATION

Having established that surfaces can considerably enhance the rate of nucleation, the question immediately arises whether the surface of the newly formed condensed protein phase can in fact act as a surface for heterogeneous nucleation itself (Fig. 3B). This question is particularly relevant for filamentous aggregated phases, where the filamentous nature stems from the fact that the affinity of the condensed phase for new monomeric building blocks is highly anisotropic Kashchiev et al. (2013), i.e. where the condensed phase grows mainly in one dimension. In a three-dimensional crystal, the formation of a new nucleus on one of the crystal faces leads usually to the spread of this nucleus across the entire face, which corresponds to the growth of the crystal along the direction normal to the face Nielsen (1984). Therefore, the number of individual crystals does not change through this type of heterogeneous nucleation. Secondary nucleation, i.e. the autocatalytic amplification of crystals has been observed Davey (2004), but has usually been attributed to attrition Tait et al. (2009). Recent computer simulations suggest, however, that pre-nucleation clusters in solution can nucleate upon contact with a crystal surface Anwar et al. (2015), and pre-nucleation clusters have indeed been observed in protein solutions Stradner et al. (2004); Sleutel and Driessche (2014).

The growth of filamentous structures, on the other hand, is typically restricted to the two ends Pinotsi et al. (2014) and hence a nucleation event on the surface of a filament, potentially followed by the detachment of the newly formed nucleus from the site of nucleation, leads to the generation of a new filament. The autocatalytic nature of this process is immediately obvious; the presence of filaments leads to the formation of more filaments (Fig. 4A). This specific type of heterogeneous nucleation is called secondary nucleation. Given the ubiquitous nature of heterogeneous nucleation discussed above, it is not surprising that secondary nucleation has been identified in some important filamentous protein systems. The first protein the condensation of which into filaments was found to involve secondary nucleation was sickle haemoglobin Ferrone et al. (1985). This protein differs by only a single amino acid from normal haemoglobin, but this small difference changes its condensation behaviour drastically. While normal haemoglobin displays the typical liquid-liquid phase separation and crystallisation transitions, deoxygenated sickle haemoglobin has a fibrillar phase as well Gunton et al. (2007). The transition from the soluble to the filamentous phase is possible within the physiological

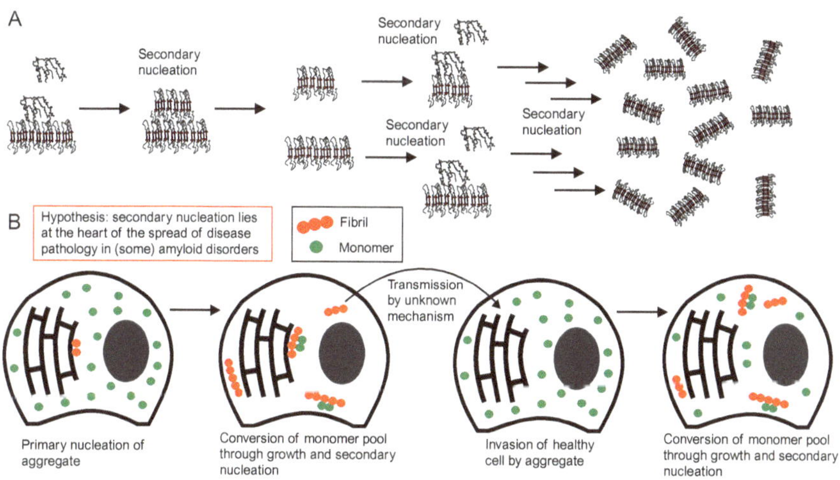

Fig. 4 The importance of secondary nucleation. (A) Secondary nucleation can lead to autocatalytic, exponential amplification of aggregate numbers. (B) The amplification of aggregates by secondary nucleation could be responsible for the spread of disease pathology in diseases, such as Parkinson's disease Peelaerts et al. (2015).

region of parameter space. Through careful experiments and data analysis, it was found that an autocatalytic secondary nucleation process needed to be included in the model in order to be able to describe the kinetic data of sickle haemoglobin polymerisation Hofrichter et al. (1974); Ferrone et al. (1985).

More recently, secondary nucleation is being identified in an increasing number of amyloid fibril forming proteins, interestingly mostly in those that are directly related to disease: a short fragment of the IAPP peptide Ruschak and Miranker (2007), the tau protein Ramachandran and Udgaonkar (2012), the Aβ peptide Cohen et al. (2013) and α-synuclein Buell et al. (2014b). It should be noted that those amyloid proteins, of which it has been proposed that proliferation occurs through fragmentation, such as mammalian Knowles et al. (2009) and yeast prions Shorter and Lindquist (2004); Wang et al. (2011), rather than secondary nucleation, are not explicitly included in this list. The existence of an autocatalytic secondary nucleation pathway is a very important feature, as it allows rapid amplification of low numbers of aggregates; indeed, in principle a single seed filament, that is introduced into a pool of metastable soluble protein, can induce the condensation into filaments of the entire pool through repeated cycles of growth and secondary nucleation of filaments (Fig. 4A). This mechanism can be hypothesised to be at the origin of the spread of disease pathologies (amyloid plaques, neurofibrillar tangles and Lewy bodies) through the brains of patients of Alzheimer's Braak and Braak (1995) and Parkinson's disease Braak et al. (2003) and other neurodegenerative disorders. In this scenario, small numbers of aggregates would diffuse across the extracellular space (e.g. in the case of extracellular pathologies, such as amyloid plaques) or be taken up by a healthy cell (in the case of intracellular pathologies, such as Lewy bodies in Parkinson's disease) and amplified from the inherent pool of soluble peptides and proteins (Fig. 4B), akin to a viral infection. Hence the crucial question to ask about an amyloid fibril forming protein might not be so much whether or not the first aggregate forms through homogeneous or heterogeneous primary nucleation, but rather whether or not secondary nucleation is able to amplify this first aggregate exponentially. It is possible that while most amino acid sequences are able to form amyloid fibrils, only a sub-set of these sequences is capable of autocatalytic replication. The exponential nature of such an autocatalytic process cannot in all cases be confined by cellular mechanisms and might ultimately lead to disease through loss of function or gain of toxic function of the protein aggregates Knowles et al. (2014).

More generally, secondary nucleation as discussed above might simply be one manifestation of the general feature that monomeric and oligomeric species of a given protein have differential affinities for the various types of surfaces of filamentous species. In some amyloid systems it is for example observed that early in the aggregation process, mainly very thin so-called 'protofilaments' are observed, whereas at later stages the dominant species are thicker, so-called 'mature' fibrils. It has been proposed that the mature fibrils form through coalescence or annealing of the protofilaments. This might be possible under conditions of elevated temperature and concentration, where entropic effects can lead to liquid crystalline ordering (Fig. 1) and hence alignment of the protofilaments Bolisetty et al. (2011). Under the commonly encountered solution conditions, a more likely scenario is in many cases that a new protofilament nucleates from monomer on the surface of an existing protofilament, and indeed this mechanism is supported by recent experimental evidence from studies of the Aβ peptide Jeong et al. (2013). Once a certain well-defined (for a given sequence and set of solution conditions) number of protofilaments has assembled in this way, a mature fibril is created, which can be described as 'surface-saturated', explaining the extremely homogeneous endpoint of some aggregation reactions Buell et al. (2014b). Further nuclei that are forming on the surface of the mature fibril are then more weakly bonded and can detach Cohen et al. (2013), leading to the prolific type of secondary nucleation discussed above.

5. EXPERIMENTAL METHODS TO STUDY NUCLEATION OF CONDENSED PROTEIN PHASES

In order to be able to study the formation of a new protein phase, it is necessary to be able to observe it. This seems like an obvious point, but is the main reason why our knowledge is at such different stages concerning the nucleation of different types of condensed protein phases. If the condensed phase is three dimensional, and can grow in all three dimensions to a length scale that is visible by conventional (diffraction-limited) optical microscopy, nucleation can be directly observed in a label-free manner, assuming that each visible domain of the new phase has originated from a single nucleation event (Fig. 5A). The most obvious example of this methodology is of course the study of protein crystallisation. Upon transfer of the protein solution to conditions under which the formation of crystals is favourable (temperature, salt concentration, crowding agents), the number of visible

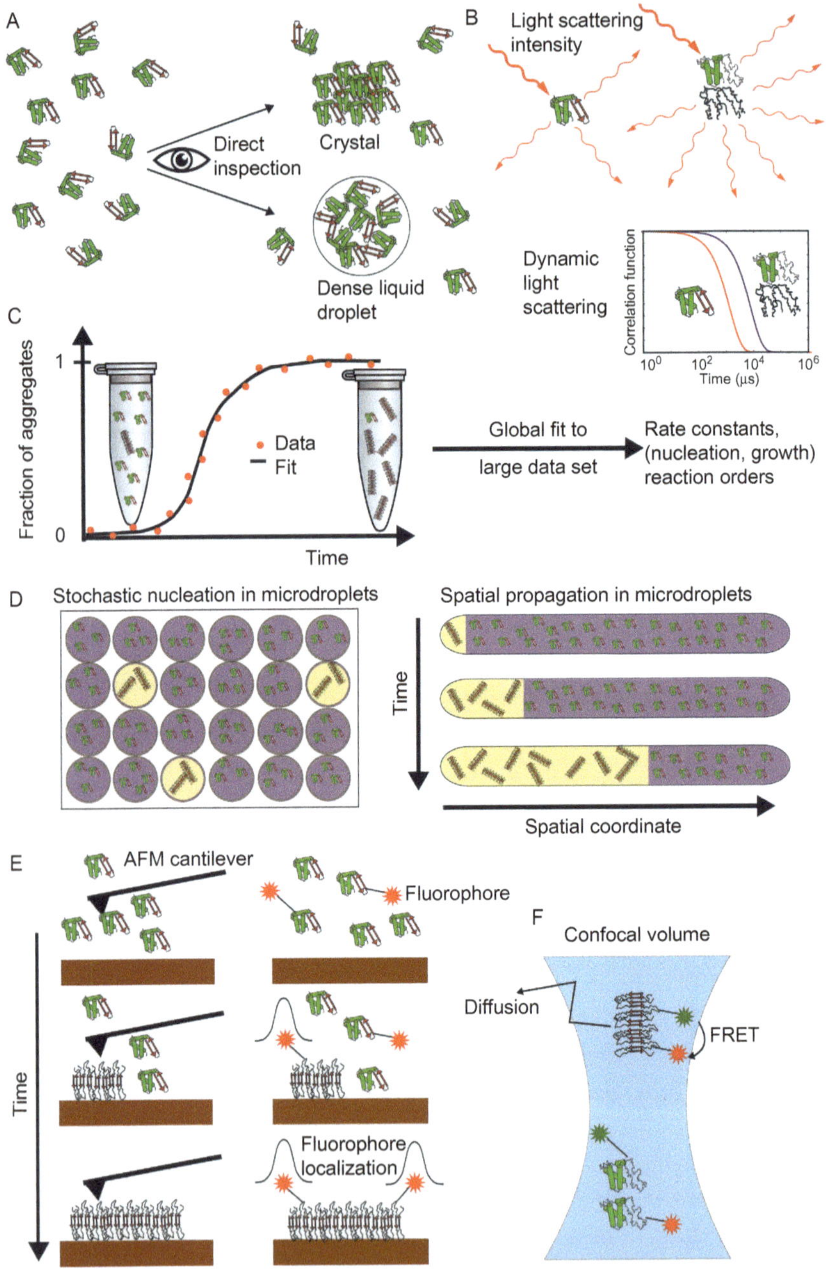

Fig. 5 Experimental methods to study nucleation processes in protein condensation phenomena. (A) The formation of macroscopic pieces of a new protein phase (e.g. crystal and dense liquid droplet) can be observed directly by light microscopic inspection. (B) Light scattering intensity and dynamic light scattering measurements are well-suited

crystallites can be counted as a function of time and together with measurements of the crystal growth rates, the rate of nucleation can be inferred Galkin and Vekilov (2001); Selimović et al. (2009). However, for most of the other condensed phases, in particular filamentous ones, until recently it has not been possible to directly observe individual structures *in situ*. This situation has, however, changed in recent years due to the advances in scanning probe microscopy, as well as optical microscopy (beyond the diffraction limit) and single molecule detection techniques. These methods are only beginning to unfold their potential for the study of protein nucleation and will be discussed below. It is interesting to note here that peptide structures of dimensions intermediate between macroscopic crystals and amyloid fibrils exist as well, most notably micro- and nanocrystals formed by short peptides Reches and Gazit (2003); Nelson et al. (2005); Liang et al. (2010). In some of these cases, it is possible to directly observe the nucleation of the crystalline phase, e.g. by fluorescence microscopy Liang et al. (2010). Such assemblies are extremely interesting, as they represent to some extent intermediates between structures of three-dimensional order (crystal) and structures of one-dimensional order (filaments) and in some cases it has been observed that both states are accessible to a given short peptide sequence Marshall et al. (2010). These system can play the role of important models for the general understanding of peptide self-assembly; however the structural and dynamic insight gained from their study cannot be projected in a straightforward manner onto their longer parent polypeptides.

Most experimental studies of protein filament formation have so-far relied on measuring the total content in the condensed phase of a given

---

to detect low concentrations of aggregates in protein samples. (C) If the global progression of a condensation reaction (e.g. amyloid fibril formation) can be precisely measured, the data can be analysed with an appropriate mathematical model and the nucleation rate can be extracted Oosawa (1975); Ferrone et al. (1985); Meisl et al. (2016). (D) Amyloid fibril formation in microdroplets. Left: If the reaction volumes are made very small, individual nucleation events can potentially be detected Knowles et al. (2011). Right: Furthermore, the spatial propagation of an aggregation reaction due to secondary nucleation processes can be studied in such microscopic reaction volumes Knowles et al. (2011); Cohen et al. (2014). (E) Scanning probe (left, Kellermayer et al. (2013); Varongchayakul et al. (2013)) and super-resolution optical (right Pinotsi et al. (2014)) microscopy can be used to investigate individual aggregates, and potentially even their formation can be studied on a single particle basis. (F) Single molecule optical techniques, based on fluorescence measurements in a confocal microscope can be used to obtain concentration and structural information of early aggregates Orte et al. (2008); Cremades et al. (2012).

macroscopic sample, i.e. the overall progression of the condensation reaction. The measurable quantity is in this case a macroscopic property of the reaction mixture that changes as the soluble protein is converted into the condensed structures, such as birefringeance Hofrichter et al. (1974), heat release Hofrichter et al. (1974), fluorescence intensity LeVine (1993); Bolognesi et al. (2010) (if a suitable probe can be found that interacts differentially with the soluble and the condensed phase) or simply the total mass of the condensed phase separated from the solution by centrifugation Wang et al. (2011). These types of measurements yield the total mass of the condensed phase; however, for the calculation of nucleation rates, the number of aggregates needs to be known as a function of time. Different solutions have been proposed for this problem. If the size distribution of aggregates can be measured in addition to the total mass, for example through light scattering Lomakin et al. (1996), sedimentation velocity analysis Mok et al. (2015) in an analytical ultracentrifuge, or, more precisely, through atomic force microscopy Xue et al. (2009), the number concentration of different species can be estimated. These different techniques have different strengths. For example, while light scattering is non-invasive and exquisitely sensitive to the presence of aggregates in otherwise monomeric solutions, it fails in very polydisperse distributions (Fig. 5B). However, the most commonly applied strategy has been to mathematically model the process under study and to fit the experimental data to a mathematical expression derived from the model, that links the macroscopic observable to the microscopic rate constants. Ideally, the fit parameters allow then to compute the rate constants and rates Fig. 5C. This is a powerful approach that has been used successfully in a range of landmark problems, most notably the polymerisation of actin Oosawa (1975) and that of sickle haemoglobin Ferrone et al. (1985). In order for this method to provide maximum insight, data of very high quality are needed, ideally spanning a wide range of initial concentrations. All the parameters that influence the rate of the condensation reaction need to be well-defined and controlled. Furthermore, all conceivable and reasonable models need to be tested against the data, in order to decide which one is best at describing the data overall. In this context the importance of global fitting should be highlighted Meisl et al. (2016). Any one progress curve of a condensation reaction in a given sample can often be fitted to a large number of different mathematical expressions, the prime example being progress curves of unseeded amyloid fibril formation reactions that can often be fitted to any sigmoidal function, of which there are a very large number of different types. This is due to the limited information content of a single progress curve, that can lead to overfitting, as the mathematical expression of sigmoidal functions contains several free parameters. Therefore, molecular rate constants can only be confidently extracted from fits to a single progress curve in very simple situations, such as when a single process dominates, as is the case in seeded aggregation reactions Buell et al. (2014b).

A global fit of multiple progress curves, on the other hand, to a model that contains significantly less free parameters than independent information in the data, can be an efficient test of whether or not the model that has been chosen represents a good description of the molecular system under study. If that is the case, the relevant rates and rate constants can be determined Cohen et al. (2013); Meisl et al. (2016). This strategy has been applied with the aim to determine the rates of the various nucleation processes that play a role in amyloid fibril formation. A particularly prominent example is the amyloid β peptide, where this method has been used to establish the existence of an autocatalytic secondary nucleation pathway. It was found that the only way to achieve a satisfying global fit was to include a nucleation process that depends both on the concentration of soluble protein and already formed fibrils Cohen et al. (2013). This example also highlights another important point, namely that in cases where a secondary process contributes strongly to the creation of new growth competent aggregates, it can be extremely challenging to accurately determine the rate of primary nucleation. It has often been proposed that the lag time of the unseeded aggregation reaction is a direct measure for the rate of primary nucleation. However, in the presence of secondary processes, this is incorrect; it can be shown that the lag time depends only very weakly (logarithmically) on the rate of primary nucleation Knowles et al. (2009); Cohen et al. (2011). Secondary processes more generally play a role in many experimental studies of amyloid fibril formation by proteins that do not display secondary nucleation on the surfaces of existing fibrils. A large part of all experimental studies of amyloid fibril formation in bulk solution is performed under conditions of mechanical agitation, which is very conducive to fibril fragmentation Xue et al. (2008); Knowles et al. (2009). Under such conditions, the primary nucleation rate cannot be determined very accurately, as the key experimental observables, the lag time and the steepest slope of the aggregation curve, depend almost exclusively on the product of the rate constants of the growth and secondary processes Cohen et al. (2011). Accurate primary nucleation rates can only be extracted from bulk measurements combined with kinetic analysis, if the dominant process responsible for the formation of new aggregates is primary nucleation (homogeneous or heterogeneous). This is, for example, the case for the polymerisation of actin Oosawa (1975), but has only been established reliably for a very small number of amyloid fibril forming peptides and proteins, such as α-synuclein in the presence of lipid vesicles Galvagnion et al. (2015). It is also a plausible scenario in the case of poly-Q peptides, where the initial aggregation behaviour appears to be in agreement with a dominance of primary nucleation Kar et al. (2011).

If an amyloid formation reaction featuring a secondary process is performed in a macroscopic volume, it is therefore difficult to determine exactly how many individual primary nucleation events were responsible for the overall progress of the reaction. One possibility to circumvent this difficulty, is to divide the macroscopic volume into many small sub-volumes and monitor the reaction progress separately in each individual sub-volume. The principal idea behind this approach is to isolate the individual primary nucleation events. For any given primary nucleation rate, there is a characteristic size of a sub-volume such that during the time scale corresponding to the bulk reaction, the number of primary nucleation events in the sub-volume will be of order one. If the sub-volumes are made even smaller, many of them will not feature a primary nucleation event during this time scale. If the given protein in addition features a sufficiently fast secondary process, allowing an aggregate from a single nucleation event to multiply into a detectable amount of aggregates, this kind of experiment can be used to count primary nucleation events. A practical realisation of this approach has been developed and consists of encapsulating monomeric protein solutions into microdroplets Meier et al. (2009); Knowles et al. (2011), followed by the detection of protein aggregates via dye fluorescence, similar to most bulk experiments. This strategy, which has been inspired by earlier work on protein crystallisation, not only allows miniaturisation of the reaction volumes, but importantly also precise control of the surface chemistry. If protein repellent surfactants are chosen, for example based on polyethylene glycol (PEG), heterogeneous nucleation can be virtually eliminated and homogeneous primary nucleation can be studied. Such experiments offer one of the very few opportunities to evaluate the intrinsically stochastic nature of the nucleation process, by measuring the distribution of nucleation times (Fig. 5D). The lack of reproducibility between seemingly identical aggregation reactions in many bulk experiments has often been interpreted as also reflecting the stochastic nucleation process. However, it has been pointed out that for known values of the nucleation and growth rate constants, the number of primary nucleation events in any macroscopic volume is usually so high that intrinsic stochasticity cannot manifest itself Buell et al. (2014a); Arosio et al. (2015). The lack of reproducibility in such bulk experiments therefore stems from differences in initial conditions, such as the protein concentration, and the presence of small amounts of pre-formed seed aggregates.

A further advantage of microvolumes as opposed to bulk systems is that the latter can also be used to study the spatial propagation of the aggregation process in a controlled manner Knowles et al. (2011), due to the absence of convection. The initial primary nucleation event occurs at a given spatial position, but the resulting aggregate, as well as the daughter aggregates that result from secondary processes can diffuse away from the initial site and can act as seeds at a different location (Fig. 5E). Within a pool of supersaturated protein, this can lead to a so-called reaction-diffusion wave that propagates in time and space Knowles et al. (2011); Cohen et al. (2014). It has been proposed that this situation might be a mimic for the spatial propagation of amyloid pathology in diseased brains.

Both bulk and microvolume approaches described above are usually reliant on the fluorescence of amyloidophilic dyes, such as Thioflavin-T (ThT) for the detection of the aggregates. In standard experimental setups, such as fluorescence platereaders or conventional microscopes, single aggregates can not normally be detected based on ThT fluorescence, due to low quantum yields and limited sensitivity. However, the detection and observation of the formation of individual aggregates is the most direct method to study the nucleation process and in addition, it does not rely on the contribution from secondary processes for aggregate amplification. As already mentioned above, in the case of crystal formation, the detection of individual crystals is relatively straightforward with conventional optical microscopy. However, filamentous aggregates usually only have one dimension comparable to, or larger than, the wavelength of visible light. Therefore, more advanced imaging technologies are required for the direct observation of amyloid fibrils, for example based on scanning probe microscopy. In situ atomic force microscopy has been used to observe the formation Varongchayakul et al. (2013) and growth Hoyer et al. (2004); Kellermayer et al. (2013) of individual fibrillar aggregates. More recently, super-resolution optical microscopy, with sub-diffraction-limited resolution, such as total internal reflection fluorescence (TIRF) microscopy Ban and Goto (2006) or stochastic optical reconstruction microscopy (STORM) Pinotsi et al. (2014) has also been used for that purpose, but has so-far mostly focused on the growth of aggregates Ban and Goto

(2006); Wördehoff et al. (2015), rather than the *de novo* formation. Such experiments usually require the aggregates to be immobilised on a surface and are therefore naturally well-adapted for the study of heterogeneous nucleation and growth processes, but are less suitable for studying homogeneous nucleation.

A strategy that has been presented as an alternative for the use in bulk solution consists of the detection of individual fluorescently labelled aggregates with a confocal microscope, where the aggregates size can be determined through absolute fluorescence intensity, two-colour coincidence Orte et al. (2008), fluorescence correlation spectroscopy Crick et al. (2006) or photobleaching Ding et al. (2009) measurements. While this strategy is able to detect small quantities of early aggregates, it cannot be used to directly observe nucleation. For reliable analysis, the samples in such experiments usually need to be diluted to concentrations far below the critical concentration for aggregation. Therefore, the aggregation reaction and the measurement need to be performed under separate conditions. Nevertheless, such approaches allow the counting of individual small aggregates and can even provide some structural information about these aggregates if combined with additional types of analyses, such as FRET efficiencies Hillger et al. (2007); Cremades et al. (2012).

Of particular interest in the context of disease-related protein aggregation is also the observation of aggregate formation within (live or fixed) biological cells. The rapid development of super-resolution optical microscopy in recent years now enables microscopic experiments with sufficient resolution to be performed Pinotsi et al. (2016), while more indirect measurements, such as fluorescence lifetime measurements Esbjörner et al. (2014) or fluorescence complementation assays Outeiro et al. (2008) were employed previously, in order to unambiguously detect aggregates within cells.

6. AMYLOID OLIGOMERS AND (CRITICAL) AMYLOID NUCLEI

One of the aims in the study of the nucleation of condensed protein phases, be they of crystalline or fibrillar nature, has been to understand the nucleation process in terms of a critical nucleus size, corresponding to the least stable species on the condensation pathway. Addition of further monomers to species smaller than the critical nucleus increases the free energy of the system, whereas the addition of a monomer to the critical nucleus decreases the free energy. The size of a critical nucleus can be defined both in the framework of a continuum model of nucleation (for example through its radius, assuming a spherical nucleus, which defines the surface-to-volume ratio), as well as through a discrete number of monomers. If the nucleus consists of a large number of monomers, the two scenarios are equivalent. Within the framework of classical nucleation theory, the critical nucleus is concentration dependent; this dependence stems from the fact that the interfacial energy (the cost of creating an interface between the solution phase and the newly formed condensed phase) and hence the energy difference between condensed phase and solution are concentration dependent. Therefore, a critical nucleus size always needs to be quoted together with the reaction conditions under which it has been determined. This concept of a critical nucleus size is well-adapted for the cases in which a given aggregate is essentially uniquely defined by the number of monomers it consists of, and potentially an additional order parameter, that describes how closely a given aggregate resembles the final structure (Price et al. (2015) and Fig. 6A). This can only be the case in systems where no internal degrees of freedom are involved in the condensation, i.e. in systems where the monomer both in solution and in the aggregate has a unique structure, which is not necessarily the same in both cases (a potential difference could be nucleotide binding, as in actin polymerisation). In systems where internal degrees of freedom participate in the condensation reaction, such as amyloid fibril formation, each species needs to be defined through its size, as well as through the internal state of all the composing monomers. In such cases a large variety of different species of equal aggregation number (and similar free energy) might exist that differ in the internal coordinates of the composing monomers. Idealised models of amyloid fibril nucleation, called 'one-step-' and 'two-step nucleation', have been proposed Auer et al. (2012), whereby the structural rearrangements occur simultaneously with the association reaction ('one step'), or else subsequently ('two step'), see Fig. 6B. Within the framework of such a simple model, critical nucleus sizes can also be defined for amyloid fibrils. However, based on the polymeric nature of aggregating proteins and the observations of large varieties of different oligomeric species in aggregating samples, a more sophisticated theoretical description might be necessary. We define the nucleation of an amyloid (proto-)fibril as the complete set of processes that leads from a monomer to a minimal fibril. The latter is defined as that structure, from which on the addition of more monomers features size-independent kinetic and thermodynamic signatures. The nucleation process defined in this way may consist of many individual events, some of which correspond to

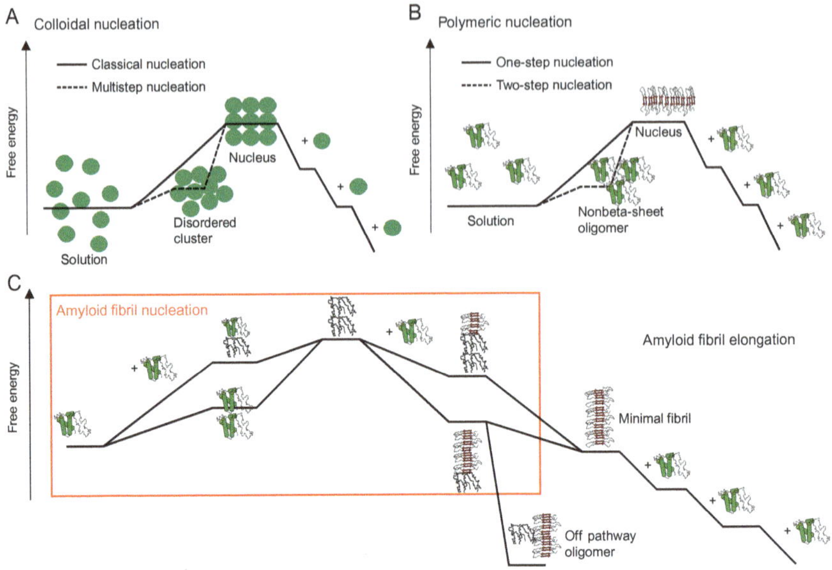

Fig. 6 Different scenarios and descriptions of nucleation processes for protein structures. (A) Classical nucleation theory is suitable for nucleation processes in which no internal degrees of freedom change ('colloidal nucleation'). In principle, the nucleus can be defined entirely by specifying the number of monomeric building blocks. However, in some cases, the introduction of an order parameter is also necessary, as the nucleation can proceed via dense, disorderd intermediate phases ('multistep nucleation'). (B) If internal coordinates change during the nucleation process ('polymeric nucleation'), this can happen simultaneously with the association reaction ('one-step nucleation') or subsequently ('two-step' nucleation Auer et al. (2012)). In this simplified picture, the nucleus is the minimal fibril. (C) We propose that in the case of amyloid fibril formation, the entire process that leads from a monomer to a minimal fibril should be referred to as 'nucleation'. In a realistic picture of this nucleation process, that accounts for the complexity and diversity of the observed species, alternative pathways might exist that lead to the nucleus. Multiple monomer addition and internal rearrangement processes might contribute to the formation of the minimal fibril. The nucleus, i.e. the structure highest in free energy, might not be identical to the minimal fibril. This scenario also accounts for side reactions that lead to off-pathway oligomers Lorenzen et al. (2014).

addition of monomer, whereas some correspond to changes in internal coordinates ('rearrangements' or 'reptations' de Gennes (1971), in the language of polymer science). Also, there may not be a unique sequence and/or order of molecular events that result in a given end product, for example a protofilament (Fig. 6). Nevertheless, there is still likely a single highest free energy state along the nucleation pathway, corresponding to the classical critical nucleus, but which requires its internal coordinates to be specified in addition to its size, in order to be well-defined.

Such a complex multi-step process is difficult to treat analytically Garcia et al. (2014), but unless an appropriate expression for the nucleation rate is available, the critical nucleus size cannot be determined from even the most accurate data. In particular, in such complex cases, it is not possible in a straightforward manner to extract critical nucleus sizes from empirical, simplifying expressions of the nucleation rate, such as $\frac{dP}{dt} = k_n m^{n_c}$, where P is the number of minimal fibrils, m is the monomer concentration and n_c is the reaction order of the nucleation process with respect to monomer Knowles et al. (2009); Cohen et al. (2011). It has been pointed out that such an expression cannot correctly describe the nucleation rate, due to the fact that the expression does not vanish at or below the critical concentration Kashchiev (2015). In addition, authors who used this type of expression successfully in the context of global fitting of aggregation kinetic data have stated that n_c in such expressions is a reaction order, and not related to a nucleus size in a straightforward manner Meisl et al. (2016).

For the amyloid β peptide, the reaction order n_c has been determined to be 2 for both primary and secondary nucleation Cohen et al. (2013), a finding which has been used by other authors to extract a critical nucleus size of 1 Ferrone (2015). Indeed, the idea of a monomeric critical nucleus size has been repeatedly proposed Kar et al. (2011), and at first sight, it seems plausible that this value should indeed be possible in a polymeric nucleation process, i.e. one that involves internal degrees of freedom (for a colloidal nucleation reaction that involves no internal degrees of freedom, a critical nucleus size of one would mark the transition from nucleation to spinodal decomposition, i.e. nucleation-less condensation). However, apart from the fact that these analyses are usually based on an oversimplified expression of the nucleation rate (see above), the very idea of a monomer as a critical nucleus is only plausible under very specific circumstances. This will be illustrated in the following.

Within the framework of an energy landscape theory of protein aggregation Buell et al. (2010), it can be easily explained why the presence of a template (i.e. a seed fibril) allows a given protein molecule to adopt a structure (β-sheet) that it never populates in its monomeric form: the presence of the template modifies the free energy landscape (both barriers and troughs) and enables new pathways and states, which are energetically inaccessible for the isolated monomer. A critical nucleus size of one then amounts to a monomer being able to act as an equally good template as a fully formed protofilament. Surely, this is an artificially created limiting scenario. It seems implausible that all of the intermolecular cooperativity is concentrated in the growth process, and none of it in the nucleation process, i.e. that the nucleation is an entirely monomolecular process. Furthermore, it has been shown that a large part of the favourable contribution to the kinetics of templating reactions in seeded growth stems from the hydrophobic effect Buell et al. (2012); this effect can be expected to also be present if two monomers interact, independently of their conformation, suggesting that nucleation usually is also (at least) a bimolecular process.

However, we do not dismiss the possibility of a monomeric nucleus altogether, in particular considering the case of homopolymeric polypeptides, with the prime example being polyglutamine (poly-Q) peptides. In those systems, it can be imagined that intramolecular contacts are very similar in nature to intermolecular contacts, an argument supported by the compact, 'self-solvating' nature of long poly-Q sequences Crick et al. (2006). Therefore, a critical nucleus of a poly-aminoacid could be equally well composed of parts of the sequences of distinct molecules or of parts of the sequence of a single molecule, and the likelihood of a monomeric nucleus can be expected to increase with poly-Q length Kar et al. (2011). These considerations notwithstanding, all the difficulties associated with the determination of critical nuclei from bulk kinetic data outlined above apply here as well; apparently well-established critical nucleus sizes should be interpreted with care, as has been pointed out before Bernacki (2009). Furthermore, it has also been pointed out that deviations from simple homogeneous nucleation through the presence of a variety of oligomeric intermediates in poly Q aggregation Crick et al. (2013) render the determination of the critical nucleus size non-straightforward Vitalis and Pappu (2011).

Potentially the most direct and convincing experimental evidence currently available that monomeric nucleus sizes might be possible has been provided by a study where individual protein molecules have been stretched with an atomic force microscope on a surface, a process that was then shown to be able to induce filament growth Varongchayakul et al. (2013). However, the nucleus in this situation is no longer in thermal equilibrium with its environment and hence classical nucleation theory cannot be applied to describe this situation.

One of the distinct features of the known amyloid forming proteins is that in many cases, a variety of aggregated states can be found to be populated during aggregation reactions Benilova et al. (2012). Not only the final product of the reaction, the amyloid fibrils, can show diversity in characteristics, such as their thickness and twist period Petkova et al. (2005), but also intermediate structures. Despite their often transient nature, it has been possible to investigate some of the more long-lived non-fibrillar intermediates in some detail Lambert et al. (1998); Campioni et al. (2010); Lorenzen et al. (2014); Chen et al. (2015). However, almost by definition, those structures that are inert enough such that they can be isolated and studied have been shown not to be direct precursors to fibrils in the cases where this has been systematically tested Lorenzen et al. (2014). In some cases non-fibrillar intermediates can have well-defined sizes and consist of several to dozens of monomers Lorenzen et al. (2014). Overall, it appears that the involvement of internal degrees of freedom upon aggregation by amyloid forming polypeptides renders a whole plethora of states accessible, some of which can even have similar thermodynamic stabilities as the mature fibrils Paslawski et al. (2014) and hence correspond to frustrated ('glassy') states that are trapped in deep secondary free energy minima (Fig. 6C). Those, however, that are direct precursors to amyloid fibrils ('on-pathway') are often the least well-characterised as they are most transient and their populations are small Cohen et al. (2013). However, the study of oligomeric aggregates that form under conditions where amyloid fibrils are observed, but that are not direct precursors to fibrils ('off-pathway') can also be important as these might be relevant in protein aggregation diseases, given that they are able to accumulate more strongly than structures on the direct pathway to fibril formation.

Overall, many open questions remain as to the process of amyloid fibril nucleation and the complete mechanism has not yet been elucidated for any amyloidogenic protein. Even some of the best-studied and most relevant peptides and proteins, such as the amyloid β-peptide, still present a range of fundamental open questions. While recent years have seen considerable progress in that it has been possible to describe large data sets globally with a relatively simple model that involves autocatalytic secondary nucleation Cohen et al. (2013), it is striking how much the results of individual experimental studies with this peptide can differ. For example, under some conditions the population of non-fibrillar oligomers during an aggregation reaction peaks during the growth phase, never exceeds a few percent and the duration of the aggregation reaction rarely exceeds 1h Cohen et al. (2013). Under other conditions, however, it was reported that a significant fraction of the peptide forms oligomers during the lagtime of the aggregation Lee et al. (2011); Garai and Frieden (2013); Brener et al. (2015), which lasts several hours, at similar Lee et al. (2011); Garai and Frieden (2013) or even much higher higher peptide concentrations Brener et al. (2015). Partly, such discrepancies can probably be explained through the different concentration ranges and solution conditions employed, as it has indeed been shown that very small changes in solution conditions can have drastic effects on the relative importance of individual molecular rate constants in amyloid fibril formation Buell et al. (2014b). However, in many cases such differences probably stem also from additional effects, such as the presence of labels Garai and Frieden (2013), or indeed of undesired impurities, in particular if the peptide has not been produced recombinantly. It is of vital importance to address and resolve these discrepancies, in order for comparability between experimental data from different laboratories to be established and a global description of these complex systems to be achieved.

7. OUTLOOK

As discussed above, from an experimental point of view, the nucleation of fibrillar protein aggregates is very challenging, due to the often indirect nature of the experiments, involving mathematical modelling and fitting procedures Meisl et al. (2016). Furthermore, no currently available experimental technique has a sufficient resolution to allow the mechanistic determination of these nucleation processes at an atomic or even just single amino acid residue level, and it is unlikely that such methods will be developed in the near future. We are currently lacking structural information even for most of the end products, the mature amyloid fibrils, which are well-defined and can be produced in large quantities. Structural information available for the heterogeneous and/or transient intermediates of the aggregation process is even more sparse. Therefore, ultimately, the mechanistic insight at the highest level of resolution will need to be provided by molecular dynamics simulations, that will need to be guided by experimental results. All atom, explicit solvent molecular dynamics simulations of an aggregation reaction Barz et al. (2014), from monomer to protofilament, will allow the nucleation process to be completely described. Unfortunately, such simulations are out of reach of currently available computers for realistic experimental concentrations and time scales. This is clearly illustrated in Fig. 7, where experimentally measured nucleation rates have been translated into waiting times for on-pathway nucleation events in sample 'boxes' of $(25\ \text{nm})^3$, corresponding to typical sizes of samples in MD simulations Urbanc et al. (2010); Barz et al. (2014). There are still many orders of magnitude difference between typical simulation times (μs) and typical waiting times for an

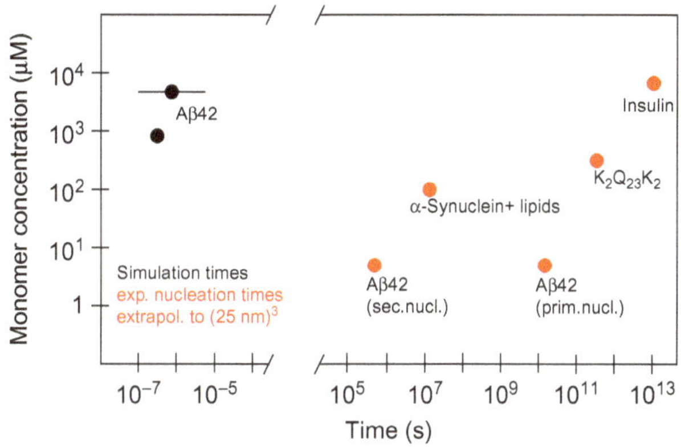

Fig. 7 Comparison of experimental nucleation times (computed as the inverse of the rate of formation) of various amyloid fibrils with simulations of the amyloid β peptide. The experimental times refer to the average time it takes for an amyloid fibril to form in a volume of $(25\ nm)^3$, at the concentration of the respective experiments. The data for the Aβ peptide Cohen et al. (2013),α-synuclein Galvagnion et al. (2015) and $K_2Q_{23}K_2$Kar et al. (2011) are based on experiments in bulk solution, whereas the data for insulin is based on the results of microdroplet experiments Knowles et al. (2011). For the secondary nucleation of Aβ, we have assumed that the $(25\ nm)^3$ box contains one fibril that consists of 10 monomers, corresponding to a nominal fibril mass concentration of ~1 mM. For the heterogeneous primary nucleation of α-synuclein, we have assumed that one wall of the $(25\ nm)^3$ box is covered with a bilayer of DMPS, corresponding to a nominal DMPS concentration of 78 mM. The points for the simulations are taken from Barz et al. (2014) and Urbanc et al. (2010). The latter point has an error bar in time dimension to reflect that in this computational method, the step size is not related to a fixed time period Urbanc et al. (2010). On-pathway nucleation events are rare and require waiting times many orders of magnitude beyond the currently possible simulation times.

"on-pathway" nucleation event, i.e. one that leads to a complete amyloid fibril, in the nanoscopic simulation volume ($>10^6$s).

Various coarse graining schemes have been developed and applied in order to circumvent the problem of the intractable computation times Pellarin and Caflisch (2006); Auer and Kashchiev (2010); Šarić et al. (2014). While being able to provide important insight into the general physical principles that govern protein nucleation and growth processes, coarse-grained simulations will not be able to provide the nucleation mechanisms of individual peptide and protein systems. Despite the generic nature of amyloid fibril formation, determining the nucleation mechanisms at the necessary level of resolution for targeted intervention necessitates inclusion of the full chemical detail of the given amino acid sequences. The route towards this ambitious aim will need to involve simulations of peptides of increasing length and complexity, that will need to be benchmarked against accurate thermodynamic and kinetic experimental data of such model systems. Such data is currently mostly lacking, due to the focus of the scientific community on the more complex, disease-related peptides and proteins. Therefore, we would like to close this paper by advocating the systematic and quantitative study of the condensation behaviour of simple model peptides for the benefit of future understanding of more complex and relevant systems.

8. CONCLUSIONS

The formation of amyloid fibrils is a special case of the general phenomenon of protein condensation into insoluble phases, and is not normally included in phase diagrams of protein condensation. The significant structural rearrangements involved in amyloid fibril formation render the application of classical colloidal models of protein condensation, which do not consider internal degrees of freedom, inappropriate and prevent the straightforward experimental determination of critical nucleus sizes, a concept more appropriate for colloidal condensation processes.

Experimental studies of amyloid fibril formation have so far mostly not been well designed for the selective study of homogeneous nucleation; rather, heterogeneous nucleation is likely to play an important role in many experimental systems. However, due to its ubiquitous nature, heterogeneous nucleation is likely to be more relevant for biological protein condensation phenomena than homogeneous nucleation. In experimental studies, the surfaces need to be well-defined and controlled in order for a detailed and quantitative analysis to be possible.

Secondary nucleation is a special case of heterogeneous nucleation in which new aggregates are nucleated on the surfaces of existing aggregates. This autocatalytic behaviour leads to rapid proliferation of aggregates and could be the defining difference between functional (and hence controlled) and deleterious, disease-related protein filament formation.

Ultimately, molecular dynamics simulations at atomic detail, guided by experimental data, will be needed to obtain insight into the kinetics, thermodynamics and structures of transient intermediates on the pathway from monomeric to fibrillar proteins.

ACKNOWLEDGMENTS

I would like to thank Dr. Céline Galvagnion and Alessia Peduzzo for helpful comments on the manuscript.

REFERENCES

Anwar, J., Khan, S., Lindfors, L., 2015. Secondary crystal nucleation: nuclei breeding factory uncovered. Angew. Chem. Int. Ed. Engl. 54, 14681–14684.

Arosio, P., Knowles, T.P.J., Linse, S., 2015. On the lag phase in amyloid fibril formation. Phys. Chem. Chem. Phys. 17, 7606–7618.

Auer, S., Frenkel, D., 2001. Prediction of absolute crystal-nucleation rate in hard-sphere colloids. Nature 409, 1020–1023.

Auer, S., Kashchiev, D., 2010. Phase diagram of alpha-helical and beta-sheet forming peptides. Phys. Rev. Lett. 104, 168105.

Auer, S., Ricchiuto, P., Kashchiev, D., 2012. Two-step nucleation of amyloid fibrils: omnipresent or not? J. Mol. Biol. 422, 723–730.

Baldwin, A.J., Knowles, T.P.J., Tartaglia, G.G., Fitzpatrick, A.W., Devlin, G.L., Shammas, S.L., Waudby, C.A., Mossuto, M.F., Meehan, S., Gras, S.L., Christodoulou, J., Anthony-Cahill, S.J., Barker, P.D., Vendruscolo, M., Dobson, C.M., 2011. Metastability of native proteins and the phenomenon of amyloid formation. J. Am. Chem. Soc. 133, 14160–14163.

Ban, T., Goto, Y., 2006. Direct observation of amyloid growth monitored by total internal reflection fluorescence microscopy. Methods Enzymol. 413, 91–102.

Barz, B., Olubiyi, O.O., Strodel, B., 2014. Early amyloid β-protein aggregation precedes conformational change. Chem. Commun. (Camb) 50, 5373–5375.

Benilova, I., Karran, E., Strooper, B.D., 2012. The toxic Aβ oligomer and Alzheimer's disease: an emperor in need of clothes. Nat. Neurosci. 15, 349–357.

Bernacki, J. P Murphy, R.M., 2009. Model discrimination and mechanistic interpretation of kinetic data in protein aggregation studies. Biophys. J. 96, 2871–2887.

Bianchi, E., Blaak, R., Likos, C.N., 2011. Patchy colloids: state of the art and perspectives. Phys. Chem. Chem. Phys. 13, 6397–6410.

Bolisetty, S., Adamcik, J., Mezzenga, R., 2011. Snapshots of fibrillation and aggregation kinetics in multistranded amyloid β-lactoglobulin fibrils. Soft Matter 7, 493–499.

Bolognesi, B., Kumita, J.R., Barros, T.P., Esbjörner, E.K., Luheshi, L.M., Crowther, D.C., Wilson, M.R., Dobson, C.M., Favrin, G., Yerbury, J.J., 2010. ANS binding reveals common features of cytotoxic amyloid species. ACS Chem. Biol. 5, 735–740.

Booth, D.R., Sunde, M., Bellotti, V., Robinson, C.V., Hutchinson, W.L., Fraser, P.E., Hawkins, P.N., Dobson, C.M., Radford, S.E., Blake, C.C., Pepys, M.B., 1997. Instability, unfolding and aggregation of human lysozyme variants underlying amyloid fibrillogenesis. Nature 385, 787–793.

Borgia, M.B., Nickson, A.A., Clarke, J., Hounslow, M.J., 2013. A mechanistic model for amorphous protein aggregation of immunoglobulin-like domains. J. Am. Chem. Soc. 135, 6456–6464.

Braak, H., Braak, E., 1995. Staging of Alzheimer's disease-related neurofibrillary changes. Neurobiol. Aging 16271-8. Discussion 278-84.

Braak, H., Tredici, K.D., Rüb, U., de Vos, R.A.I., Steur, E.N.H.J., Braak, E., 2003. Staging of brain pathology related to sporadic Parkinson's disease. Neurobiol. Aging 24, 197–211.

Brangwynne, C.P., Tompa, P., Pappu, R.V., 2015. Polymer physics of intracellular phase transitions. Nat. Phys. 11, 899–904.

Brant, D.A., Flory, P.J., 1965. The configuration of random polypeptide chains. II. Theory. J. Am. Chem. Soc. 87 (13), 2791–2800.

Brener, O., Dunkelmann, T., Gremer, L., van Groen, T., Mirecka, E.A., Kadish, I., Willuweit, A., Kutzsche, J., Jürgens, D., Rudolph, S., Tusche, M., Bongen, P., Pietruszka, J., Oesterhelt, F., Langen, K.-J., Demuth, H.-U., Janssen, A., Hoyer, W., Funke, S.A., Nagel-Steger, L., Willbold, D., 2015. QIAD assay for quantitating a compound's efficacy in elimination of toxic Aβ oligomers. Sci. Rep. 5, 13222.

Buell, A.K., Blundell, J.R., Dobson, C.M., Welland, M.E., Terentjev, E.M., Knowles, T.P.J., 2010. Frequency factors in a landscape model of filamentous protein aggregation. Phys. Rev. Lett. 104, 228101.

Buell, A.K., Dhulesia, A., Mossuto, M.F., Cremades, N., Kumita, J.R., Dumoulin, M., Welland, M.E., Knowles, T.P.J., Salvatella, X., Dobson, C.M., 2011. Population of nonnative states of lysozyme variants drives amyloid fibril formation. J. Am. Chem. Soc. 133, 7737–7743.

Buell, A.K., Dhulesia, A., White, D.A., Knowles, T.P.J., Dobson, C.M., Welland, M.E., 2012. Detailed analysis of the energy barriers for amyloid fibril growth. Angew. Chem. Int. Ed. Engl. 51, 5247–5251.

Buell, A.K., Dobson, C.M., Knowles, T.P.J., 2014a. The physical chemistry of the amyloid phenomenon: thermodynamics and kinetics of filamentous protein aggregation. Essays Biochem. 56, 11–39.

Buell, A.K., Galvagnion, C., Gaspar, R., Sparr, E., Vendruscolo, M., Knowles, T.P.J., Linse, S., Dobson, C.M., 2014b. Solution conditions determine the relative importance of nucleation and growth processes in α-synuclein aggregation. Proc. Natl. Acad. Sci. USA 111 (21), 7671–7676.

Cabaleiro-Lago, C., Quinlan-Pluck, F., Lynch, I., Dawson, K.A., Linse, S., 2010. Dual effect of amino modified polystyrene nanoparticles on amyloid β protein fibrillation. ACS Chem. Neurosci. 1, 279–287.

Campioni, S., Mannini, B., Zampagni, M., Pensalfini, A., Parrini, C., Evangelisti, E., Relini, A., Stefani, M., Dobson, C.M., Cecchi, C., Chiti, F., 2010. A causative link between the structure of aberrant protein oligomers and their toxicity. Nat. Chem. Biol. 6, 140–147.

Campioni, S., Carret, G., Jordens, S., Nicoud, L., Mezzenga, R., Riek, R., 2014. The presence of an air-water interface affects formation and elongation of alpha-synuclein fibrils. J. Am. Chem. Soc. 136, 2866–2875.

Chayen, N.E., Saridakis, E., Sear, R.P., 2006. Experiment and theory for heterogeneous nucleation of protein crystals in a porous medium. Proc. Natl. Acad. Sci. USA 103, 597–601.

Chen, S.W., Drakulic, S., Deas, E., Ouberai, M., Aprile, F.A., Arranz, R., Ness, S., Roodveldt, C., Guilliams, T., De-Genst, E.J., Klenerman, D., Wood, N.W., Knowles, T.P.J., Alfonso, C., Rivas, G., Abramov, A.Y., Valpuesta, J.M., Dobson, C.M., Cremades, N., 2015. Structural characterization of toxic oligomers that are kinetically trapped during α-synuclein fibril formation. Proc. Natl. Acad. Sci. USA 112, E1994–E2003.

Chiti, F., Dobson, C.M., 2009. Amyloid formation by globular proteins under native conditions. Nat. Chem. Biol. 5, 15–22.

Cohen, S.I.A., Vendruscolo, M., Welland, M.E., Dobson, C.M., Terentjev, E.M., Knowles, T.P.J., 2011. Nucleated polymerization with secondary pathways.I. Time evolution of the principal moments. J. Chem. Phys. 135, 065105.

Cohen, S.I.A., Linse, S., Luheshi, L.M., Hellstrand, E., White, D.A., Rajah, L., Otzen, D.E., Vendruscolo, M., Dobson, C.M., Knowles, T.P.J., 2013. Proliferation of amyloid-β 42 aggregates occurs through a secondary nucleation mechanism. Proc. Natl. Acad. Sci. USA 110, 9758–9763.

Cohen, S.I.A., Rajah, L., Yoon, C.H., Buell, A.K., White, D.A., Sperling, R.A., Vendruscolo, M., Terentjev, E.M., Dobson, C.M., Weitz, D.A., Knowles, T.P.J., 2014. Spatial propagation of protein polymerization. Phys. Rev. Lett. 112, 098101.

Cremades, N., Cohen, S.I.A., Deas, E., Abramov, A.Y., Chen, A.Y., Orte, A., Sandal, M., Clarke, R.W., Dunne, P., Aprile, F.A., Bertoncini, C.W., Wood, N.W., Knowles, T.P.J., Dobson, C.M., Klenerman, D., 2012. Direct observation of the interconversion of normal and toxic forms of α-synuclein. Cell 149, 1048–1059.

Crick, S.L., Jayaraman, M., Frieden, C., Wetzel, R., Pappu, R.V., 2006. Fluorescence correlation spectroscopy shows that monomeric polyglutamine molecules form collapsed structures in aqueous solutions. Proc. Natl. Acad. Sci. USA 103, 16764–16769.

Crick, S.L., Ruff, K.M., Garai, K., Frieden, C., Pappu, R.V., 2013. Unmasking the roles of N- and C-terminal flanking sequences from exon 1 of huntingtin as modulators of polyglutamine aggregation. Proc. Natl. Acad. Sci. USA 110, 20075–20080.

Damodaran, S., 2005. Protein stabilization of emulsions and foams. J. Food Sci. 70 (3), 54–66.

Davey, R.J., 2004. Crystallization: how come you look so good? Nature 428, 374–375.

de Gennes, P.G., 1971. Reptation of a polymer chain in the presence of fixed obstacles. J. Chem. Phys. 55, 572.

Derjaguin, B., Landau, L., 1941. Theory of the stability of strongly charged lyophobic sols and of the adhesion of strongly charged particles in solutions of electrolytes. Acta Phys. Chem. URSS 14, 633.

Ding, H., Wong, P.T., Lee, E.L., Gafni, A., Steel, D.G., 2009. Determination of the oligomer size of amyloidogenic protein beta-amyloid(1-40) by single-molecule spectroscopy. Biophys. J. 97, 912–921.

Dobrynin, A.V., Rubinstein, M., 2005. Theory of polyelectrolytes in solutions and at surfaces. Prog. Polym. Sci. 30, 1049–1118.

Domike, K.R., Donald, A.M., 2007. Thermal dependence of thermally induced protein spherulite formation and growth: kinetics of beta-lactoglobulin and insulin. Biomacromolecules 8, 3930–3937.

Dumetz, A.C., Chockla, A.M., Kaler, E.W., Lenhoff, A.M., 2008. Protein phase behavior in aqueous solutions: crystallization, liquid-liquid phase separation, gels, and aggregates. Biophys. J. 94, 570–583.

Dunker, A.K., Lawson, J.D., Brown, C.J., Williams, R.M., Romero, P., Oh, J.S., Oldfield, C.J., Campen, A.M., Ratliff, C.M., Hipps, K.W., Ausio, J., Nissen, M.S., Reeves, R., Kang, C., Kissinger, C.R., Bailey, R.W., Griswold, M.D., Chiu, W., Garner, E.C., Obradovic, Z., 2001. Intrinsically disordered protein. J. Mol. Graph Model 19, 26–59.

Ellis, R.J., 2001. Macromolecular crowding: obvious but underappreciated. Trends Biochem. Sci. 26, 597–604.

Esbjörner, E.K., Chan, F., Rees, E., Erdelyi, M., Luheshi, L.M., Bertoncini, C.W., Kaminski, C.F., Dobson, C.M., Schierle, G.S.K., 2014. Direct observations of amyloid β self-assembly in live cells provide insights into differences in the kinetics of aβ(1-40) and aβ(1-42) aggregation. Chem. Biol. 21, 732–742.

Ferrone, F.A., 2015. Assembly of Aβ proceeds via monomeric nuclei. J. Mol. Biol. 427, 287–290.

Ferrone, F.A., Hofrichter, J., Eaton, W.A., 1985. Kinetics of sickle hemoglobin polymerization. II. A double nucleation mechanism. J. Mol. Biol. 183, 611–631.

Firat-Karalar, E. N Welch, M.D., 2011. New mechanisms and functions of actin nucleation. Curr. Opin. Cell Biol. 23, 4–13.

Flory, P., 1953. Principles of Polymer Chemistry. Cornell Univ. Press, Ithaca, NY.

Galkin, O., Vekilov, P.G., 2001. Nucleation of protein crystals: critical nuclei, phase behavior, and control pathways. J. Cryst. Growth 232, 63–76.

Galvagnion, C., Buell, A.K., Meisl, G., Michaels, T.C.T., Vendruscolo, M., Knowles, T.P.J., Dobson, C.M., 2015. Lipid vesicles trigger α-synuclein aggregation by stimulating primary nucleation. Nat. Chem. Biol. 11, 229–234.

Garai, K., Frieden, C., 2013. Quantitative analysis of the time course of Aβ oligomerization and subsequent growth steps using tetramethylrhodamine-labeled Aβ. Proc. Natl. Acad. Sci. USA 110, 3321–3326.

Garcia, G.A., Cohen, S.I.A., Dobson, C.M., Knowles, T.P.J., 2014. Nucleation-conversion-polymerization reactions of biological macromolecules with prenucleation clusters. Phys. Rev. E 89 (89), 032712.

Giehm, L., Oliveira, C.L.P., Christiansen, G., Pedersen, J.S., Otzen, D.E., 2010. SDS-induced fibrillation of alpha-synuclein: an alternative fibrillation pathway. J. Mol. Biol. 401, 115–133.
Gitlin, I., Carbeck, J.D., Whitesides, G.M., 2006. Why are proteins charged? Networks of charge-charge interactions in proteins measured by charge ladders and capillary electrophoresis. Angew. Chem. Int. Ed. Engl. 45, 3022–3060.
Grey, M., Dunning, C.J., Gaspar, R., Grey, C., Brundin, P., Sparr, E., Linse, S., 2015. Acceleration of α-synuclein aggregation by exosomes. J. Biol. Chem. 290, 2969–2982.
Gunton, J.D., Shirayev, A., Pagan, D.L., 2007. Protein condensation–Kinetic pathways to crystallisation and disease. Cambridge University Press, Cambridge.
Hanson, J., Lowy, J., 1963. The structure of F-actin and of actin filaments isolated from muscle. J. Mol. Biol. 6, 46–60.
Hasler, L., Heymann, J.B., Engel, A., Kistler, J., Walz, T., 1998. 2D crystallization of membrane proteins: rationales and examples. J. Struct. Biol. 121, 162–171.
Hellstrand, E., Nowacka, A., Topgaard, D., Linse, S., Sparr, E., 2013. Membrane lipid co-aggregation with α-synuclein fibrils. PLoS One 8, e77235.
Hillger, F., Nettels, D., Dorsch, S., Schuler, B., 2007. Detection and analysis of protein aggregation with confocal single molecule fluorescence spectroscopy. J. Fluoresc. 17, 759–765.
Hirota-Nakaoka, N., Hasegawa, K., Naiki, H., Goto, Y., 2003. Dissolution of beta2-microglobulin amyloid fibrils by dimethylsulfoxide. J. Biochem. 134, 159–164.
Hofrichter, J., Ross, P.D., Eaton, W.A., 1974. Kinetics and mechanism of deoxyhemoglobin s gelation: a new approach to understanding sickle cell disease. Proc. Natl. Acad. Sci. USA 71, 4864–4868.
Hoover, W., Ree, F., 1968. Melting transition and communal entropy for hard spheres. J. Chem. Phys. 49, 3609–3617.
Hoyer, W., Cherny, D., Subramaniam, V., Jovin, T.M., 2004. Rapid self-assembly of alpha-synuclein observed by in situ atomic force microscopy. J. Mol. Biol. 340, 127–139.
Israelachvili, J., 1992. Intermolecular and Surface Forces. Academic Press, Cambridge, MA.
Iyer, A., Petersen, N.O., Claessens, M.M.A.E., Subramaniam, V., 2014. Amyloids of alpha-synuclein affect the structure and dynamics of supported lipid bilayers. Biophys. J. 106, 2585–2594.
Jean, L., Lee, C.F., Vaux, D.J., 2012. Enrichment of amyloidogenesis at an air-water interface. Biophys. J. 102, 1154–1162.
Jeong, J.S., Ansaloni, A., Mezzenga, R., Lashuel, H.A., Dietler, G., 2013. Novel mechanistic insight into the molecular basis of amyloid polymorphism and secondary nucleation during amyloid formation. J. Mol. Biol. 425, 1765–1781.
Jimenez, J.L., Nettleton, E.J., Bouchard, M., Robinson, C.V., Dobson, C.M., Saibil, H.R., 2002. The protofilament structure of insulin amyloid fibrils. Proc. Natl. Acad. Sci. USA 99, 9196–9201.
Job, D., Valiron, O., Oakley, B., 2003. Microtubule nucleation. Curr. Opin. Cell Biol. 15, 111–117.
Jung, J.-M., Mezzenga, R., 2010. Liquid crystalline phase behavior of protein fibers in water: experiments versus theory. Langmuir 26, 504–514.
Kar, K., Jayaraman, M., Sahoo, B., Kodali, R., Wetzel, R., 2011. Critical nucleus size for disease-related polyglutamine aggregation is repeat-length dependent. Nat. Struct. Mol. Biol. 18, 328–336.
Kashchiev, D., 2015. Protein polymerization into fibrils from the viewpoint of nucleation theory. Biophys. J. 109, 2126–2136.
Kashchiev, D., Cabriolu, R., Auer, S., 2013. Confounding the paradigm: peculiarities of amyloid fibril nucleation. J. Am. Chem. Soc. 135, 1531–1539.

Kellermayer, M.S.Z., Murvai, U., Horváth, A., Lászlóffi, E., Soós, K., Penke, B., 2013. Epitaxial assembly dynamics of mutant amyloid β25-35 N27C fibrils explored with time-resolved scanning force microscopy. Biophys. Chem. 184C, 54–61.

Knowles, T.P.J., Waudby, C.A., Devlin, G.L., Cohen, S.I.A., Aguzzi, A., Vendruscolo, M., Terentjev, E.M., Welland, M.E., Dobson, C.M., 2009. An analytical solution to the kinetics of breakable filament assembly. Science 326, 1533–1537.

Knowles, T.P.J., White, D.A., Abate, A.R., Agresti, J.J., Cohen, S.I.A., Sperling, R.A., Genst, E.J.D., Dobson, C.M., Weitz, D.A., 2011. Observation of spatial propagation of amyloid assembly from single nuclei. Proc. Natl. Acad. Sci. USA 108, 14746–14751.

Knowles, T.P.J., Vendruscolo, M., Dobson, C.M., 2014. The amyloid state and its association with protein misfolding diseases. Nat. Rev. Mol. Cell Biol. 15, 384–396.

Krapivsky, P.L., Redner, S., Ben-Naim, E., 2010. A Kinetic View of Statistical Physics. Cambridge University Press, Cambridge.

Lambert, M.P., Barlow, A.K., Chromy, B.A., Edwards, C., Freed, R., Liosatos, M., Morgan, T.E., Rozovsky, I., Trommer, B., Viola, K.L., Wals, P., Zhang, C., Finch, C.E., Krafft, G.A., Klein, W.L., 1998. Diffusible, nonfibrillar ligands derived from abeta1-42 are potent central nervous system neurotoxins. Proc. Natl. Acad. Sci. USA 95, 6448–6453.

Lee, J., Culyba, E.K., Powers, E.T., Kelly, J.W., 2011. Amyloid-β forms fibrils by nucleated conformational conversion of oligomers. Nat. Chem. Biol. 7, 602–609.

Levin, A., Mason, T.O., Adler-Abramovich, L., Buell, A.K., Meisl, G., Galvagnion, C., Bram, Y., Stratford, S.A., Dobson, C.M., Knowles, T.P.J., Gazit, E., 2014. Ostwald's rule of stages governs structural transitions and morphology of dipeptide supramolecular polymers. Nat. Commun. 5, 5219.

LeVine, H., 1993. Thioflavine T interaction with synthetic Alzheimer's disease beta-amyloid peptides: detection of amyloid aggregation in solution. Protein Sci. 2, 404–410.

Liang, Y., Lynn, D.G., Berland, K.M., 2010. Direct observation of nucleation and growth in amyloid self-assembly. J. Am. Chem. Soc. 132, 6306–6308.

Lin, Y.-C., Petersson, E.J., Fakhraai, Z., 2014. Surface effects mediate self-assembly of amyloid-β peptides. ACS Nano 8, 10178–10186.

Linse, S., Cabaleiro-Lago, C., Xue, W.-F., Lynch, I., Lindman, S., Thulin, E., Radford, S.E., Dawson, K.A., 2007. Nucleation of protein fibrillation by nanoparticles. Proc. Natl. Acad. Sci. USA 104, 8691–8696.

Lomakin, A., Chung, D.S., Benedek, G.B., Kirschner, D.A., Teplow, D.B., 1996. On the nucleation and growth of amyloid beta-protein fibrils: detection of nuclei and quantitation of rate constants. Proc. Natl. Acad. Sci. USA 93, 1125–1129.

Lorenzen, N., Nielsen, S.B., Buell, A.K., Kaspersen, J.D., Arosio, P., Vad, B.S., Paslawski, W., Christiansen, G., Valnickova-Hansen, Z., Andreasen, M., Enghild, J.J., Pedersen, J.S., Dobson, C.M., Knowles, T.P.J., Otzen, D.E., 2014. The role of stable α-synuclein oligomers in the molecular events underlying amyloid formation. J. Am. Chem. Soc. 136, 3859–3868.

Mackay, J.P., Matthews, J.M., Winefield, R.D., Mackay, L.G., Haverkamp, R.G., Templeton, M.D., 2001. The hydrophobin EAS is largely unstructured in solution and functions by forming amyloid-like structures. Structure 9, 83–91.

Marshall, K.E., Hicks, M.R., Williams, T.L., Hoffmann, S.V., Rodger, A., Dafforn, T.R., Serpell, L.C., 2010. Characterizing the assembly of the sup35 yeast prion fragment, gnnqqny: structural changes accompany a fiber-to-crystal switch. Biophys. J. 98, 330–338.

Meehan, S., Berry, Y., Luisi, B., Dobson, C.M., Carver, J.A., MacPhee, C.E., 2004. Amyloid fibril formation by lens crystallin proteins and its implications for cataract formation. J. Biol. Chem. 279, 3413–3419.

Meier, M., Kennedy-Darling, J., Choi, S.H., Norstrom, E.M., Sisodia, S.S., Ismagilov, R.F., 2009. Plug-based microfluidics with defined surface chemistry to miniaturize and control aggregation of amyloidogenic peptides. Angew. Chem. Int. Ed. Engl. 48, 1487–1489.

Meisl, G., Yang, X., Hellstrand, E., Frohm, B., Kirkegaard, J.B., Cohen, S.I.A., Dobson, C.M., Linse, S., Knowles, T.P.J., 2014. Differences in nucleation behavior underlie the contrasting aggregation kinetics of the Aβ40 and Aβ42 peptides. Proc. Natl. Acad. Sci. USA 111, 9384–9389.

Meisl, G., Kirkegaard, J.B., Arosio, P., Michaels, T.C.T., Vendruscolo, M., Dobson, C.M., Linse, S., Knowles, T.P.J., 2016. Molecular mechanisms of protein aggregation from global fitting of kinetic models. Nat. Protoc. 11, 252–272.

Mezzenga, R., Fischer, P., 2013. The self-assembly, aggregation and phase transitions of food protein systems in one, two and three dimensions. Rep. Prog. Phys. 76, 046601.

Mok, Y.-F., Howlett, G.J., Griffin, M.D.W., 2015. Sedimentation velocity analysis of the size distribution of amyloid oligomers and fibrils. Methods Enzymol. 562, 241–256.

Morel, B., Varela, L., Conejero-Lara, F., 2010. The thermodynamic stability of amyloid fibrils studied by differential scanning calorimetry. J. Phys. Chem. B 114, 4010–4019.

Nayak, A., Dutta, A.K., Belfort, G., 2008. Surface-enhanced nucleation of insulin amyloid fibrillation. Biochem. Biophys. Res. Commun. 369, 303–307.

Nelson, R., Sawaya, M.R., Balbirnie, M., Madsen, A.O., Riekel, C., Grothe, R., Eisenberg, D., 2005. Structure of the cross-beta spine of amyloid-like fibrils. Nature 435, 773–778.

Nielsen, A.E., 1984. Electrolyte crystal growth mechanisms. J. Cryst. Growth 67 (2), 289–310.

Oosawa, F., 1975. Thermodynamics of the Polymerization of Protein. Academic Press Inc, Cambridge, MA.

Orte, A., Birkett, N.R., Clarke, R.W., Devlin, G.L., Dobson, C.M., Klenerman, D., 2008. Direct characterization of amyloidogenic oligomers by single-molecule fluorescence. Proc. Natl. Acad. Sci. USA 105, 14424–14429.

Outeiro, T.F., Putcha, P., Tetzlaff, J.E., Spoelgen, R., Koker, M., Carvalho, F., Hyman, B.T., McLean, P.J., 2008. Formation of toxic oligomeric alpha-synuclein species in living cells. PLoS One 3, e1867.

Parmar, A.S., Gottschall, P.E., Muschol, M., 2007. Pre-assembled clusters distort crystal nucleation kinetics in supersaturated lysozyme solutions. Biophys. Chem. 129, 224–234.

Paslawski, W., Andreasen, M., Nielsen, S.B., Lorenzen, N., Thomsen, K., Kaspersen, J.D., Pedersen, J.S., Otzen, D.E., 2014. High stability and cooperative unfolding of α-synuclein oligomers. Biochemistry 53, 6252–6263.

Patel, A., Lee, H.O., Jawerth, L., Maharana, S., Jahnel, M., Hein, M.Y., Stoynov, S., Mahamid, J., Saha, S., Franzmann, T.M., Pozniakovski, A., Poser, I., Maghelli, N., Royer, L.A., Weigert, M., Myers, E.W., Grill, S., Drechsel, D., Hyman, A.A., Alberti, S., 2015. A liquid-to-solid phase transition of the ALS protein FUS accelerated by disease mutation. Cell 162, 1066–1077.

Peelaerts, W., Bousset, L., der Perren, A.V., Moskalyuk, A., Pulizzi, R., Giugliano, M., den Haute, C.V., Melki, R., Baekelandt, V., 2015. α-synuclein strains cause distinct synucleinopathies after local and systemic administration. Nature 522, 340–344.

Pellarin, R., Caflisch, A., 2006. Interpreting the aggregation kinetics of amyloid peptides. J. Mol. Biol. 360, 882–892.

Petkova, A.T., Leapman, R.D., Guo, Z., Yau, W.-M., Mattson, M.P., Tycko, R., 2005. Self-propagating, molecular-level polymorphism in Alzheimer's beta-amyloid fibrils. Science 307, 262–265.

Pinotsi, D., Buell, A.K., Galvagnion, C., Dobson, C.M., Schierle, G.S.K., Kaminski, C.F., 2014. Direct observation of heterogeneous amyloid fibril growth kinetics via two-color super-resolution microscopy. Nano Lett. 14, 339–345.

Pinotsi, D., Michel, C.H., Buell, A.K., Dobson, C.M., Kaminski, C.F., Schierle, G.S.K., 2016. Nanoscopic insights into seeding mechanisms and toxicity of α-synuclein species in neurons. Proc. Natl. Acad. Sci. USA 113, 3815–3819.

Price, S., Veesler, S., Pan, H., Lewtas, K., Smets, M., Rimez, B., Myerson, A., Hughes, C., Hare, A., Zhang, F., Meekes, H., Mazzotti, M., Rosbottom, I., Khamar, D., van den Ende, J., Fabian, L., Black, S., Taulelle, F., Gich, M., Vekilov, P., Toroz, D., Bertran, C.A., Sefcik, J., Schroeder, S., Booth, S., Rasmuson, A., Breynaert, E., Simone, E., Hammond, R., Sear, R., de Yoreo, J., Davey, R., Anwar, J., Ristic, R., Corzo, D.M.C., Roberts, K., Harris, K., Colfen, H., Turner, T., 2015. Molecular self-assembly and clustering in nucleation processes: general discussion. Faraday Discuss. 179, 155–197.

Pusey, P., van Megen, W., 1986. Phase behaviour of concentrated suspensions of nearly hard colloidal spheres. Nature 320, 340–342.

Ramachandran, G., Udgaonkar, J.B., 2011. Understanding the kinetic roles of the inducer heparin and of rod-like protofibrils during amyloid fibril formation by tau protein. J. Biol. Chem. 286, 38948–38959.

Ramachandran, G., Udgaonkar, J.B., 2012. Evidence for the existence of a secondary pathway for fibril growth during the aggregation of tau. J. Mol. Biol. 421, 296–314.

Reches, M., Gazit, E., 2003. Casting metal nanowires within discrete self-assembled peptide nanotubes. Science 300, 625–627.

Roach, L., Song, H., Ismagilov, R., 2005. Controlling nonspecific protein adsorption in a plug-based microfluidic system by controlling interfacial chemistry using fluorous-phase surfactants. Anal Chem. 77 (3), 785–796.

Ruschak, A.M., Miranker, A.D., 2007. Fiber-dependent amyloid formation as catalysis of an existing reaction pathway. Proc. Natl. Acad. Sci. USA 104, 12341–12346.

Schladitz, C., Vieira, E.P., Hermel, H., Möhwald, H., 1999. Amyloid-beta-sheet formation at the air-water interface. Biophys. J. 77, 3305–3310.

Selimović, S., Jia, Y., Fraden, S., 2009. Measuring the nucleation rate of lysozyme using microfluidics. Cryst. Growth Des. 9, 1806–1810.

Shammas, S.L., Knowles, T.P.J., Baldwin, A.J., Macphee, C.E., Welland, M.E., Dobson, C.M., Devlin, G.L., 2011. Perturbation of the stability of amyloid fibrils through alteration of electrostatic interactions. Biophys. J. 100, 2783–2791.

Shorter, J., Lindquist, S., 2004. Hsp104 catalyzes formation and elimination of self-replicating Sup35 prion conformers. Science 304, 1793–1797.

Sleutel, M., Driessche, A.E.S.V., 2014. Role of clusters in nonclassical nucleation and growth of protein crystals. Proc. Natl. Acad. Sci. USA 111, E546–E553.

Stradner, A., Sedgwick, H., Cardinaux, F., Poon, W.C.K., Egelhaaf, S.U., Schurtenberger, P., 2004. Equilibrium cluster formation in concentrated protein solutions and colloids. Nature 432, 492–495.

Sunde, M., Serpell, L.C., Bartlam, M., Fraser, P.E., Pepys, M.B., Blake, C.C., 1997. Common core structure of amyloid fibrils by synchrotron X-ray diffraction. J. Mol. Biol. 273, 729–739.

Tait, S., White, E.T., Litster, J.D., 2009. A study on nucleation for protein crystallization in mixed vessels. Cryst. Growth Des. 9 (5), 2198–2206.

Terakawa, M.S., Yagi, H., Adachi, M., Lee, Y.-H., Goto, Y., 2015. Small liposomes accelerate the fibrillation of amyloid β (1-40). J. Biol. Chem. 290, 815–826.

Thomson, J.A., Schurtenberger, P., Thurston, G.M., Benedek, G.B., 1987. Binary liquid phase separation and critical phenomena in a protein/water solution. Proc. Natl. Acad. Sci. USA 84, 7079–7083.

Tjernberg, L., Hosia, W., Bark, N., Thyberg, J., Johansson, J., 2002. Charge attraction and beta propensity are necessary for amyloid fibril formation from tetrapeptides. J. Biol. Chem. 277, 43243–43246.

Tosi, G., Fermani, S., Falini, G., Gavira, J.A., Ruiz, J.M.G., 2011. Hetero- vs homogeneous nucleation of protein crystals discriminated by supersaturation. Cryst. Growth Des. 11 (5), 1542–1548.
Trigg, B.J., Lee, C.F., Vaux, D.J., Jean, L., 2013. The air-water interface determines the outcome of seeding during amyloidogenesis. Biochem. J. 456, 67–80.
Tsekova, D., Dimitrova, S., Nanev, C.N., 1999. Heterogeneous nucleation (and adhesion) of lysozyme crystals. J. Cryst. Growth 196, 226–233.
Urbanc, B., Betnel, M., Cruz, L., Bitan, G., Teplow, D.B., 2010. Elucidation of amyloid beta-protein oligomerization mechanisms: discrete molecular dynamics study. J. Am. Chem. Soc. 132, 4266–4280.
Vácha, R., Linse, S., Lund, M., 2014. Surface effects on aggregation kinetics of amyloidogenic peptides. J. Am. Chem. Soc. 136, 11776–11782.
Varongchayakul, N., Johnson, S., Quabili, T., Cappello, J., Ghandehari, H., Solares, S.D.J., Hwang, W., Seog, J., 2013. Direct observation of amyloid nucleation under nanomechanical stretching. ACS Nano 7, 7734–7743.
Verwey, E., Overbeek, J., 1948. Theory of the Stability of Lyophobic Colloids. Elsevier, Amsterdam.
Vitalis, A., Pappu, R.V., 2011. Assessing the contribution of heterogeneous distributions of oligomers to aggregation mechanisms of polyglutamine peptides. Biophys. Chem. 159, 14–23.
Šarić, A., Chebaro, Y.C., Knowles, T.P.J., Frenkel, D., 2014. Crucial role of nonspecific interactions in amyloid nucleation. Proc. Natl. Acad. Sci. USA 111, 17869–17874.
Wang, X., Hammer, N.D., Chapman, M.R., 2008. The molecular basis of functional bacterial amyloid polymerization and nucleation. J. Biol. Chem. 283, 21530–21539.
Wang, C., Shah, N., Thakur, G., Zhou, F., Leblanc, R.M., 2010. Alpha-synuclein in alpha-helical conformation at air-water interface: implication of conformation and orientation changes during its accumulation/aggregation. Chem. Commun. (Camb) 46, 6702–6704.
Wang, Y.-Q., Buell, A.K., Wang, X.-Y., Welland, M.E., Dobson, C.M., Knowles, T.P.J., Perrett, S., 2011. Relationship between prion propensity and the rates of individual molecular steps of fibril assembly. J. Biol. Chem. 286, 12101–12107.
Wördehoff, M.M., Bannach, O., Shaykhalishahi, H., Kulawik, A., Schiefer, S., Willbold, D., Hoyer, W., Birkmann, E., 2015. Single fibril growth kinetics of α-synuclein. J. Mol. Biol. 427, 1428–1435.
Xue, W.-F., Homans, S.W., Radford, S.E., 2008. Systematic analysis of nucleation-dependent polymerization reveals new insights into the mechanism of amyloid self-assembly. Proc. Natl. Acad. Sci. USA 105, 8926–8931.
Xue, W.-F., Homans, S.W., Radford, S.E., 2009. Amyloid fibril length distribution quantified by atomic force microscopy single-particle image analysis. Protein Eng. Des. Sel. 22, 489–496.
Yoshimura, Y., Lin, Y., Yagi, H., Lee, Y.-H., Kitayama, H., Sakurai, K., So, M., Ogi, H., Naiki, H., Goto, Y., 2012. Distinguishing crystal-like amyloid fibrils and glass-like amorphous aggregates from their kinetics of formation. Proc. Natl. Acad. Sci. USA 109, 14446–14451.
Zhu, M., Li, J., Fink, A.L., 2003. The association of alpha-synuclein with membranes affects bilayer structure, stability, and fibril formation. J. Biol. Chem. 278, 40186–40197.
Zurdo, J., Guijarro, J.I., Jimenez, J.L., Saibil, H.R., Dobson, C.M., 2001. Dependence on solution conditions of aggregation and amyloid formation by an SH3 domain. J. Mol. Biol. 311, 325–340.

CHAPTER SIX

What Makes a Prion: Infectious Proteins From Animals to Yeast

K.S. MacLea[1]
University of New Hampshire, Manchester, NH, United States
[1]Corresponding author: e-mail address: kyle.maclea@unh.edu

Contents

International Review of Cell and Molecular Biology, Volume 329
ISSN 1937-6448
http://dx.doi.org/10.1016/bs.ircmb.2016.08.012

Abstract

While philosophers in ancient times had many ideas for the cause of contagion, the modern study of infective agents began with Fracastoro's 1546 proposal that invisible "spores" spread infectious disease. However, firm categorization of the pathogens of the natural world would need to await a mature germ theory that would not arise for 300 years. In the 19th century, the earliest pathogens described were bacteria and other cellular microbes. By the close of that century, the work of Ivanovsky and Beijerinck introduced the concept of a virus, an infective particle smaller than any known cell. Extending into the early–mid-20th century there was an explosive growth in pathogenic microbiology, with a cellular or viral cause identified for nearly every transmissible disease. A few occult pathogens remained to be discovered, including the infectious proteins (prions) proposed by Prusiner in 1982. This review discusses the prions identified in mammals, yeasts, and other organisms, focusing on the amyloid-based prions. I discuss the essential biochemical properties of these agents and the application of this knowledge to diseases of protein misfolding and aggregation, as well as the utility of yeast as a model organism to study prion and amyloid proteins that affect human and animal health. Further, I summarize the ideas emerging out of these studies that the prion concept may go beyond proteinaceous infectious particles and that prions may be a subset of proteins having general nucleating or seeding functions involved in noninfectious as well as infectious pathogenic protein aggregation.

ABBREVIATIONS

ALS amyotrophic lateral sclerosis
BSE bovine spongiform encephalopathy
CJD Creutzfeldt–Jakob disease
CWD chronic wasting disease
FFI fatal familial insomnia
FTLD frontotemporal lobar degeneration
GFP green fluorescent protein
HMM hidden Markov model
MBM meat and bone meal
MSA multiple system atrophy
ND nucleation domain
ORD oligopeptide repeat domain
ORF open reading frame
PFD prion-forming domain
PrLD prion-like domain

PrP prion protein
TMV tobacco mosaic virus
TSE transmissible spongiform encephalopathy

1. INTRODUCTION

As long as there have been humans, curing and preventing illness in humankind has been a goal that crosses all cultural and geographic boundaries. Key to any real understanding of how to heal the sick was careful study of illness, identification of true causes of diverse types of sickness, and experiments to assess methods of cure and prevention. The first section explores the historical development of infectious disease etiology (Section 2) culminating in the proposal of a purely protein-based infectious agent, the prion. Scientific evidence for the existence of infectious prions in animals and in yeasts and other species is presented in Section 3. While a subset of proteins were identified with this unusual pathogenicity and transmissibility, the essential question of why only some proteins displayed this behavior was the next big question, addressed in Section 4. Some answers of what makes a protein a prion grew out of basic structural characterization of prions, examining their amyloid structure, and further experiments in animals and yeasts have begun to fine-tune that understanding. Finally, this growing understanding of prions has had implications for noninfectious protein aggregation diseases in humans and animals and has led to an enlargement of the prion concept, discussed in Section 5.

2. PATHOGENS AND THE EMERGENCE OF THE PRION HYPOTHESIS

2.1 The Causative Agents of Infectious Disease

Diseases of antiquity such as leprosy and plague left indelible marks on cultures and civilizations but also had no known and agreed-upon cause. Some blamed supernatural forces, others vapors and miasmas, and still others diet, living conditions, and atmospheric climate. The ancient Greek physician Galen, working in the 2nd century CE from the medical principles of Hippocrates and others, was the primary proponent of the idea of diseases caused by miasma (pollution) or poor quality air. In 1546, Girolamo Fracastoro, the eminent Venetian physician, published his work *De Contagione et Contagiosis Morbis* promulgating the idea of "spores," directly transmitted (*contagion*) and also distantly transmitted, and fomites "not themselves corrupt" indirectly spreading these seeds of disease. This work was published during the time he was serving as the elected physician of the Council of Trent and proved to be an influential counterpoint to the prevailing notion of miasmas. However, Galen's miasma theory of disease would not be fully supplanted in the minds of physicians and scientists until the last years of the 19th century with the advent of the germ theory of disease (Table 1).

2.2 Cellular Causes of Infectious Diseases

A medieval Dutch draper who wanted to see his threads better, Antonie van Leeuwenhoek, became the celebrated lens and microscope maker that

Table 1 Prevailing Notions of Natural Causes of Disease With Notable Milestones

Time Frame	Agent	Advocate(s)	Physical Basis
Ancient until 19th century	Miasma	Galen of Pergamon, Indian, and Chinese philosophers	Bad airs
Ancient until 19th century	Contagion	Fracastoro and others	Direct contact with sick people
1836	Living germ or seed	Bassi	Fungal pathogen, no microscopic evidence
1865–1870	Microbe	Pasteur	Fungal pathogen
1876	Bacterium	Koch	Anthrax bacillus
1898	Virus	Beijerinck, Loeffler, and Frosch	Tobacco mosaic virus (TMV) and aphthovirus
1942	Virus	Cohen and Stanley	TMV composed of nucleic acid and protein
20th century	Slow virus	Many	Virus composed of nucleic acid and protein with long incubation period
1982	Prion	Prusiner	Animal disease caused by protein only (no nucleic acid)
1994	Prion	Wickner	Yeast infectious protein (no nucleic acid) explains unusual genetics of [*PSI*$^+$], [URE3] traits

introduced the world to the first observations of microscopic organisms. Beginning in 1673, van Leeuwenhoek's 190 letters to the Royal Society described observations of the first cells that he termed *animalculum* (very small animals). In the course of his work, van Leeuwenhoek noted not only the first unicellular organisms (protists) but also the first bacteria and subcellular structures. The English scientist Robert Hooke coined the term *cell* in his 1665 book *Micrographia* to describe the individual compartments in cork and living plants that were analogous to the *animalcules* of van Leeuwenhoek.

Although microscopic cells and microbes were known from the 17th century, for nearly 200 years after van Leeuwenhoek and Hooke doctors and scientists saw no connection between the cellular microbes and disease, even in some cases postulating that organisms found in diseased tissues were the effect, rather than the cause, of injury. A "germ theory" arose in the 19th century, connecting the presence of infectious organisms with disease. Agostino Bassi (1838, silkworm disease) gained rapid acceptance for his work but Ignaz Semmelweis (1847–61, childbed or puerperal fever) met with substantial resistance for a germ theory of disease.

The French chemist Louis Pasteur firmly established the germ theory of disease with his experiments demonstrating a microbial cause for fermentation, disproving spontaneous generation, developing "pasteurization," and linking particular silkworm diseases to microbes (1857–70). German scientist Ferdinand Cohn soon formally described and classified the *Bacteria* (1875). Visiting Cohn at Breslau, physician Robert Koch demonstrated the use of pure cultures of anthrax bacilli to cause the illness in previously healthy animals (1876 with refinements continuing in later years). While developing his famous postulates for connecting specific microorganisms with specific diseases, Koch in the 1880s made several other connections between disease-causing or pathogenic organisms and their specific organic diseases, notably cholera and tuberculosis. Many other scientists and physicians contributed their observations to the growing body of evidence that supported the germ theory of disease.

2.3 Noncellular Causes of Disease in Animals

Building on the work of Pasteur, Koch, and others in the mid–late 19th century, the microbiological agents responsible for the great diseases of antiquity were, one after another, systematically identified. As described, the first pathogenic agents identified were those in which the organisms in question could be readily observed under the microscope, such as Pasteur's discovery of a microsporidian parasite as the cause of the pébrine disease of silkworms and Koch's discovery of the bacterium *Bacillus anthracis* as the cause of anthrax.

However, some diseases stymied the efforts of even the giants of the new fields of bacteriology and microbiology. Although Pasteur successfully developed a rabies vaccine in 1886, he could not identify the causative agent, speculating that it was too small to be visible through the use of the microscope. Another French microbiologist, Charles Chamberland, developed a special porcelain filter that excluded anything as large as the known bacteria (1884). The Chamberland Filter proved important for extending the germ theory of disease beyond the cellular parasites, protists, and bacteria. Russian scientist Dmitri Ivanovsky used a Chamberland Filter to remove bacteria and isolate the tobacco mosaic virus (1892), although it was not initially perceived to be anything other than a bacterial toxin. The Dutch microbiologist Martinus Beijerinck in 1898 realized that Ivanovsky's filtrate actually contained a new infectious agent that he referred to both as a *contagium vivum fluidum* (living fluid germ) and as a *virus* (slimy poison liquid). In the same year, Friedrich Loeffler and Paul Frosch discovered the first animal virus (aphthovirus for foot-and-mouth disease) using a similar filter.

The composition of viruses was not immediately understood. American virologist Wendell Stanley, working with Ivanovsky's filtered agent, now known as tobacco mosaic virus (TMV), successfully crystallized it, proving it was not a liquid as Beijerinck has proposed. However, Stanley initially believed that TMV contained only protein and only later realized the concomitant presence of a nucleic acid (Cohen and Stanley, 1942; Stanley, 1935). The scientific community had not yet firmly settled on nucleic acid as the particle of heredity by this time, but evidence was accumulating.

Since Friedrich Miescher's 1869 discovery of the *nuclein* or nucleic acid found in nuclei of eukaryotic cells, scientists had been probing its structure. Phoebus Levene's 1919 tetranucleotide hypothesis of nucleic acid structure (Levene, 1919) held sway in the scientific community for decades, suggesting nucleic acid would be a poor informational molecule and that therefore protein would be a superior basis for the particles of heredity. When Frederick Griffith's 1928 pneumococcal "transforming principle" (molecule of heredity) (Griffith, 1928) was proven to be nucleic acid (Avery et al., 1944), the composition and structure of viral genetic information also became a point of intense interest. It was Alfred Hershey and Martha Chase, working with bacteriophage (bacterial virus) T2, who demonstrated that the nucleic acid portion of the virus was its hereditary material as well (Hershey and Chase, 1952).

By this time, a host of viruses had been identified as the causative agents of plant and animal diseases, complementing the many cellular pathogens identified in the 19th and early 20th centuries. By the mid-20th century, the majority of the pathogenic agents causing known infectious diseases had been identified (Brachman, 2003). All of these agents were cellular or viral in nature.

2.4 Unusual Disease Traits in Animals

Despite success with identifying many cellular and viral pathogens, the cause of a few rare diseases remained stubbornly difficult to pinpoint.

One of these diseases was a condition known as scrapie observed in Merino sheep in Spain in 1732 (Table 2, top). This disease, in which sheep obsessively scrape themselves against trees, fence posts, and other obstacles, also manifests a variety of symptoms affecting the nervous system: altered gait, lip smacking, and convulsions. Although clearly infectious within flocks, long and variable incubation periods made determination of etiology difficult. No virus or cellular cause had been identified as a cause of scrapie, but it had been hypothesized that the disease was caused by a "slow virus," an exceptionally slow-to-propagate virus with a long incubation period (Cuille and Chelle, 1938a; Sigurðsson, 1954).

Human diseases of unknown etiology were found with similarities to scrapie (Table 2, bottom). A human neurological disorder that would come to be known as Creutzfeldt–Jakob disease (CJD) was identified in 1920 (Creutzfeldt, 1920; Jakob, 1921). Another human disease found among the Fore tribe of Papua New Guinea, called kuru or the "laughing disease," was brought to the attention of the scientific community in 1959 (Gajdusek and Zigas, 1959; Klatzo et al., 1959). Immediately, the similarities in these diseases were noted (Hadlow, 1959; Klatzo et al., 1959), and it was postulated that all of them were infectious (like scrapie) and due to a slow virus. Later experiments proved their transmissible nature, and these diseases came to be known as transmissible spongiform encephalopathies (TSEs) on the basis of their essential neuroanatomic effect of producing tiny holes in the brain cortex of affected individuals (Fig. 1).

2.5 Non-Mendelian Inheritance of Characters in the Baker's Yeast

In 1965, yeast geneticist Brian Cox traced and described an unusual trait he called [ψ^+] (now written as [*PSI*$^+$]) in the baker's yeast *Saccharomyces cerevisiae*. The [*PSI*$^+$] trait was a suppressor of a super-suppressor of stop

Table 2 Prion Diseases in Nonhuman Mammals and Humans

Animal Disease	Mechanism	Animal(s)
Scrapie	Somatic mutation in *Prnp* gene or spontaneous conversion of normal PrP^C to abnormal PrP^{Sc} or infection from other infected animals	Sheep and goats
Bovine spongiform encephalopathy (BSE)	Infection or sporadic	Cattle
Transmissible mink encephalopathy (TME)	Infection from sheep or cattle	Mink
Chronic wasting disease (CWD)	Infection or possibly sporadic	Cervids (deer and elk)
Exotic ungulate encephalopathy	Infection with prion-contaminated meat and bone meal (MBM)	Ungulates (oryx, nyala, greater kudu, etc.)
Feline spongiform encephalopathy (FSE)	Infection with prion-contaminated meat or MBM	Domestic cats, various wild cats
Proposed canine spongiform encephalopathy	Unknown, based on a single case report	Domestic dogs
Human Disease	**Mechanism**	**Specific Hosts**
Kuru (extinct?)	Ritual funerary cannibalism	Fore tribe of Papua New Guinea
Sporadic Creutzfeldt–Jakob Disease (sCJD)	Somatic mutation in *PNRP* gene or spontaneous conversion of normal PrP^C to abnormal PrP^{Sc}	All humans
Familial CJD	Germline mutation in *PNRP* gene	Humans from CJD families
Variant CJD (vCJD)	Infection from consumption of meat from BSE cattle	All humans
Iatrogenic CJD (iCJD)	Infection from contaminated medicines or medical equipment	All humans
GSS	Germline mutation in *PNRP* gene	Humans from GSS families

Table 2 Prion Diseases in Nonhuman Mammals and Humans—cont'd

Human Disease	Mechanism	Specific Hosts
Fatal familial insomnia (FFI)	Germline mutation in *PNRP* gene	Humans from FFI families
Sporadic fatal insomnia (sFI)	Somatic mutation in *PNRP* gene or spontaneous conversion of normal PrP^{C} to abnormal PrP^{Sc}	All humans
Multiple system atrophy	Mutant alpha-synuclein infection in mice/cultured cells (artificial model) (reviewed in Supattapone, 2015)	Unknown
Other diseases	Growing recognition of prion-like and amyloid proteins in disease and other pathological changes in protein conformation	Unknown

After Colby, D.W., Prusiner, S.B., 2011a. De novo generation of prion strains. Nat. Rev. Microbiol. 9 (11), 771–7. Available at: http://www.pubmedcentral.nih.gov/articlerender.fcgi?artid=3924856&tool=pmcentrez&rendertype=abstract (accessed April 21, 2016); Colby, D.W., Prusiner, S.B., 2011b. Prions. Cold Spring Harb. Perspect. Biol. 3 (1), a006833. Available at: http://www.pubmedcentral.nih.gov/articlerender.fcgi?artid=3003464&tool=pmcentrez&rendertype=abstract (accessed February 12, 2016)

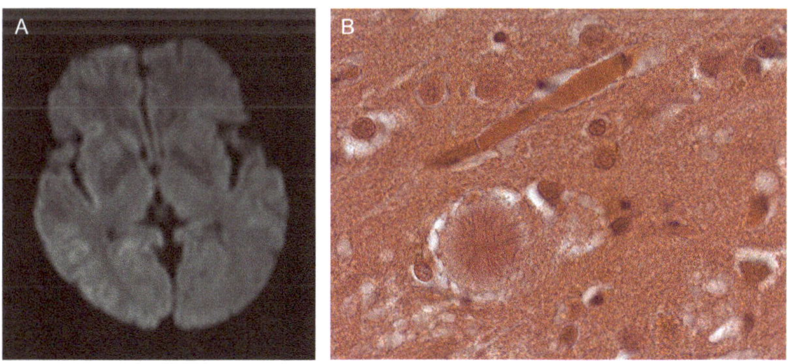

Fig. 1 Brain effects of CJD, a transmissible spongiform encephalopathy, in humans. (A) Diffusion-weighted magnetic resonance (MRI) image of a patient who presented with a rapidly progressive dementia, with initial hallucinations and behavioral change that progressed to a mute, akinetic state with myoclonus. Right cortical and striatal high signal is consistent with a diagnosis of sporadic-type Creutzfeldt–Jakob disease (sCJD). (B) Hematoxylin–eosin stained cortex of patient with variant Creutzfeldt–Jakob (vCJD) disease with florid plaques. *Panel (A) Photo courtesy of Dr. Laughlin Dawes and Wikimedia user Filip em, 2008. Panel (B) Photo is in the public domain.*

codons, a gene now known as *SUP35*. What made the trait more puzzling was that in Cox's meticulous studies of inheritance, [*PSI*$^{+}$] did not obey Mendelian principles of inheritance (Cox, 1965; reviewed in Tuite et al., 2015). Cox identified (correctly) what he referred to as a "self-replicating particle" in the cytoplasm that was involved in the inheritance of the trait. In yeast, there were three known principle cytoplasmic components that were inherited: mitochondrial DNA, yeast killer dsRNA plasmids, and 2-μm circle plasmids. The [*PSI*$^{+}$] trait was none of these, although its identity would remain a mystery for almost 30 years.

Another strangely inherited trait in yeast was identified by Lacroute (1971). In this case the gene involved was called *URE2* and the trait [URE3]. Lacroute hypothesized that the trait was mitochondrially inherited, although several features would have been very unusual for a mitochondrial trait. Lacroute also proposed an alternative to that idea, proposing that [URE3] was a "nonmitochondrial cytoplasmic replicon" of unknown nature (Lacroute, 1971). Akin to [*PSI*$^{+}$], the biochemical and genetic basis of [URE3] was not understood until the prion hypothesis had been in formulated. Connection of these traits to the prion hypothesis will be described in Section 3.7.

2.6 The Prion Hypothesis

In the animal TSEs, the hypothesis of a slow virus etiology was widely accepted, but data began to accumulate that put that etiology into question. CJD in humans was clearly hereditary. The scrapie agent was not inactivated by formalin or by UV radiation, which both inactivated known viruses (Alper et al., 1967; Pattison and Jones, 1967). Decades of struggle to find any nucleic acid in the scrapie agent continued to prove fruitless and several investigators suspected a purely proteinaceous infective nature for scrapie (Cho, 1980; Griffith, 1967; Hadlow et al., 1980; Hunter et al., 1969; Merz et al., 1983; Prusiner et al., 1978a,b, 1980a,b, 1981).

Despite the lack of evidence for nucleic acid playing a role in transmission for the TSEs, the scientists working in the field still had a healthy regard for the Central Dogma and were not ready to assume a protein-only inheritance for these diseases. However, one scientist, Stanley Prusiner, was willing to push ahead with a formal hypothesis of a fully protein infective agent, something he called the "proteinaceous infectious particle" or "prion" (Prusiner, 1982). This bold hypothesis, for which Prusiner would be awarded the Nobel Prize in Physiology or Medicine in 1997, was not proven overnight, and many lines of evidence were required to convince a skeptical scientific community. This hypothesis would later be more widely applied to the inheritance of the unusual non-Mendelian characters in yeast and what was learned in the study of prion diseases would prove applicable to the more general problem of human protein-misfolding diseases that were of a noninfectious nature as well.

3. EVIDENCE FOUND: IDENTIFICATION OF ANIMAL, YEAST, AND OTHER PRIONS

3.1 Scrapie in Sheep and Goats

TSEs have been found in a number of mammals, including humans (Table 2) with the longest studied being scrapie. Sheep and goats affected with the neurological pathology of scrapie had been the subject of scientific investigation for centuries, with the first verified report published in Germany in 1750 (Leopoldt, 1750), although cases were cited in other reports going back to 1732 in Spain and in England. Leopoldt's initial report postulates an infectious cause for scrapie, although other scientists would debate whether hereditary or other causes were more likely for many years to come (reviewed in Schneider et al., 2008). Experiments to prove transmissibility were undertaken many times but had various deficiencies leading to continued disagreement. Finally, beginning in 1936, Cuille and Chelle proved transmissibility by inoculating healthy animals with material from the central nervous systems of sick animals (Cuille and Chelle, 1936, 1938a,b,c, 1939). Small wild sheep called mouflons are also susceptible to scrapie (Wood, Lund and Done, 1992a), as are goats (Cuille and Chelle, 1939; Wood et al., 1992b).

Cuille and Chelle proposed a viral etiology for scrapie in their 1930s research, although other causes were still postulated by others. A particular designation as a "slow virus" disease (Sigurðsson, 1954) became the common way to group this disease with CJD and Kuru as they were discovered. As mentioned earlier, a protein-only transmission was also proposed by Griffith but did not immediately attract the support of the scrapie research community (Griffith, 1967). One difficulty in conducting this research was the long incubation in sheep, which was overcome by conducting experiments in mice (Chandler, 1961). Although mice remained a workhorse in studying scrapie for decades, a later hamster model was also developed which dropped the incubation period from years in sheep to 150 days in mice to 60 days in hamsters (Kimberlin and Walker, 1977).

The prion protein was identified and called PrP, with the gene being called *Prnp* in sheep and goats. Two forms were described: PrP^{Sc} (scrapie form) and PrP^{C} (cellular normal form). Many strains of scrapie were identified, mutations in the genes were identified, and it was found that some strains/mutations delayed onset of disease and others shortened the time to disease progression.

Scrapie modes of transmission have been debated for many years. Although experimental transmission can take several forms, the natural transmission of scrapie horizontally between individuals occurs through direct contact between animals and through contact with environmental contamination (reviewed in Schneider et al., 2008). Scrapie is predominantly acquired through the oral route, and the placenta and amniotic fluid are the most common sources of oral infection, although fetal parts, feces, and milk have all shown infectivity (see Schneider et al., 2008).

3.2 Bovine Spongiform Encephalopathy

With the substantial neuropathological understanding of scrapie going back decades, veterinarians and scientists in the United Kingdom quickly noticed the arrival of a new, related disease. Bovine spongiform encephalopathy (BSE) in cattle was identified in 1987 (Wells et al., 1987). BSE was noted for the classic neurological symptoms typical of spongiform encephalopathies: ataxia (contributing to "downer cattle" that cannot stand well), behavioral changes, anorexia, and death. The practice of using rendered meat and bone meal (MBM) product (which contains nervous tissue) from sheep and cattle to increase protein in animal feed was immediately suspected as a potential epidemiological cause of the BSE outbreak (Matthews, 1990; Taylor, 1989) and UK and other government inquiries agreed with that stance, leading to changes in feeding practices across the globe. It is still debated whether BSE may have arisen from sporadic BSE entering the MBM food chain or whether it may have been scrapie in slaughtered sheep in the MBM (with a subsequent rare evasion of the species barrier) that led to the widespread BSE outbreak in the United Kingdom. It was quickly recognized, however, that since a scrapie origin to the BSE outbreak was plausible, the possibility that BSE might also cross the species barrier into humans was equally plausible (Matthews, 1990; Taylor, 1989). This prediction proved prescient, with the discovery of an unusual cluster of younger Creutzfeldt–Jakob patients ("variant" CJD) in the United Kingdom only a few years later in 1996 (see Section 3.3).

3.3 Kuru, CJD, Other Prion Diseases in Humans

The first description of a human TSE disease (Table 2, bottom) was CJD in 1920–21 (Creutzfeldt, 1920; Jakob, 1921). This rare, neurodegenerative disease (CJD) was characterized in people by loss of memory and judgment and increasing dementia, concomitant with loss of muscular coordination, significant personality changes, and impaired vision. The proximate cause of these neurological deficits was death of neurons (as seen in MRI, Fig. 1A) and holes in brain tissue with concomitant buildup of plaques (as shown in histologic section, Fig. 1B). CJD was found to occur in families, but most cases were not associated with heredity and were termed sporadic CJD (sCJD). sCJD is the most common human prion disease with ~85% of all cases, with the balance made up of familial CJD and other diseases (Prusiner, 1989).

Kuru (Gajdusek and Zigas, 1959; Klatzo et al., 1959) bore many of the same neurological features as CJD and scrapie when it was identified among the Fore people of the Eastern Highlands of Papua New Guinea. Originating from a Fore word meaning "to shake," kuru was also known among the Fore as the "laughing sickness." The Fore engaged in a practice of mortuary or funerary cannibalism wherein the internal organs, including the brain, of the dead would be consumed by living relatives for spiritual purposes (Alpers, 1968). When Australian colonial administrators and Christian missionaries suppressed the practice of cannibalism, the epidemic levels of kuru observed in the 1950s rapidly declined, although because of the long and variable incubation period seen in many TSEs the last sufferer of kuru is reported to have died in 2005 (Alpers, 2008; Anon, 2009; Lindenbaum, 2008).

Beginning in the 1990s, it was recognized that human disease caused by prions went beyond the sporadic or familial forms of CJD and the exotic and largely extinct kuru. Variant CJD (vCJD) was noted in the United Kingdom in 1996, with features consistent with a CJD diagnosis, but an earlier average age of onset (Will et al., 1996). It was rapidly shown that the cause of the vCJD outbreak was consumption of food products from cattle infected with the BSE agent (Bruce et al., 1997).

Iatrogenic CJD (iCJD) has been recognized since the 1980s. In this form of CJD, improperly disinfected medical equipment, especially instruments used in brain surgeries, and also improperly prepared medicines, e.g., human growth hormone, have resulted in cases of CJD (Marzewski et al., 1988; Mocsny, 1991; Rappaport, 1987).

Finally, a few other distinctive human diseases with a prion basis are recognized. Fatal insomnia is a disease characterized by thalamic degeneration, progressive loss of neurological characteristics required for sleep, motor abnormalities, and hyperactivation of the autonomic nervous system (Lugaresi et al., 1986). First identified was a familial form of this disorder referred to as fatal familial insomnia (FFI) (Lugaresi et al., 1986), although later work found evidence of sporadic cases (sFI) as well (Barash, 2009; Montagna et al., 2003; Moody et al., 2011). Gerstmann–Sträussler–Scheinker (GSS) syndrome (reviewed in Liberski, 2012) is a very rare hereditary disease inherited in autosomal dominant fashion originally noted over 100 years ago in Austria (Dimitz, 1913) and more fully described in the 1920s and 1930s (Gerstmann, 1928; Gerstmann et al., 1936). GSS features dysarthria, ataxia, and progressive dementia, and its causative mutations in the human *PRNP* gene were identified in 1989 (Hsiao et al., 1989). The disease effects were experimentally recreated in mice shortly thereafter (Hsiao et al., 1990). Other variations in *PRNP* associated with disease in human families have been reported in unrelated groups around the world (e.g., Dlouhy et al., 1992; Hsiao et al., 1991).

3.4 Prion Diseases in Other Mammals

Other mammalian prion diseases have been described (Table 2, top) (reviewed in Greenlee and Greenlee, 2015). An infectious encephalopathy affecting ranched mink appeared as early as 1947 in the United States with a formal description in 1965 (Barlow, 1972; Burger and Hartsough, 1965; Hartsough and Burger, 1965; Marsh and Hanson, 1969). A disease of abnormal behavior, severe anorexia, and rapid death was observed 1967–79 in cervids (elk and deer) in Colorado and Wyoming (Williams and Young, 1980). Because of the substantial wasting caused by the anorexia in these animals, it was named chronic wasting disease (CWD). Despite its different name, it was immediately recognized, based on distinctive histopathology, as a spongiform encephalopathy in the same line as scrapie. Feline spongiform encephalopathy (FSE) was identified in domestic cats (Pearson et al., 1991, 1992; Wyatt et al., 1991) and later in many wild cats including lions, puma, ocelot, and cheetah (e.g., Eiden et al., 2010). An abstract from the Prion 2012 meeting in Amsterdam reported the case of a 9-week-old Rottweiler with canine spongiform encephalopathy (David and Tayebi, 2012). However, no further reports on canine spongiform encephalopathy have been published. Even though the list of species with documented cases (Table 2) is small, it remains likely that yet-undiscovered spongiform encephalopathies exist in all mammals.

3.5 Prions in Other Eukaryotes

Prion-based TSEs have only been reported in mammals. However, homologs of the PrP-encoding gene have been identified in birds, reptiles, amphibians, and fish (reviewed in Málaga-Trillo et al., 2011; Schätzl, 2007). It is unknown whether the variant PrP sequences in these species (which have several divergent features depending on taxonomic grouping) can form *bona fide* prions, amyloids, or whether TSE-like disease is present in these animals.

A protein with prion characteristics, when expressed in the yeast system, was also recently found in *Arabidopsis*, making it the first potential plant prion-like protein (Chakrabortee et al., 2016; discussed in Chernoff, 2016).

3.6 Evidence in Support of the Prion Hypothesis in Mammalian Disease

The proposal of a fully proteinaceous infectious agent and the coining of the term prion for that agent (Prusiner, 1982) did not coincide with irrefutable proof of the prion hypothesis, and certainly did not immediately satisfy all criticisms with the hypothesis. Instead, the formal statement of the prion hypothesis as the causative agent of scrapie built upon the steady framework of evidence from earlier studies (Cho, 1980; Griffith, 1967; Hadlow et al., 1980; Hunter et al., 1969; Merz et al., 1983; Prusiner et al., 1978a,b, 1980a,b, 1981) and provided a scaffold upon which to place further empirical data to support or refute it. Some of the major lines of support are provided here, although other texts provide a more complete picture of the supporting arguments (Colby and Prusiner, 2011b; Hörnlimann and Riesner, 2007; Zabel and Reid, 2015).

The laboratories of Charles Weissmann, Stanley Prusiner, and Leroy Hood, together published the identification of the gene responsible for scrapie, which encoded a protein in sheep for which several normal functions have since been determined, but no single well-determined role has been pinpointed. The gene, *Prnp* in animals and *PRNP* in humans, encoded the PrP (prion) protein (Oesch et al., 1985). The *Prnp* gene in mice was found to be colocated with a previously identified marker of mouse scrapie called *Sinc* (Dickinson et al., 1968), which provided evidence that a normal cellular (nonviral) gene locus was associated with the disease protein (Carlson et al., 1986, 1988; Hunter et al., 1987). Mice that were devoid of the PrP gene proved to be resistant to scrapie (Büeler et al., 1993). Mice that were modified to express their *Prnp* gene with the mutation corresponding to human FFI were spontaneously stricken with prion disease (Jackson et al., 2009). Prions can be made in bacteria and cause disease in mice (Legname et al., 2004). Reconstitution of the prion using a cyclic amplification technique was possible with both partially purified substrates (Deleault et al., 2005) and with infectious particles created in vitro (Barria et al., 2009). Further studies building on this theme show that it is possible to make recombinant infectious particles de novo in bacteria and without amplification in a clean laboratory that has never seen prions (Zhang et al., 2013).

The prion hypothesis holds that a natively folded cellular protein can assume an abnormal, infectious, and pathological shape that can be propagated between cells and between organisms without the need for any nucleic acid or viral structures. Although some scientists remain doubtful (Bastian et al., 2007; Manuelidis, 2007, 2013; Manuelidis et al., 2009; Somerville and Gentles, 2011), with the evidence above and other lines of evidence, most scientists are now convinced of the validity of the prion hypothesis in mammals (and, as seen below, in yeast).

3.7 Reed Wickner's Keen Observations in Yeast

The yeast traits (discussed in Section 2.5) that resulted from Cox and Lacroute's mysterious nonmitochondrial cytoplasmic particles in the baker's yeast *S. cerevisiae* (Cox, 1965; Lacroute, 1971) had long been on the mind of Reed Wickner, yeast geneticist and virologist. He began studies in 1989 (Wickner, 2012) to see if Prusiner's proposed framework of protein-only inheritance (Prusiner, 1982) could be applied to the [URE3] trait.

In 1994, Reed Wickner published this work of careful and keen observation, showing that [URE3] trait resulted from a heritable conformation of the Ure2 protein, wherein it took on a prion form that was passed to daughter cells (Wickner, 1994). This elegant hypothesis accounted for all of the unusual features of the non-Mendelian cytoplasmic inheritance of [URE3] that had vexed scientists for 30 years and immediately also suggested a mechanism for the inheritance of [PSI^+] as well (Wickner, 1994; reviewed in Tuite et al., 2015). [PSI^+] proved to be a heritable prion state of the Sup35 protein in yeast (Doel et al., 1994; Patino et al., 1996; Paushkin et al., 1996; Ter-Avanesyan et al., 1994).

In establishing the prion hypothesis for yeast proteins, Wickner had laid out three genetic criteria for a prion that should readily distinguish them from agents containing nucleic acid, such as viruses (Wickner, 1994, 2012): (a) the infection should be curable but reversible, (b) the overproduction of the relevant cellular gene should increase the frequency of prion formation, and (c) the prion-positive phenotype, inactivating a cellular protein's normal function, should match that of the loss-of-function mutant form of the same protein. All three of these criteria are met in [URE3] and

[PSI^+], where, first, low concentrations of guanidine HCl can cure prions (Ferreira et al., 2001; Lund and Cox, 1981; Tuite et al., 1981), but prions can then arise de novo in cured strains because the normal protein is still present. (Viruses would need to have nucleic acid reintroduced from outside the cell.) Second, overproduction of prion proteins increases the concentration of these proteins in the cell resulting in more prion formation (Chernoff et al., 1993; Derkatch et al., 1996; Wickner, 1994), presumably due to an increase in the probability of the misfolding event that initiates prion or oligomer formation. Finally, the *URE2* and *SUP35* genes, respectively, are necessary for the formation of the [URE3] and [PSI^+] prions, and the prion phenotype is the same as that of loss-of-function mutations for each gene (Aigle and Lacroute, 1975; Cox et al., 1988; Wickner, 1994).

With these criteria satisfied, further characterization of the nature of these prion proteins could begin. Through the work of Wickner's laboratory and the labs of Michael Ter-Avanesyan, Susan Lindquist, and Susan Liebman, and others, [URE3] and [PSI^+] began to reveal their secrets. Comparisons with the structures of animal prions would show many commonalities.

3.8 Other Fungal and Invertebrate Prions

Although they are not further discussed in this review, prions in other fungi and invertebrates have also been identified, which differ in some way from the known yeast and animal prions. For example, there is another fungal prion that differs somewhat in structure from the well-characterized yeast prions: [Het-s] the prion form of the HET-s protein in *Podospora anserina* (Baxa et al., 2007; Coustou et al., 1997; Mathur et al., 2012; Wan and Stubbs, 2014; Wickner et al., 2016). Enzymatic and nonamyloid prions have also been identified, e.g., the yeast protease B (Jones, 1991; Roberts and Wickner, 2003) and the poly-A-binding protein CPEB in *Aplysia californica* (Si and Kandel, 2016; Si et al., 2003a,b, 2010; Stephan et al., 2015).

4. WHAT MAKES A PRION: FEATURES THAT DEFINE PRIONS

4.1 Defining Features of Prions

In the course of finding evidence for the prion hypothesis in animals and fungi (see Section 3), many other characteristics about their biochemical and biophysical nature were also noted.

The primary physical characteristic of prions found in prion diseases is that these diseases exhibit amyloid deposits in nervous tissue (detailed later). In the course of early studies of these diseases, the amyloid deposits were found to be stainable with agents such as Congo red. After the identity of amyloid as protein rather than either carbohydrate or lipid, amyloid proteins were also found to be insoluble, protease and detergent resistant, beta-sheet rich, and prone to assemble into aggregate and fibril structures.

In this section, I detail the work that uncovered the overall amyloid structures of the animal (Section 4.2) and yeast (Section 4.3) prions. Knowledge of the essential structural and functional nature of prions (PrP and the yeast prions, chiefly) has logically led to the search for other prions in mammals and in yeasts (Section 4.4), although the success rate for finding new prions has been much greater in yeast. Other characteristics that define prions have also been noted over years of study (Section 4.5), and these characteristics are leading to insight into prion, amyloid, and similar diseases and their pathophysiologies.

4.2 Structural Features of Animal Prions

Animal prions are characterized by certain structural and biochemical features. The well-characterized mammalian PrP prion is known to form amyloid fibrils. Amyloids (misidentified by Rudolf Vircow in 1854 as related to starch—*amylum*—because amyloid is stained by iodine like starch) were found in nervous tissue and associated with all of the prion diseases above as well as with other amyloidoses including Alzheimer's disease (Sipe and Cohen, 2000). Amyloids were found to be different from starch under light microscopy on the basis of a green/yellow/orange birefringence when stained with Congo red dye and illuminated under polarized light (Howie, 2015). In 1959 the first electron micrographs of amyloids showed fibrils of 80–100 Å in width and of variable length (Sipe and Cohen, 2000). Amyloids were resistant to protease treatment (Kitamoto et al., 1986; Manuelidis et al., 1985; McKinley et al., 1983; Oesch et al., 1985) and detergent treatment (Glenner et al., 1969; Prusiner et al., 1987).

Native PrP protein has been crystallized (Antonyuk et al., 2009) and solved by NMR (James et al., 1997; Riek et al., 1996, 1998; Zahn et al., 2000), but working with nonnative and insoluble amyloid forms of proteins is problematic for traditional structural techniques. The secondary conformations found in amyloids were first elucidated in the 1960s and showed a beta-sheet rich structure with the beta-sheet axes perpendicular to the long axis of each fibril (the so-called cross-beta structure) (Eanes and Glenner, 1968). Many subsequent studies have borne out the basic conclusion for different animal amyloid and prion proteins (Groveman et al., 2014; Harper et al., 1997; Lyubchenko et al., 2012; Sunde et al., 1997; Tycko and Wickner, 2013) with the latter papers clarifying a parallel in-register intermolecular beta-sheet structure for the amyloid forms of these proteins.

Amyloid proteins self-assemble into large, complex aggregates and fibrils on the basis of their unusual beta-sheet rich tertiary conformations (Fig. 2). The process of fibril formation has a number of steps (Chiti and Dobson, 2006; Dobson, 2003; Eisenberg and Jucker, 2012; Gregersen et al., 2005; Knowles et al., 2014; Maji et al., 2009; Naeem and Fazili, 2011; Tanaka et al., 2006). One model is presented here, although other models have been proposed (Colby and Prusiner, 2011b). In this model, conversion of native to amyloid form is a rare event (Fig. 2A), where the misfolded proteins can associate and cause conformational conversion of other natively folded proteins (Fig. 2B). Through this process, oligomers are formed (Fig. 2C) that eventually assemble into longer fibrils (Fig. 2D). Chaperone proteins and other proteins may be involved in cleaving long fibrils into smaller pieces (Fig. 2D–C). It has been noted that the amyloid oligomer stage (Fig. 2C) is likely the most toxic to cells and tissues (reviewed in Kayed and Lasagna-Reeves, 2013; Verma et al., 2015). It is also worth noting that while amyloid formation is clearly a process that involves cytotoxicity and histotoxicity, production of rod-type and other nonamyloid aggregates is also possible with PrP and disease can still result (Wille et al., 2000).

The *Prnp/PRNP* genes in animals and humans encode the PrP protein (Basler et al., 1986; Oesch et al., 1985), and the domain structure of the translated PrP protein (Fig. 3A) has been long studied and dissected for interesting and notable features (reviewed in Colby and Prusiner, 2011a,b). The mammalian prion protein, PrP, as shown in Fig. 3A, contains five octarepeats (consensus sequence: PHGGGWGQ; Brown et al., 1997). The similar length of each repeat and number of repeats found in each protein is suggestive of some important function. The importance of the repeats in PrP is underscored because PrP repeat expansion is associated with dom inant inherited prion disease (Prusiner et al., 1998; Wadsworth et al., 2003) and removal of the repeats in a mouse model of disease slows progression (Flechsig et al., 2000). The profile of the repeat structures in PrP rose further when it was noted that there are compositional similarities between the repeats in PrP and in the yeast prion Sup35 (Fig. 3B, with similar prevalence to PrP of the amino acids proline, glycine, and glutamine in the repeats,

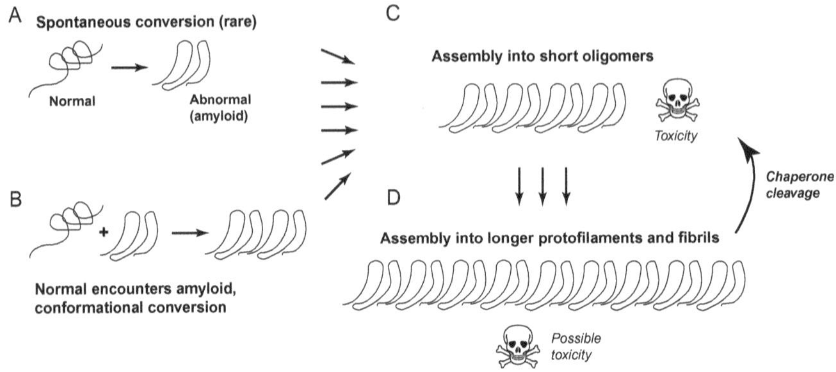

Fig. 2 Process of assembly of toxic oligomers, protofilaments, and fibrils in amyloid-based diseases, including prion diseases. (A) Spontaneous conversion between a native and normally folded protein state into an abnormal or amyloid state (beta-sheet rich) is very rare. Both forms are stable states. (B) Once an abnormal amyloid form of a protein is present in a cell, when it encounters a natively folded protein, it is capable of causing a conformational change in which the native protein assumes an amyloid structure. (C) When amyloid-structured proteins encounter each other, they have a tendency to aggregate and form, initially, short stretches of dimers, trimers, and oligomers. Evidence suggests these oligomers are more toxic to the cell than monomers or larger filaments (e.g., Simoneau et al., 2007; reviewed in, e.g., Verma et al., 2015). (D) Oligomers that pick up additional monomers or oligomers may assemble into larger protofilaments and then fibrils that can be extremely large. These fibrils are often hallmarks of amyloidoses and can be visualized in histopathologic sections with various straining and imaging techniques. Chaperones (such as Hsp104 in yeast) are capable of cleaving larger fibrils into shorter pieces, which appears to be required for proper maintenance of the prion during cell division.

for example, as detailed in Section 4.3). Indeed, in the context of yeast Sup35, its oligopeptide repeat domain (ORD) repeats can even be functionally replaced with PrP repeats, and propagation is unimpaired (Parham et al., 2001). And in a result analogous to the in vivo repeat expansion experiment, Sup35 aggregates with increasing numbers of PrP repeats have reduced times to fiber formation in vitro (Kalastavadi and True, 2008). Given the similarity between Sup35 and PrP repeats and the presence of repeat elements in other yeast prion domains—Rnq1 and New1 (Osherovich et al., 2004; Vitrenko et al., 2007)—primary sequence effects could be an important consideration for propagation of prions. However, as discovered in yeast prions (Section 4.3), primary sequence elements like repeats may instead represent a convenient genetic method of rapidly expanding amino acid compositional biases that lead to prion formation.

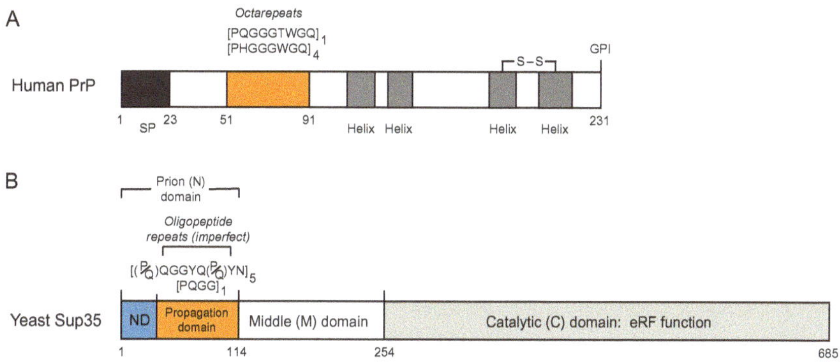

Fig. 3 Domain structures of canonical mammalian and fungal prions. Repeat domains are noted with single-letter amino acid abbreviations for repeat structures in the protein sequences. (A) Human prion protein (PrP), which can interconvert between normal PrP^{C} and abnormal PrP^{Sc} protein variants. *SP*, signal peptide; *S–S*, disulfide bridge; *GPI*, glycophosphatidylinositol anchor. (B) Yeast prion protein Sup35 (eRF3) which can give rise to the [*PSI*$^{+}$] prion. *C domain*, catalytic domain; *M domain*, middle domain; *N-domain*, prion domain; *ND*, nucleation domain region of the N-domain; *ORD*, oligopeptide repeat domain region of the N-domain.

Other structural features have been noted for PrP as well (Fig. 3A). It is doubly-glycosylated near the cysteines involved in a disulfide bridge and has a GPI-anchor for cell membrane attachment. Unlike the repeat structures noted earlier, these features have not been generally noted in the yeast prions and so may represent less commonly found domains or characteristics of prion proteins.

4.3 Structural Characterization of Yeast Prions

Although the non-Mendelian cytoplasmic characters [URE3] and [*PSI*$^{+}$] from yeast were shown to be prions in 1994, many aspects of their fundamental biology remained to be worked out. Though Wickner had shown a protein-only inheritance in the yeast prions consistent with that previously proposed in mammalian PrP, whether the yeast prions would share the basic protein structure of an abnormal amyloid fold was not known. The amyloid structure would first be noted for [*PSI*$^{+}$] (King et al., 1997) and [URE3] (Taylor et al., 1999) and the predicted (Ross et al., 2005b) parallel in-register beta-sheet structure observed for PrP would be noted for [URE3] (Baxa et al., 2007), [*PSI*$^{+}$] (Chen et al., 2009; Shewmaker et al., 2009; Wickner et al., 2008), and others (Chen et al., 2009; Engel et al., 2011). Yeast prions, found to generally form amyloid structures, were also protease and detergent resistant (Masison and Wickner, 1995).

The full history of yeast prion characterization is outside of the scope of this review (for a fuller discussion, see Wickner, 2012), but I will discuss several key structural and biochemical features of yeast prions beyond amyloid structure in this section.

Shortly after Wickner's (1994) paper, it was rapidly noted by Yury Chernoff in Susan Liebman's lab in collaboration with Susan Lindquist's lab that the chaperone protein Hsp104 was involved in propagating the [PSI^+] prion to daughter cells and cells that mate with [PSI^+] cells (Chernoff et al., 1995; Lindquist et al., 1995), and this process would be mediated by Hsp104's ability to cleave fibrils into smaller pieces (reviewed in Sweeny and Shorter, 2016, see also the arrow from Fig. 2D–C).

The function of yeast prions is a matter of some debate. Unlike the TSEs which greatly hamper neurologic function and are uniformly fatal when symptoms begin, prions in yeast, due to short generation time and rapid growth, could be beneficial (Suzuki and Tanaka, 2013; True and Lindquist, 2000) or harmful (McGlinchey et al., 2011; Nakayashiki et al., 2005; Wickner et al., 2011). In fact, there is no reason to expect that prions could not be both sometimes beneficial and sometimes harmful to the cell.

The normal function of each host protein, Sup35 and Ure2, was exploited as assays for the detection of prion activity as well. Detection of [URE3] relies on growth characteristic of the cells in the presence of a good nitrogen source. [URE3] cells in this circumstance would be able to take up ureidosuccinate, an intermediate compound in uracil biosynthesis, while cells without the [URE3] prion cannot uptake ureidosuccinate (Lacroute, 1971). This ability has been used to assay for the presence of the [URE3] prion, but it can be a difficult assay to work with (Brachmann et al., 2006). Assaying for [PSI^+] is a much easier-to-interpret test. Because Sup35 is an "omnipotent suppressor" that can read-through stop codons (Ter-Avanesyan et al., 1994), in a cellular background containing an *ade2-1* (or similar) mutant with a premature stop codon, suppression by the eRF3 function of Sup35 will lead to read-through in prion-containing cells and no read-through in prion-negative cells (Fig. 4A). Because the *ade2* mutant is nonfunctional without read-through, oxidized *P*-ribosylaminoimidazole in the adenine biosynthetic pathway will accumulate and the cells will be red in color when plated on limiting adenine (Fig. 4B, right). If the prion state removes active Sup35 from the cell by sequestering it in fibrils, read-through will occur and the cell will remain wildtype in color (Fig. 4B, left).

Unusually, both [URE3] and [PSI^+] were found in genetic screens where, uncommonly, a loss-of-function event for either protein was

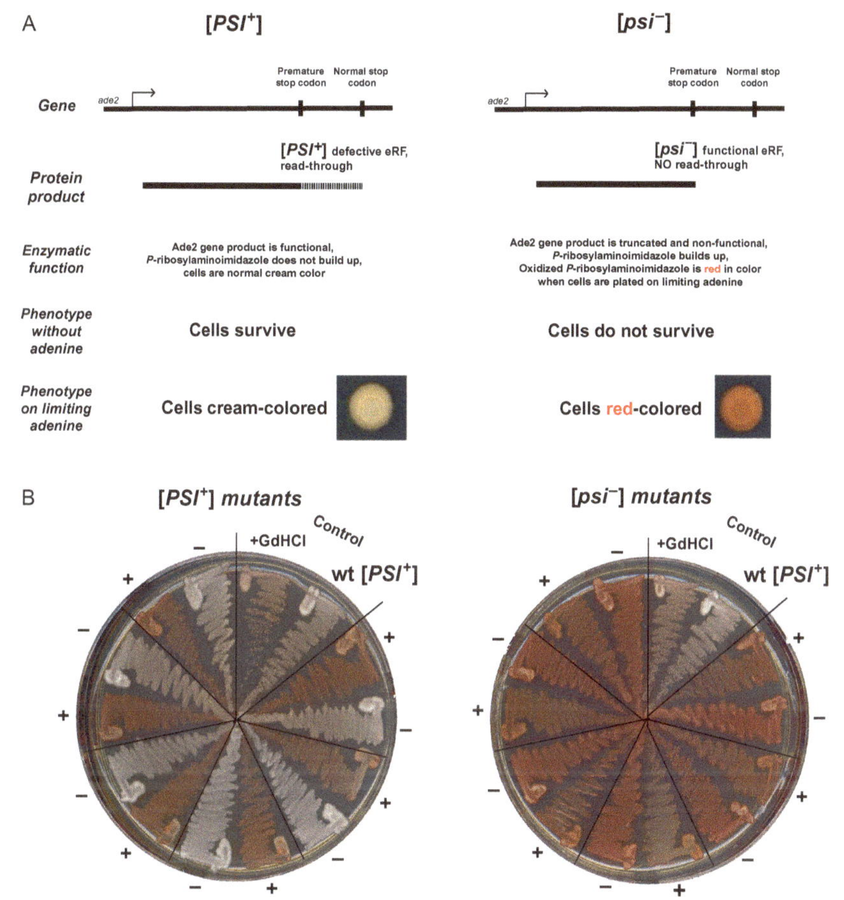

Fig. 4 Assay for presence of the yeast [*PSI*$^+$] prion using the *ade2-1* mutant nonsense suppression (eRF3) function of Sup35. (A) Schematic diagram for *ade2-1* generation of color phenotypes in the presence or absence of the [*PSI*$^+$] prion. (B) Examples of *red/white color* selection using the *ade2-1* assay. *Left*, mutant forms of Sup35 that are [*PSI*$^+$] in this assay are compared with the control wild-type [*PSI*$^+$] prion, plus or minus curing with guanidine hydrochloride (GdHCl). *Right*, mutant forms of Sup35 that are [*psi*$^-$] (nonprion) are shown.

advantageous to the cell (Cox, 1965; Lacroute, 1971). In most cases, detecting such a rare loss-of-function event would be extremely difficult. However, structural studies of [URE3] and [*PSI*$^+$] revealed an exploitable feature of these proteins that could help identify other, similar, prions.

Sup35, the protein that forms the [*PSI*$^+$] prion, features three domains (Fig. 3B): an N-terminal (N) domain that is responsible for prion formation (also called a prion-forming domain (PFD) or prion-like domain (PrLD)), a charged middle domain (M) and a C-terminal catalytic domain (C) responsible for the nonsense suppression (eRF3) function of Sup35 (Ter-Avanesyan et al., 1993). The N domain is rich in glutamine and asparagine (Q/N) amino acid residues. Within the N domain, the nucleation domain (ND), the first 39 amino acids, is more Q/N-rich than the portion of the N domain immediately after DePace et al. (1998). This section, the ORD, is also enriched in glutamine and asparagine but is primarily noted for having a series of 5½ imperfect repeats (Fig. 3B) (Osherovich et al., 2004; Shkundina et al., 2006). Ure2 also has a substantial Q/N-tract that is required for prion formation (Masison and Wickner, 1995). What made these Q/N-rich domains of even greater interest was that these domains were modular (the compact Q/N-rich portion of the protein enabled the protein to assume an amyloid shape without contribution from the rest of the three-dimensional structure) and also transferrable (that amyloid/prion-forming ability could be fused to many other proteins and cause them to also become amyloid/prion forming) (Baxa et al., 2002; Li and Lindquist, 2000). In both the Sup35 and Ure2 yeast prion proteins, the prion domain was also dispensable and could be deleted without affecting catalytic functions (domains reviewed in Ross et al., 2005a,b).

The prion domains of the [URE3] and [*PSI*$^+$] prions have a curious conformational property as well. For almost all known proteins, three-dimensional structure and function are inextricably linked to the primary sequence, the ordered series of amino acids. In the beta-sheet rich [URE3] and [*PSI*$^+$] prions, it is possible to actually scramble the order of the amino acids in each PFD (using a random number generator) and retain both the amyloid structure and the prion function/effects in the cell (Ross et al., 2004, 2005a,b; Shewmaker et al., 2006).

The ability to scramble amino acid order while retaining structure and function is an especially curious property given that, as detailed in Section 4.2, Sup35 has been utilized as a model for examining the role of prion protein repeats in formation and propagation of aggregates (Dong et al., 2007; Kalastavadi and True, 2008; Parham et al., 2001; Tank et al., 2007), and the mammalian PrP repeats have been repeatedly suggested to be important for disease (Flechsig et al., 2000; Prusiner et al., 1998; Wadsworth et al., 2003).

In the case of [*PSI*$^+$], the two portions of the PFD (the N-terminal ND region and the C-terminal ORD region) have distinct amino acid compositions (Toombs et al., 2011). The distinct compositions seem to relate to different functions of each subdomain: the ND is required for nucleation or formation of the prion, and the ORD is required to propagate or maintain the prion (DePace et al., 1998; Osherovich et al., 2004; Shkundina et al., 2006). The ability to scramble prion primary sequence and still generate functional prions led to important experiments, discussed later, useful in understanding yeast prions and in identifying new candidate prions.

4.4 Making Predictions: Using Biochemical Knowledge of Known Prions to Identify Other Prions and Understand the Prion Structure–Function Relationship

Given the longer history of study of the animal prions, it might be expected that after Prusiner's prion hypothesis (Prusiner, 1982) gained traction, other animal prions would be rapidly discovered. That has not been the case, although some (bottom part of Table 2), including the alpha-synucleinopathies, appear to form *bona fide* infectious prions. Alpha-synuclein, which has no sequence similarity to PrP, has recently been reported using mouse animal and cell culture models of human multiple system atrophy (MSA) as a prion (Prusiner et al., 2015; Watts et al., 2013; Woerman et al., 2015; reviewed in Supattapone, 2015). Alpha-synucleinopathies aggregate alpha-synuclein with other proteins in pathological structures called Lewy bodies (Mezey et al., 1998; Spillantini et al., 1997) that are found in Parkinson's disease, MSA, Lewy body dementia, and some cases of Alzheimer's disease (Yokota et al., 2002). It is likely that other human prion or prion-like diseases may still await discovery. True infectious prions in mammals have not been easily found, but as noted in Section 5, the enlargement of the prion concept may instead show that other prion-like diseases have been hiding, perhaps, in plain sight.

Despite difficulties in identifying new animal prions, a whole host of new candidate and verified yeast prions have been found since Wickner's (1994) recognition of the prion hypothesis in *Saccharomyces*. The ease of genetic screens and manipulation in yeast has made a host of different approaches possible. These studies in turn have led to greater structural insights, and each new observation has improved methods for identifying other prions, resulting in more discoveries. The current list of likely yeast prions is ~18 in *S. cerevisiae* alone. And because prions are a subset of aggregative proteins that form a major new class of human diseases and the proteins responsible for these human diseases share characteristics with yeast prions, identifying new prions in yeast (reviewed in MacLea and Ross, 2011) is a topic of considerable interest with applications in human disease. Several techniques have been used or proposed to identify new prions in yeast: (1) Prion–prion interactions; (2) Q/N-content or other composition; and (3) Other bioinformatics and proteomics methods.

4.4.1 Prion–Prion Interactions Help Reveal New Prions

Prions interact frequently with other prions in yeast, and these interactions can have variable effects on prion formation and propagation (Gonzalez Nelson and Ross, 2011). The [*PIN*$^+$]/[*RNQ*$^+$] prion has been most well-studied in its effects on other prions, particularly its ability to promote formation of the [*PSI*$^+$] prion (Derkatch et al., 1997, 2000, 2001). The identification of [*PIN*$^+$]/[*RNQ*$^+$], described later, allowed Irina Derkatch to perform a genetic screen to identify factors that could substitute for [*PIN*$^+$] in allowing [*PSI*$^+$] formation (Derkatch et al., 2001). This method identified 11 candidate prions, of which one was shown to be prion-like in certain assays but has not been shown to form prions in its native state (New1), and two were identified as likely prions (Swi1 and Cyc8) (Derkatch et al., 2001; Du et al., 2008; Patel et al., 2009). This genetic screen was unique to [*PIN*$^+$] and given that little is known about the seeding or other mechanism responsible for the behavior of [*PIN*$^+$] in the cell, this method has not been used in additional screens.

4.4.2 Q/N or Other Amino Acid Composition as a Tool for Prion Identification

[*PSI*$^+$], encoded by the *SUP35* gene in yeast, has a PFD that is both modular and transferable and has an extremely easy-to-use and robust assay for prion formation (Fig. 4 and see earlier), making it the ideal platform on which to test other candidate prions. A classical experimental scheme using Sup35 in this manner involves replacing the N domain (PFD) of Sup35 (see Fig. 3B) with any candidate open reading frame (ORF) and then assessing its function in the *ade2-1* assay conventionally used to monitor [*PSI*$^+$] function (Fig. 4). Using this scheme, additional prions would soon be identified in yeast, including [*NU*$^+$] encoded by New1 (Michelitsch and Weissman, 2000) and [*PIN*$^+$]/[*RNQ*$^+$] encoded by Rnq1 (Derkatch et al., 2001; Santoso et al., 2000; Sondheimer and Lindquist, 2000). The PFDs of New1 and Rnq1 were also Q/N-rich and also transferrable, conferring the ability to aggregate even on the green fluorescent protein (GFP) in the absence of Sup35 (Osherovich and Weissman, 2001; Osherovich et al., 2004; Sondheimer and Lindquist, 2000). The New1 PFD has additional similarities to Sup35, including separation of the formation and propagation functions within the PFD (Osherovich et al., 2004; discussed later for Sup35).

When New1 and Rnq1 were identified and shown to have similar Q/N-content and characteristics to Sup35 and Ure2, two large-scale bioinformatics screens looking for Q/N-rich predicted prions in the yeast proteome were undertaken, in Jonathan Weissman's lab (Michelitsch and Weissman, 2000) and by Harrison and Gerstein (2003). Melissa Michelitsch found 107 candidate yeast prion proteins, including most (8/11) found by Irina Derkatch, all four of the previously identified prions (Ure2, Sup35, New1, Rnq1) and four that were later shown to be *bona fide* prions (Swi1, Cyc8, Mot3, Sfp1) (Alberti et al., 2009; Du et al., 2008; Michelitsch and Weissman, 2000; Patel et al., 2009; Rogoza et al., 2010). Paul Harrison found 172 prion candidates of which 101/172 were found by Michelitsch and 9/11 of the proteins found by Irina Derkatch in her genetic screen (Harrison and Gerstein, 2003). All 8 of the proven/likely prions found above were also found in this study (Ure2, Sup35, Rnq1, Swi1, Cyc8, Mot3, Sfp1). Michelitsch and Harrison both identified a large number of candidate prion proteins, but determining which of these candidates to examine further was not obvious given the methods used. A combination of the bioinformatics screen with an experimental approach was necessary.

The method of fusing prospective candidate PFDs to Sup35 to test prionogenicity and three other aggregation assays was used in a major study out of Susan Lindquist's lab to address this central criticism of previous bioinformatics screens. In this study (Alberti et al., 2009), a computational tool called a hidden Markov model (HMM) was first used to identify the 100 most-similar proteins to Ure2, Sup35, Rnq1, and New1. In a mammoth experiment, each of those 100 ORFs was then tested in four different tests of prion-like activity, and 23 proteins were found that could induce prion formation in the context of Sup35 (Alberti et al., 2009). This method did not identify all potential prions since two known prion proteins, Cyc8 and Mot3, did not show prion activity in this assay. Showing the utility of this combined bioinformatics/empirical approach, although 67/100 of the ORFs had been previously implicated by Michelitsch and Harrison (Harrison and Gerstein, 2003; Michelitsch and Weissman, 2000), most did not have prion activity in one, two, three, or four of the prion candidate testing methods (Alberti et al., 2009).

The enormous combined screen of Simon Alberti and Randal Halfmann in Susan Lindquist's lab (Alberti et al., 2009) provided a data set of immense value, adding in the experimental results for all four assays of aggregative/prion activity to the computational screens previously conducted. Still, within the data set generated, there was found to be no substantial relationship between the degree of similarity of each of the 100 ORFs to previously known prion sequences with their results in the four assays (Alberti et al., 2009; Ross and Toombs, 2010; Toombs et al., 2010). While at first blush this suggests that amino acid composition may not be the main determinant of prion propensity, the incompleteness of previous knowledge on what made a prion and the small sample size likely meant that the algorithm was not optimized for this situation. What was needed was an experiment that would give scoring values for each amino acid so that an increase or decrease in propensity to form prions could be calculated, without relying on previously discovered yeast prions.

In Eric Ross's laboratory, Trey Toombs used a scrambled version of Sup35 and replaced two short segments with a random sequence to generate two libraries of mutants (Ross and Toombs, 2010; Toombs et al., 2010). For each library, different regions of the Sup35 protein ND were modified and he then compared (in each library) the amino acid composition for a naïve subset of clones (with no selection) with a subset that could form prions and generated a prion-propensity score for each amino acid. This allowed regions and whole ORFs and proteomes to be scanned and scored to evaluate overall predicted prion propensities. Using another algorithm, FoldIndex, that measures order/disorder propensity (Prilusky et al., 2005), Toombs found that known yeast PFDs had extended disordered regions with only modest prion propensities (Ross and Toombs, 2010; Toombs et al., 2010). Although not a perfect predictor, this method did improve (Toombs et al., 2010) on the blind HMM method used in Lindquist's lab and was reasonably effective at predicting prion propensities for the proteins examined in the four assays of aggregative/prion function (Alberti et al., 2009). The resulting algorithm for screening yeast proteins for prion propensity was named PAPA (Ross and Toombs, 2010; Ross et al., 2013; Toombs et al., 2010).

The Toombs experiment measured, by its design, the combined processes of prion formation and prion propagation or maintenance. A follow-up study showed that the two subdomains within the PFD of Sup35 had amino acid compositions that were not identical. That is, the composition of the ND (responsible for formation) and the ORD (responsible for maintenance) of Sup35 was different, and therefore propagation of prions to daughter cells had slightly different compositional requirements than nucleation (Toombs et al., 2011). Further work addressed this compositional bias and allowed calculation of separate prion maintenance propensities (MacLea et al., 2015), which may in the future allow these processes to be better dissected and lead to more accurate prediction algorithms for fully functional prions.

4.4.3 Other Bioinformatics and Proteomics Methods for Prion Identification

Numerous algorithms have been developed to predict protein aggregation propensity, chiefly using the mammalian amyloids as a basis. Algorithms including TANGO (Fernandez-Escamilla et al., 2004), Zyggregator (Tartaglia et al., 2008), BETASCAN (Bryan et al., 2009), Waltz (Maurer-Stroh et al., 2010), and ZipperDB (Goldschmidt et al., 2010) have been somewhat successful at finding known amyloids in mammalian databases but have had less utility in identifying yeast prions. Although there is probably more to the story, the amyloidogenesis in both systems is thought to be rather different. Mammalian amyloids appear to require a shorter, highly amyloidogenic stretch, while yeast prions appear to require longer stretches of modest prion propensity with intrinsic disorder as estimated by FoldIndex (Esteras-Chopo et al., 2005; Prilusky et al., 2005; Ross and Toombs, 2010). Newer algorithms focused on yeast prions, such as ArchCandy, which incorporates three-dimensional modeling, may prove useful as well (Bondarev et al., 2013) but at the moment no verified new prions have been identified using these methods.

Simulations of molecular dynamics for short-peptide stretches found commonly in mammalian prions were used in the creation of some of the algorithms above and have shed some light on how the conformational conversion process from native to amyloid shape may occur at the molecular level. Similar simulations for the Q/N-rich prions have also been undertaken (Berryman et al., 2011; Halfmann et al., 2011). Proteomics methods including two-dimensional gels and mass spectrometry have been proposed and used in small studies, but the insolubility of the amyloidogenic proteins makes these kinds of techniques very tricky to interpret. Other methods may prove useful in the future for identification of more amyloid and prion proteins. Any such method developed will need to work around difficult intrinsic properties of these proteins, including insolubility, protease and detergent resistance, and more. Methods that are not biased in the same ways as earlier studies (looking only at Q/N-rich proteins, relying on fusion to Sup35 for an assay, etc.) will likely yield the most fruit in years to come. One such study that exploits the difficult intrinsic properties of prion and amyloid proteins was recently published (Kryndushkin et al., 2013) and may be a useful template for future proteomics experiments to identify new prions or similar proteins.

4.5 Strains

In Section 4, overall physical structures of animal (Section 4.2) and yeast (Section 4.3) prions have been examined, showing key features of these proteins, e.g., amyloid structure, staining properties, protease and detergent resistance, domain structures, repeat sequences, and amino acid compositions. These properties of "what makes a prion" were the initial seeds upon which further studies have been built. In learning to identify new prions, chiefly in yeast (Section 4.4), new features of both yeast and animal prions and amyloids have been noted, further expanding the field's knowledge of the essential characteristics and diversity of prions and amyloids. One key, but unusual, feature of prions has not yet been discussed: distinct prion strains.

Like other pathogens, prions have strain differences, and these strain differences are propagated when the prions are transmitted. This was first noted in scrapie (Dickinson and Meikle, 1969; Fraser and Dickinson, 1973). Animal prion strains appear to be caused by conformational diversity (different stable forms with tertiary conformational variability) being inherited more or less faithfully (Bessen and Marsh, 1994; Colby and Prusiner, 2011a; Collinge et al., 1996; Peretz et al., 2001; Telling et al., 1994). Yeast prions have widely appreciated strain differences as well (Huang et al., 2013; King and Diaz-Avalos, 2004; Marcelino-Cruz et al., 2011; Tanaka et al., 2004, 2006) that appear to be passed vertically and can be passed ex vivo cell to cell using traditional experimental techniques as well. Because prions are not easily passed horizontally in yeast, it is unclear whether strains can be naturally transmitted this way.

5. THE ENLARGING PRION CONCEPT IN DISEASE AND BEYOND

5.1 Introduction

Prion diseases such as the TSEs were ultimately identified and set apart from other diseases on the basis of their etiology by a "proteinaceous infectious particle" or prion. While this was a useful designation in the early years of prion studies, when scientific consensus on the existence of prions was far from sure, it is now becoming clear that the segregation of prions from other agents of pathological protein aggregation is inappropriate. For example, noninfective amyloids such as amyloid precursor protein and tau, when injected directly into the central nervous system of other animals, appear to be able to cause disease (Clavaguera et al., 2009; Haass et al., 1995). Human patients have also acquired Lewy body-type pathologic inclusions from brain grafts (Kordower et al., 2008). From these and other observations (e.g., Eisenberg and Jucker, 2012; Jucker and Walker, 2011), it appears clear that the line separating the infectious prions from the noninfectious amyloids or pathologic aggregates is thinner than previously thought. As a result, the consensus is that the prion concept itself is enlarging to encompass other diseases of aberrant protein aggregation as well (Colby and Prusiner, 2011b; Walker and Jucker, 2015).

5.2 Developing a Definition of a General Category of Prion-Like Conformational States

It was recently proposed that a new category of prion and prion-like diseases should together share certain essential characteristics (Colby and Prusiner, 2011b). (1) A posttranslational conformational change occurs in a native protein to a form with high beta-sheet content; (2) Oligomers are formed from the high beta-sheet protein forms and are toxic to cells; (3) Polymerization into fibrils results in reduced toxicity of the high beta-sheet forms; (4) "Plaques," "tangles," or "bodies" result from sequestration of the fibrils inside and outside of cells, in the central nervous system; and (5) Mutations in these proteins may cause familial heritability of these traits.

5.3 Prion-Like Proteins, Quasi-Prions, and Prionoids

A growing awareness of the broad swath of prion-like phenomena has necessitated some new terms to distinguish these categories. Paul Harrison's lab has suggested the categories of prion and prion-like proteins, with the latter category made up of quasi-prions and prionoids (Harbi and Harrison, 2014). Briefly, prions have firm evidence of prion behavior, with fully infective particles made in vitro (strongest evidence, e.g., Sup35) or not (weaker, e.g., Cyc8). Quasi-prions behave similarly to prions but do not meet the infection requirements of a prion but can still pass the quasi-prion to progeny (for example, the likely prionogenic proteins from the Alberti et al., 2009 study or RepA-WH1 in bacteria). Prionoids have been shown to propagate between cells in multicellular organisms (for example, Tau in Alzheimer's disease). Regardless of the specific nomenclature, the rising realization in the aggregation and prion communities that there is overlap and cross talk between the fields that may allow leaps in one area to rapidly crosspollinate to another area across these categories make an understanding of the relatedness of the concepts especially apt and timely. For example, in Section 5.4, the application of discoveries in the yeast realm to studies of familial human diseases illustrate that these prion-like phenomena clearly share a biochemical and cellular basis.

5.4 The Intersection of Animals and Yeast: Studies of Yeast Prions Have Lead to Understanding of Human Amyloid Diseases

Yeast prions have helped us to find amyloid proteins in humans. Although PrP is by far the most well-studied human prion protein, Q/N-rich proteins are overrepresented in the human proteome (Harrison and Gerstein, 2003; Michelitsch and Weissman, 2000), and study of these proteins in the context of yeast has been useful for identifying aggregating proteins in humans (reviewed in Cascarina and Ross, 2014). All of the following suspect amyloid proteins were tested in the yeast prion model. For example, amyloidogenic proteins generated from mutant TDP-43 alleles were linked with amyotrophic lateral sclerosis (ALS), frontotemporal lobar degeneration (FTLD), Alzheimer's, and Parkinson's diseases (Da Cruz and Cleveland, 2011; Johnson et al., 2008, 2009; Lagier-Tourenne et al., 2010; Neumann et al., 2006). Mutations in FUS/TLS, EWSR1, and hnRNPA1 and hnRNPA2B1 were shown to cause ALS in some families (Couthouis et al., 2012; Daigle et al., 2013; Kim et al., 2013; Kwiatkowski et al., 2009; Sun et al., 2011; Vance et al., 2009). Additional human amyloid proteins have been found in this way as well (reviewed in Cascarina and Ross, 2014), and it is extremely likely that additional discoveries will be made in the coming years by fusing advanced genetic and pedigree analysis of humans with the experimental virtues of the simple, well-worn yeast prion analysis system. In undertaking studies such as these, it is interesting to note that these human proteins, in large part, share more sequence/structure characteristics with the yeast prions than they do with PrP, demonstrating that fundamental biology is at work, probably for all eukaryotic cells and perhaps for all cells.

5.5 What Ties Together Prion-Like Phenomena

Abnormal accumulation of disease-specific protein aggregates is a hallmark of most neurodegenerative disorders. These include Parkinson's disease (PD), ALS, MSA, FTLD, and others. The proteins implicated in these disorders are numerous (reviewed in Walker and Jucker, 2015), but they all involve aggregation-prone proteins, many with PrLDs, ability to form beta-sheet rich secondary conformations, and the ability to spread locally within brain regions and form plaques or similar deposits with concomitant toxicities. In short, they meet the requirements set above for prion-like behavior

(Section 5.2) (Colby and Prusiner, 2011b). What all of these disease-causing proteins fundamentally share is that they are based on seeded aggregation of proteins. As the field moves forward, grouping the diseases together that are caused by seeded abnormal protein aggregation is perhaps the best starting place for a new understanding of the prion concept. What Walker and Jucker have referred to as a "proteinaceous nucleating particle" (Walker and Jucker, 2015) brings the prion diseases and the nonprion amyloid diseases together with yet-to-be-discovered variants under the umbrella term "prion." While this term has not yet been widely used to encompass infectious and noninfectious aggregating proteins (and indeed whether the term is ever used in that fashion), the enlargement of the prion concept and the acknowledgement that there is relatively little difference between prions and noninfectious amyloids has already begun.

6. CONCLUDING REMARKS

In this review, I have discussed the history of the discovery of prions in mammals and the resulting recognition that previously discovered but unexplained non-Mendelian traits in the baker's yeast *S. cerevisiae* represented prions as well. The essential genetic, biochemical, and biophysical features of the mammalian prions and amyloids, and the yeast prions and prion-like molecules, while broadly similar, show significant differences as well. Despite this, understanding of the simple yeast prion system has allowed for major health and basic science discoveries in the mammalian context and insights from mammals have informed the studies of prion proteins in yeast. The collective discoveries in this area have grown larger through a recognition that aggregative proteins form a larger constellation of related phenomena (including many diseases). Because of this, the scientists and physicians studying aggregating proteins responsible for human and animal disease, whether infective or not, would do well to familiarize themselves with the literature across the whole gamut of prion, prion-like, and amyloid proteins, because these phenomena clearly demonstrate fundamental similarity at the cellular level that can be exploited to solve problems in all parts of the field.

REFERENCES

Aigle, M., Lacroute, F., 1975. Genetical aspects of [URE3], a non-mitochondrial, cytoplasmically inherited mutation in yeast. Mol. Gen. Genet. 136 (4), 327–335. Available at: http://www.ncbi.nlm.nih.gov/pubmed/16095000 (accessed May 25, 2016).

Alberti, S., et al., 2009. A systematic survey identifies prions and illuminates sequence features of prionogenic proteins. Cell 137 (1), 146–158. Available at: http://www.pubmedcentral.nih.

gov/articlerender.fcgi?artid=2683788&tool=pmcentrez&rendertype=abstract (accessed June 13, 2011).
Alper, T., et al., 1967. Does the agent of scrapie replicate without nucleic acid? Nature 214 (5090), 764–766. Available at: http://www.ncbi.nlm.nih.gov/pubmed/4963878 (accessed May 14, 2016).
Alpers, M.P., 1968. Kuru: implications of its transmissibility for the interpretation of its changing epidemiologic pattern. In: Bailey, O.T., Smith, D.E. (Eds.), The Central Nervous System, Some Experimental Models of Neurological Diseases. Proceedings of the Fifty-Sixth Annual Meeting of the International Academy of Pathology, Washington, DC, 12–15 March 1967. In: International Academy of Pathology Monograph, vol. 9. Williams and Wilkins, Baltimore, MD, pp. 234–251.
Alpers, M.P., 2008. Review. The epidemiology of kuru: monitoring the epidemic from its peak to its end. Philos. Trans. R. Soc. Lond. B Biol. Sci. 363 (1510), 3707–3713. Available at: http://www.pubmedcentral.nih.gov/articlerender.fcgi?artid=2577135&tool=pmcentrez&rendertype=abstract (accessed May 1, 2016).
Anon, 2009. A Life of Determination. Available at: http://www.med.monash.edu.au/news/2009/michael-alpers-biography.html.
Antonyuk, S.V., et al., 2009. Crystal structure of human prion protein bound to a therapeutic antibody. Proc. Natl. Acad. Sci. U.S.A. 106 (8), 2554–2558. Available at: http://www.ncbi.nlm.nih.gov/pubmed/19204296 (accessed July 21, 2016).
Avery, O.T., Macleod, C.M., McCarty, M., 1944. Studies on the chemical nature of the substance inducing transformation of pneumococcal types: induction of transformation by a desoxyribonucleic acid fraction isolated from pneumococcus type III. J. Exp. Med. 79 (2), 137–158. Available at: http://www.pubmedcentral.nih.gov/articlerender.fcgi?artid=2135445&tool=pmcentrez&rendertype=abstract (accessed March 18, 2016).
Barash, J.A., 2009. Clinical features of sporadic fatal insomnia. Rev. Neurol. Dis. 6 (3), E87–E93. Available at: http://www.ncbi.nlm.nih.gov/pubmed/19898273 (accessed May 19, 2016).
Barlow, R.M., 1972. Transmissible mink encephalopathy: pathogenesis and nature of the aetiological agent. J. Clin. Pathol. Suppl. (R. Coll. Pathol.) 6, 102–109. Available at: http://www.pubmedcentral.nih.gov/articlerender.fcgi?artid=1347258&tool=pmcentrez&rendertype=abstract (accessed May 18, 2016).
Barria, M.A., et al., 2009. De novo generation of infectious prions in vitro produces a new disease phenotype. PLoS Pathog. 5 (5), e1000421. Available at: http://journals.plos.org/plospathogens/article?id=10.1371/journal.ppat.1000421 (accessed May 19, 2016).
Basler, K., et al., 1986. Scrapie and cellular PrP isoforms are encoded by the same chromosomal gene. Cell 46 (3), 417–428. Available at: http://linkinghub.elsevier.com/retrieve/pii/0092867486906628 (accessed July 21, 2016).
Bastian, F.O., et al., 2007. Spiroplasma spp. from transmissible spongiform encephalopathy brains or ticks induce spongiform encephalopathy in ruminants. J. Med. Microbiol. 56 (Pt. 9), 1235–1242. Available at: http://www.ncbi.nlm.nih.gov/pubmed/17761489. accessed May 19, 2016).
Baxa, U., et al., 2002. Mechanism of inactivation on prion conversion of the Saccharomyces cerevisiae Ure2 protein. Proc. Natl. Acad. Sci. U.S.A. 99 (8), 5253–5260. Available at: http://www.ncbi.nlm.nih.gov/pubmed/17761489 (accessed May 25, 2016).
Baxa, U., et al., 2007. Characterization of beta-sheet structure in Ure2p1-89 yeast prion fibrils by solid-state nuclear magnetic resonance. Biochemistry 46 (45), 13149–13162. Available at: http://dx.doi.org/10.1021/bi700826b (accessed May 24, 2016).
Berryman, J.T., Radford, S.E., Harris, S.A., 2011. Systematic examination of polymorphism in amyloid fibrils by molecular-dynamics simulation. Biophys. J. 100 (9), 2234–2242. Available at: http://www.pubmedcentral.nih.gov/articlerender.fcgi?artid=3149254&tool=pmcentrez&rendertype=abstract (accessed May 26, 2016).

Bessen, R.A., Marsh, R.F., 1994. Distinct PrP properties suggest the molecular basis of strain variation in transmissible mink encephalopathy. J. Virol. 68 (12), 7859–7868. Available at: http://www.ncbi.nlm.nih.gov/pubmed/7966576 (accessed July 21, 2016).

Bondarev, S.A., et al., 2013. Effect of charged residues in the N-domain of Sup35 protein on prion [PSI+] stability and propagation. J. Biol. Chem. 288 (40), 28503–28513. Available at: http://www.ncbi.nlm.nih.gov/pubmed/23965990 (accessed January 7, 2014).

Brachman, P.S., 2003. Infectious diseases—past, present, and future. Int. J. Epidemiol. 32 (5), 684–686. Available at: http://ije.oxfordjournals.org/content/32/5/684.full (accessed December 10, 2015).

Brachmann, A., Toombs, J.A., Ross, E.D., 2006. Reporter assay systems for [URE3] detection and analysis. Methods (San Diego, Calif.) 39 (1), 35–42. Available at: http://www.sciencedirect.com/science/article/pii/S1046202306000545 (accessed May 24, 2016).

Brown, D.R., et al., 1997. The cellular prion protein binds copper in vivo. Nature 390 (6661), 684–687. Available at: http://www.ncbi.nlm.nih.gov/pubmed/9414160 (accessed October 20, 2011).

Bruce, M.E., et al., 1997. Transmissions to mice indicate that "new variant" CJD is caused by the BSE agent. Nature 389 (6650), 498–501. Available at: http://www.ncbi.nlm.nih.gov/pubmed/9333239 (accessed May 18, 2016).

Bryan, A.W., et al., 2009. BETASCAN: probable beta-amyloids identified by pairwise probabilistic analysis. PLoS Comput. Biol. 5 (3), e1000333. Available at: http://www.pubmedcentral.nih.gov/articlerender.fcgi?artid=2653728&tool=pmcentrez&rendertype=abstract (accessed May 2, 2016).

Büeler, H., et al., 1993. Mice devoid of PrP are resistant to scrapie. Cell 73 (7), 1339–1347. Available at: http://www.ncbi.nlm.nih.gov/pubmed/8100741 (accessed March 3, 2016).

Burger, D., Hartsough, G.R., 1965. Encephalopathy of mink. II. Experimental and natural transmission. J. Infect. Dis. 115 (4), 393–399. Available at: http://www.ncbi.nlm.nih.gov/pubmed/5837893 (accessed May 18, 2016).

Carlson, G.A., et al., 1986. Linkage of prion protein and scrapie incubation time genes. Cell 46 (4), 503–511. Available at: http://www.ncbi.nlm.nih.gov/pubmed/3015416 (accessed May 19, 2016).

Carlson, G.A., et al., 1988. Genetic control of prion incubation period in mice. Ciba Found. Symp. 135, 84–99. Available at: http://www.ncbi.nlm.nih.gov/pubmed/2900721 (accessed May 19, 2016).

Cascarina, S.M., Ross, E.D., 2014. Yeast prions and human prion-like proteins: sequence features and prediction methods. Cell. Mol. Life Sci. 71 (11), 2047–2063. Available at: http://www.ncbi.nlm.nih.gov/pubmed/24390581 (accessed June 24, 2014).

Chakrabortee, S., et al., 2016. Luminidependens (LD) is an Arabidopsis protein with prion behavior. Proc. Natl. Acad. Sci. U.S.A. 113 (21), 6065–6070. Available at: http://www.pnas.org/content/113/21/6065.full (accessed May 10, 2016).

Chandler, R.L., 1961. Encephalopathy in mice produced by inoculation with scrapie brain material. Lancet (London, England) 1 (7191), 1378–1379. Available at: http://www.ncbi.nlm.nih.gov/pubmed/13692303 (accessed May 16, 2016).

Chen, B., et al., 2009. Measurement of amyloid fibril mass-per-length by tilted-beam transmission electron microscopy. Proc. Natl. Acad. Sci. U.S.A. 106 (34), 14339–14344. Available at: http://www.pubmedcentral.nih.gov/articlerender.fcgi?artid=2732815&tool=pmcentrez&rendertype=abstract (accessed May 24, 2016).

Chernoff, Y.O., 2016. Are there prions in plants? Proc. Natl. Acad. Sci. U.S.A. 113 (22), 6097–6099. http://dx.doi.org/10.1073/pnas.1605671113. Available at: http://www.ncbi.nlm.nih.gov/pubmed/27217577 (accessed May 25, 2016).

Chernoff, Y.O., Derkach, I.L., Inge-Vechtomov, S.G., 1993. Multicopy SUP35 gene induces de-novo appearance of psi-like factors in the yeast Saccharomyces cerevisiae.

Curr. Genet. 24 (3), 268–270. Available at: http://www.ncbi.nlm.nih.gov/pubmed/8221937 (accessed May 25, 2016).
Chernoff, Y.O., et al., 1995. Role of the chaperone protein Hsp104 in propagation of the yeast prion-like factor [psi+]. Science (New York, N.Y.) 268 (5212), 880–884. Available at: http://www.ncbi.nlm.nih.gov/pubmed/7754373 (accessed October 19, 2011).
Chiti, F., Dobson, C.M., 2006. Protein misfolding, functional amyloid, and human disease. Annu. Rev. Biochem. 75, 333–366. Available at: http://www.annualreviews.org/doi/full/10.1146/annurev.biochem.75.101304.123901 (accessed July 15, 2011).
Cho, H.J., 1980. Requirement of a protein component for scrapie infectivity. Intervirology 14 (3–4), 213–216. Available at: http://www.ncbi.nlm.nih.gov/pubmed/6786997 (accessed May 14, 2016).
Clavaguera, F., et al., 2009. Transmission and spreading of tauopathy in transgenic mouse brain. Nat. Cell Biol. 11 (7), 909–913. Available at: http://www.nature.com/doifinder/10.1038/ncb1901 (accessed July 21, 2016).
Cohen, S.S., Stanley, W.M., 1942. The molecular size and shape of the nucleic acid of tobacco mosaic virus. J. Biol. Chem. 144, 589–598. Available at: http://www.jbc.org/content/144/3/589.full.pdf (accessed May 12, 2016).
Colby, D.W., Prusiner, S.B., 2011a. De novo generation of prion strains. Nature Rev. Microbiol. 9 (11), 771–777. Available at: http://www.pubmedcentral.nih.gov/articlerender.fcgi?artid=3924856&tool=pmcentrez&rendertype=abstract (accessed April 21, 2016).
Colby, D.W., Prusiner, S.B., 2011b. Prions. Cold Spring Harb. Perspect. Biol. 3 (1), a006833. Available at: http://www.pubmedcentral.nih.gov/articlerender.fcgi?artid=3003464&tool=pmcentrez&rendertype=abstract (accessed February 12, 2016).
Collinge, J., et al., 1996. Molecular analysis of prion strain variation and the aetiology of "new variant" CJD. Nature 383 (6602), 685–690. Available at: http://www.ncbi.nlm.nih.gov/pubmed/8878476 (accessed July 21, 2016).
Coustou, V., et al., 1997. The protein product of the het-s heterokaryon incompatibility gene of the fungus Podospora anserina behaves as a prion analog. Proc. Natl. Acad. Sci. U.S.A. 94 (18), 9773–9778. Available at: http://www.pubmedcentral.nih.gov/articlerender.fcgi?artid=23266&tool=pmcentrez&rendertype=abstract (accessed May 3, 2016).
Couthouis, J., et al., 2012. Evaluating the role of the FUS/TLS-related gene EWSR1 in amyotrophic lateral sclerosis. Hum. Mol. Genet. 21 (13), 2899–2911. Available at: http://hmg.oxfordjournals.org/content/21/13/2899 (accessed May 26, 2016).
Cox, B.S., 1965. Ψ, A cytoplasmic suppressor of super-suppressor in yeast. Heredity 20 (4), 505–521. Available at: http://dx.doi.org/10.1038/hdy.1965.65 (accessed June 9, 2014).
Cox, B.S., Tuite, M.F., McLaughlin, C.S., 1988. The psi factor of yeast: a problem in inheritance. Yeast (Chichester, England) 4 (3), 159–178. Available at: http://www.ncbi.nlm.nih.gov/pubmed/3059716 (accessed May 25, 2016).
Creutzfeldt, H.G., 1920. Über eine eigenartige herdförmige Erkrankung des Zentralnervensystems. Z Gesamte Neurol Psychiatr 57, 1–18.
Cuille, J., Chelle, P.L., 1936. Pathologie animale—La maladie dite tremblante du mouton est-elle inoculable? C. R. Hebd. Seances Acad. Sci. 203, 1552–1554.
Cuille, J., Chelle, P.L., 1938a. Investigations of scrapie in sheep. Vet. Med. 34, 417–418.
Cuille, J., Chelle, P.L., 1938b. La tremblante du mouton est-elle determinee par un virus filtrable? C. R. Acad. Sci. Hebd. Seances Acad. Sci. 206, 1687–1688.
Cuille, J., Chelle, P.L., 1938c. Le tremblante du mouton est bien inoculable. C. R. Acad. Sci. Hebd. Seances Acad. Sci. 206, 78–79.
Cuille, J., Chelle, P.L., 1939. Transmission experimentale de la tremblante a la chevre. C. R. Acad. Sci. Hebd. Seances Acad. Sci. 208, 1058–1060.

Da Cruz, S., Cleveland, D.W., 2011. Understanding the role of TDP-43 and FUS/TLS in ALS and beyond. Curr. Opin. Neurobiol. 21 (6), 904–919. Available at: http://www.sciencedirect.com/science/article/pii/S0959438811000973 (accessed May 26, 2016).

Daigle, J.G., et al., 2013. RNA-binding ability of FUS regulates neurodegeneration, cytoplasmic mislocalization and incorporation into stress granules associated with FUS carrying ALS-linked mutations. Hum. Mol. Genet. 22 (6), 1193–1205. Available at: http://hmg.oxfordjournals.org/content/22/6/1193 (accessed May 26, 2016).

David, M., Tayebi, M., 2012. Canine spongiform encephalopathy—a new form of animal prion disease. Prion 6, 6.

Deleault, N.R., et al., 2005. Protease-resistant prion protein amplification reconstituted with partially purified substrates and synthetic polyanions. J. Biol. Chem. 280 (29), 26873–26879. Available at: http://www.ncbi.nlm.nih.gov/pubmed/15917229 (accessed May 19, 2016).

DePace, A.H., et al., 1998. A critical role for amino-terminal glutamine/asparagine repeats in the formation and propagation of a yeast prion. Cell 93 (7), 1241–1252. Available at: http://www.ncbi.nlm.nih.gov/pubmed/9657156.

Derkatch, I.L., et al., 1996. Genesis and variability of [PSI] prion factors in Saccharomyces cerevisiae. Genetics 144 (4), 1375–1386. Available at: http://www.pubmedcentral.nih.gov/articlerender.fcgi?artid=1207691&tool=pmcentrez&rendertype=abstract (accessed May 25, 2016).

Derkatch, I.L., et al., 1997. Genetic and environmental factors affecting the de novo appearance of the [PSI+] prion in Saccharomyces cerevisiae. Genetics 147 (2), 507–519. Available at: http://www.pubmedcentral.nih.gov/articlerender.fcgi?artid=1208174&tool=pmcentrez&rendertype=abstract (accessed May 26, 2016).

Derkatch, I.L., et al., 2000. Dependence and independence of [PSI(+)] and [PIN(+)]: a two-prion system in yeast? EMBO J. 19 (9), 1942–1952. Available at: http://www.pubmedcentral.nih.gov/articlerender.fcgi?artid=305693&tool=pmcentrez&rendertype=abstract (accessed May 26, 2016).

Derkatch, I.L., et al., 2001. Prions affect the appearance of other prions: the story of [PIN(+)]. Cell 106 (2), 171–182. Available at: http://www.ncbi.nlm.nih.gov/pubmed/11511345 (accessed September 14, 2011).

Dickinson, A.G., Meikle, V.M.H., 1969. A comparison of some biological characteristics of the mouse-passaged scrapie agents, 22A and ME7. Genet. Res. 13 (2), 213. Available at: http://www.journals.cambridge.org/abstract_S0016672300002895 (accessed July 21, 2016).

Dickinson, A.G., Meikle, V.M., Fraser, H., 1968. Identification of a gene which controls the incubation period of some strains of scrapie agent in mice. J. Comp. Pathol. 78 (3), 293–299. Available at: http://www.ncbi.nlm.nih.gov/pubmed/4970191 (accessed May 19, 2016).

Dimitz, L., 1913. Bericht der Vereines fur Psychiatrie und Neurologie in Wien (Vereinsjahr 1912/1913), Sitzung vom 11 Juni 1912. Jahrb. Psychiatr. Neurol. 34, 384.

Dlouhy, S.R., et al., 1992. Linkage of the Indiana kindred of Gerstmann-Sträussler-Scheinker disease to the prion protein gene. Nat. Genet. 1 (1), 64–67. Available at: http://www.ncbi.nlm.nih.gov/pubmed/1363809 (accessed May 19, 2016).

Dobson, C.M., 2003. Protein folding and misfolding. Nature 426 (6968), 884–890. Available at: http://www.ncbi.nlm.nih.gov/pubmed/14685248 (accessed October 14, 2014).

Doel, S.M., et al., 1994. The dominant PNM2- mutation which eliminates the psi factor of Saccharomyces cerevisiae is the result of a missense mutation in the SUP35 gene. Genetics 137 (3), 659–670. Available at: http://www.pubmedcentral.nih.gov/articlerender.fcgi?artid=1206025&tool=pmcentrez&rendertype=abstract (accessed February 15, 2012).

Dong, J., et al., 2007. Probing the role of PrP repeats in conformational conversion and amyloid assembly of chimeric yeast prions. J. Biol. Chem. 282 (47), 34204–34212. Available at: http://www.pubmedcentral.nih.gov/articlerender.fcgi?artid=2262835&tool=pmcentrez&rendertype=abstract (accessed October 28, 2011).

Du, Z., et al., 2008. Newly identified prion linked to the chromatin-remodeling factor Swi1 in Saccharomyces cerevisiae. Nat. Genet. 40 (4), 460–465. Available at: http://www.pubmedcentral.nih.gov/articlerender.fcgi?artid=2633598&tool=pmcentrez&rendertype=abstract (accessed May 26, 2016).

Eanes, E.D., Glenner, G.G., 1968. X-ray diffraction studies on amyloid filaments. J. Histochem. Cytochem. 16 (11), 673–677. Available at: http://www.ncbi.nlm.nih.gov/pubmed/5723775 (accessed May 24, 2016).

Eiden, M., et al., 2010. Biochemical and immunohistochemical characterization of feline spongiform encephalopathy in a German captive cheetah. J. Gen. Virol. 91 (Pt. 11), 2874–2883. Available at: http://jgv.microbiologyresearch.org/content/journal/jgv/10.1099/vir.0.022103-0#tab2 (accessed May 16, 2016).

Eisenberg, D., Jucker, M., 2012. The amyloid state of proteins in human diseases. Cell 148 (6), 1188–1203. Available at: http://www.pubmedcentral.nih.gov/articlerender.fcgi?artid=3353745&tool=pmcentrez&rendertype=abstract (accessed February 15, 2016).

Engel, A., et al., 2011. Amyloid of the Candida albicans Ure2p prion domain is infectious and has an in-register parallel β-sheet structure. Biochemistry 50 (27), 5971–5978. Available at: http://www.pubmedcentral.nih.gov/articlerender.fcgi?artid=3144561&tool=pmcentrez&rendertype=abstract (accessed May 24, 2016).

Esteras-Chopo, A., Serrano, L., López de la Paz, M., 2005. The amyloid stretch hypothesis: recruiting proteins toward the dark side. Proc. Natl. Acad. Sci. U.S.A. 102 (46), 16672–16677. Available at: http://www.pubmedcentral.nih.gov/articlerender.fcgi?artid=1283810&tool=pmcentrez&rendertype=abstract (accessed May 26, 2016).

Fernandez-Escamilla, A.-M., et al., 2004. Prediction of sequence-dependent and mutational effects on the aggregation of peptides and proteins. Nat. Biotechnol. 22 (10), 1302–1306. Available at: http://www.ncbi.nlm.nih.gov/pubmed/15361882 (accessed May 26, 2016).

Ferreira, P.C., et al., 2001. The elimination of the yeast [PSI+] prion by guanidine hydrochloride is the result of Hsp104 inactivation. Mol. Microbiol. 40 (6), 1357–1369. Available at: http://www.ncbi.nlm.nih.gov/pubmed/11442834 (accessed October 20, 2011).

Flechsig, E., et al., 2000. Prion protein devoid of the octapeptide repeat region restores susceptibility to scrapie in PrP knockout mice. Neuron 27 (2), 399–408. Available at: http://www.ncbi.nlm.nih.gov/pubmed/10985358 (accessed October 28, 2011).

Fraser, H., Dickinson, A.G., 1973. Scrapie in mice. J. Comp. Pathol. 83 (1), 29–40. Available at: http://linkinghub.elsevier.com/retrieve/pii/0021997573900248 (accessed July 21, 2016).

Gajdusek, D.C., Zigas, V., 1959. Kuru. Am. J. Med. 26, 442–469.

Gerstmann, J., 1928. über ein noch nicht beschriebenes Reflexphanomen beieiner Erkrankung des zerebellaren Systems. Wien Medizin Wochenschr 78, 906–908.

Gerstmann, J., Sträussler, E., Scheinker, I., 1936. Über eine eigenartige hereditär- familiäre Erkrankung des Zentralnervensystems. Zeitschrift für die gesamte Neurologie und Psychiatrie 154 (1), 736–762. Available at: http://link.springer.com/10.1007/BF02865827 (accessed May 19, 2016).

Glenner, G.G., et al., 1969. Physical and chemical properties of amyloid fibers. II. Isolation of a unique protein constituting the major component from human splenic amyloid fibril concentrates. J. Histochem. Cytochem. 17 (12), 769–780. Available at: http://www.ncbi.nlm.nih.gov/pubmed/4983715 (accessed May 24, 2016).

Goldschmidt, L., et al., 2010. Identifying the amylome, proteins capable of forming amyloid-like fibrils. Proc. Natl. Acad. Sci. U.S.A. 107 (8), 3487–3492. Available at: http://www.

pubmedcentral.nih.gov/articlerender.fcgi?artid=2840437&tool=pmcentrez&rendertype=abstract (accessed March 28, 2016).
Gonzalez Nelson, A.C., Ross, E.D., 2011. Interactions between non-identical prion proteins. Semin. Cell Dev. Biol., 22 (5), 437–443. http://dx.doi.org/10.1016/j.semcdb.2011.02.022. Available at: http://www.ncbi.nlm.nih.gov/pubmed/21354317.
Greenlee, J.J., Greenlee, M.H.W., 2015. The transmissible spongiform encephalopathies of livestock. ILAR J. 56 (1), 7–25. Available at: http://www.ncbi.nlm.nih.gov/pubmed/25991695 (accessed May 18, 2016).
Gregersen, N., Bolund, L., Bross, P., 2005. Protein misfolding, aggregation, and degradation in disease. Mol. Biotechnol. 31 (2), 141–150. Available at: http://www.ncbi.nlm.nih.gov/pubmed/16170215 (accessed May 24, 2016).
Griffith, F., 1928. The significance of pneumococcal types. J. Hyg. 27 (2), 113–159. Available at: http://www.pubmedcentral.nih.gov/articlerender.fcgi?artid=2167760&tool=pmcentrez&rendertype=abstract (accessed January 14, 2015).
Griffith, J.S., 1967. Self-replication and scrapie. Nature 215 (5105), 1043–1044. Available at: http://www.ncbi.nlm.nih.gov/pubmed/4964084 (accessed May 14, 2016).
Groveman, B.R., et al., 2014. Parallel in-register intermolecular β-sheet architectures for prion-seeded prion protein (PrP) amyloids. J. Biol. Chem. 289 (35), 24129–24142. Available at: http://www.pubmedcentral.nih.gov/articlerender.fcgi?artid=4148845&tool=pmcentrez&rendertype=abstract (accessed May 24, 2016).
Haass, C., et al., 1995. The Swedish mutation causes early-onset Alzheimer's disease by β-secretase cleavage within the secretory pathway. Nat. Med. 1 (12), 1291–1296. Available at: http://www.nature.com/doifinder/10.1038/nm1295-1291 (accessed July 21, 2016).
Hadlow, W.J., 1959. Scrapie and kuru. Lancet 274 (7097), 289–290. Available at: http://www.thelancet.com/article/S0140673659920811/fulltext (accessed May 14, 2016).
Hadlow, W.J., et al., 1980. Brain tissue from persons dying of Creutzfeldt-Jakob disease causes scrapie-like encephalopathy in goats. Ann. Neurol. 8 (6), 628–632. Available at: http://www.ncbi.nlm.nih.gov/pubmed/7011169 (accessed May 14, 2016).
Halfmann, R., et al., 2011. Opposing effects of glutamine and asparagine govern prion formation by intrinsically disordered proteins. Mol. Cell 43 (1), 72–84. Available at: http://www.pubmedcentral.nih.gov/articlerender.fcgi?artid=3132398&tool=pmcentrez&rendertype=abstract (accessed April 1, 2016).
Harbi, D., Harrison, P.M., 2014. Classifying prion and prion-like phenomena. Prion 8 (2), 161–165. http://dx.doi.org/10.4161/pri.27960. Available at: http://www.pubmedcentral.nih.gov/articlerender.fcgi?artid=4189883&tool=pmcentrez&rendertype=abstract (accessed May 18, 2016).
Harper, J.D., et al., 1997. Observation of metastable Abeta amyloid protofibrils by atomic force microscopy. Chem. Biol. 4 (2), 119–125. Available at: http://www.ncbi.nlm.nih.gov/pubmed/9190286 (accessed May 24, 2016).
Harrison, P.M., Gerstein, M., 2003. A method to assess compositional bias in biological sequences and its application to prion-like glutamine/asparagine-rich domains in eukaryotic proteomes. Genome Biol. 4 (6), R40. Available at: http://www.pubmedcentral.nih.gov/articlerender.fcgi?artid=193619&tool=pmcentrez&rendertype=abstract.
Hartsough, G.R., Burger, D., 1965. Encephalopathy of mink. I. Epizootiologic and clinical observations. J. Infect. Dis. 115 (4), 387–392. Available at: http://www.ncbi.nlm.nih.gov/pubmed/5891240 (accessed May 18, 2016).
Hershey, A.D., Chase, M., 1952. Independent functions of viral protein and nucleic acid in growth of bacteriophage. J. Gen. Physiol. 36 (1), 39–56. Available at: http://www.pubmedcentral.nih.gov/articlerender.fcgi?artid=2147348&tool=pmcentrez&rendertype=abstract (accessed April 25, 2016).

Hörnlimann, B., Riesner, D., 2007. Prions in Humans and Animals. Walter de Gruyter. Available at: https://books.google.com/books?id=Nh7E_gUV7bAC&pgis=1 (accessed May 18, 2016).

Howie, A.J., 2015. "Green (or apple-green) birefringence" of Congo red-stained amyloid. Amyloid, 22 (3), 205–206. http://dx.doi.org/10.3109/13506129.2015.1054026. Available at: http://www.tandfonline.com/doi/full/10.3109/13506129.2015.1054026 (accessed May 24, 2016).

Hsiao, K., et al., 1989. Linkage of a prion protein missense variant to Gerstmann-Sträussler syndrome. Nature 338 (6213), 342–345. Available at: http://www.nature.com/nature/journal/v338/n6213/pdf/338342a0.pdf (accessed April 5, 2016).

Hsiao, K.K., et al., 1990. Spontaneous neurodegeneration in transgenic mice with mutant prion protein. Science (New York, N.Y.) 250 (4987), 1587–1590. Available at: http://www.ncbi.nlm.nih.gov/pubmed/1980379 (accessed May 19, 2016).

Hsiao, K.K., et al., 1991. A prion protein variant in a family with the telencephalic form of Gerstmann-Sträussler-Scheinker syndrome. Neurology 41 (5), 681–684. Available at: http://www.ncbi.nlm.nih.gov/pubmed/1674116 (accessed May 19, 2016).

Huang, V.J., Stein, K.C., True, H.L., 2013. Spontaneous variants of the [RNQ+] prion in yeast demonstrate the extensive conformational diversity possible with prion proteins. J. Ma, Ed PLoS One 8 (10), e79582. Available at: http://dx.plos.org/10.1371/journal.pone.0079582 (accessed November 11, 2013).

Hunter, G.D., et al., 1969. Further studies of the infectivity and stability of extracts and homogenates derived from scrapie affected mouse brains. J. Comp. Pathol. 79 (1), 101–108. Available at: http://www.ncbi.nlm.nih.gov/pubmed/4304706 (accessed May 14, 2016).

Hunter, N., et al., 1987. Linkage of the scrapie-associated fibril protein (PrP) gene and Sinc using congenic mice and restriction fragment length polymorphism analysis. J. Gen. Virol. 68 (Pt. 10), 2711–2716. Available at: http://www.ncbi.nlm.nih.gov/pubmed/2889794 (accessed May 19, 2016).

Jackson, W.S., et al., 2009. Spontaneous generation of prion infectivity in fatal familial insomnia knockin mice. Neuron 63 (4), 438–450. Available at: http://www.pubmedcentral.nih.gov/articlerender.fcgi?artid=2775465&tool=pmcentrez&rendertype=abstract (accessed May 19, 2016).

Jakob, A., 1921. Über eigenartige Erkrankungen des Zentralnervensystems mit bemerkenswertem anatomischem Befunde (Spastische Pseudosklerose-Encephalomyelopathie mit disseminierten Degenerationsherden). Z. Gesamte Neurol. Psychiatr. 64, 147–228.

James, T.L., et al., 1997. Solution structure of a 142-residue recombinant prion protein corresponding to the infectious fragment of the scrapie isoform. Proc. Natl. Acad. Sci. U.S.A. 94 (19), 10086–10091. Available at: http://www.ncbi.nlm.nih.gov/pubmed/9294167 (accessed July 21, 2016).

Johnson, B.S., et al., 2008. A yeast TDP-43 proteinopathy model: exploring the molecular determinants of TDP-43 aggregation and cellular toxicity. Proc. Natl. Acad. Sci. U.S.A. 105 (17), 6439–6444. Available at: http://www.pnas.org/content/105/17/6439 (accessed April 13, 2016).

Johnson, B.S., et al., 2009. TDP-43 is intrinsically aggregation-prone, and amyotrophic lateral sclerosis-linked mutations accelerate aggregation and increase toxicity. J. Biol. Chem. 284 (30), 20329–20339. Available at: http://www.jbc.org/content/284/30/20329 (accessed May 26, 2016).

Jones, E.W., 1991. Three proteolytic systems in the yeast saccharomyces cerevisiae. J. Biol. Chem. 13, 7963–7966. Available at: http://www.ncbi.nlm.nih.gov/pubmed/2022624 (accessed May 24, 2016).

Jucker, M., Walker, L.C., 2011. Pathogenic protein seeding in Alzheimer disease and other neurodegenerative disorders. Ann. Neurol. 70 (4), 532–540. Available at: http://www.ncbi.nlm.nih.gov/pubmed/22028219 (accessed July 20, 2016).

Kalastavadi, T., True, H.L., 2008. Prion protein insertional mutations increase aggregation propensity but not fiber stability. BMC Biochem. 9, 7. Available at: http://www.pubmedcentral.nih.gov/articlerender.fcgi?artid=2276218&tool=pmcentrez&rendertype=abstract (accessed September 28, 2011).

Kayed, R., Lasagna-Reeves, C.A., 2013. Molecular mechanisms of amyloid oligomers toxicity. J. Alzheimer's Dis. 33 (Suppl. 1(s1)), S67–S78. Available at: http://content.iospress.com/articles/journal-of-alzheimers-disease/jad129001 (accessed May 17, 2016).

Kim, H.J., et al., 2013. Mutations in prion-like domains in hnRNPA2B1 and hnRNPA1 cause multisystem proteinopathy and ALS. Nature. 495, 467–473. Available at: http://dx.doi.org/10.1038/nature11922 (accessed March 5, 2013).

Kimberlin, R.H., Walker, C., 1977. Characteristics of a short incubation model of scrapie in the golden hamster. J. Gen. Virol. 34 (2), 295–304. Available at: http://www.ncbi.nlm.nih.gov/pubmed/402439 (accessed May 16, 2016).

King, C.-Y., Diaz-Avalos, R., 2004. Protein-only transmission of three yeast prion strains. Nature 428 (6980), 319–323. Available at: http://www.ncbi.nlm.nih.gov/pubmed/15029195 (accessed July 21, 2016).

King, C.Y., et al., 1997. Prion-inducing domain 2–114 of yeast Sup35 protein transforms in vitro into amyloid-like filaments. Proc. Natl. Acad. Sci. U.S.A. 94 (13), 6618–6622. Available at: http://www.pubmedcentral.nih.gov/articlerender.fcgi?artid=21207&tool=pmcentrez&rendertype=abstract (accessed May 24, 2016).

Kitamoto, T., et al., 1986. Scrapie-associated fibrils (SAF) purification method yields amyloid proteins from systemic and cerebral amyloidosis. Biosci. Rep. 6 (5), 459–465. Available at: http://www.ncbi.nlm.nih.gov/pubmed/2874846 (accessed May 24, 2016).

Klatzo, I., Gajdusek, D.C., Zigas, V., 1959. Pathology of kuru. Lab. Invest. 4, 799–847.

Knowles, T.P.J., Vendruscolo, M., Dobson, C.M., 2014. The amyloid state and its association with protein misfolding diseases. Nat. Rev. Mol. Cell Biol. 15 (6), 384–396. Available at: http://dx.doi.org/10.1038/nrm3810 (accessed July 11, 2014).

Kordower, J.H., et al., 2008. Lewy body-like pathology in long-term embryonic nigral transplants in Parkinson's disease. Nat. Med. 14 (5), 504–506. Available at: http://www.nature.com/doifinder/10.1038/nm1747 (accessed July 21, 2016).

Kryndushkin, D., et al., 2013. Non-targeted identification of prions and amyloid-forming proteins from yeast and mammalian cells. J. Biol. Chem. 288 (38), 27100–27111. Available at: http://www.ncbi.nlm.nih.gov/pubmed/23926098 (accessed December 3, 2013).

Kwiatkowski, T.J., et al., 2009. Mutations in the FUS/TLS gene on chromosome 16 cause familial amyotrophic lateral sclerosis. Science (New York, N.Y.) 323 (5918), 1205–1208. Available at: http://science.sciencemag.org/content/323/5918/1205.abstract (accessed February 4, 2016).

Lacroute, F., 1971. Non-Mendelian mutation allowing ureidosuccinic acid uptake in yeast. J. Bacteriol. 106 (2), 519–522. Available at: http://www.pubmedcentral.nih.gov/articlerender.fcgi?artid=285125&tool=pmcentrez&rendertype=abstract.

Lagier-Tourenne, C., Polymenidou, M., Cleveland, D.W., 2010. TDP-43 and FUS/TLS: emerging roles in RNA processing and neurodegeneration. Hum. Mol. Genet. 19 (R1), R46–R64. Available at: http://hmg.oxfordjournals.org/content/19/R1/R46 (accessed May 26, 2016).

Legname, G., et al., 2004. Synthetic mammalian prions. Science (New York, N.Y.) 305 (5684), 673–676. Available at: http://www.ncbi.nlm.nih.gov/pubmed/15286374 (accessed May 19, 2016).

Leopoldt, J.G., 1750. Nützliche und auf die Erfahrung gegründete Einleitung zu der LandWirthschafft.

Levene, P.A., 1919. The structure of yeast nucleic acid. IV. Ammonia hydrolysis. J. Biol. Chem. 40 (2), 415–424. Available at: http://www.jbc.org/content/40/2/415.citation (accessed May 12, 2016).

Li, L., Lindquist, S., 2000. Creating a protein-based element of inheritance. Science (New York, N.Y.) 287 (5453), 661–664. Available at: http://www.ncbi.nlm.nih.gov/pubmed/10650001 (accessed May 25, 2016).

Liberski, P.P., 2012. Gerstmann-Sträussler-Scheinker disease. Adv. Exp. Med. Biol. 724, 128–137. Available at: http://www.ncbi.nlm.nih.gov/pubmed/22411239 (accessed May 19, 2016).

Lindenbaum, S., 2008. Review. Understanding kuru: the contribution of anthropology and medicine. Philos. Trans. R. Soc. Lond. Ser. B: Biol. Sci. 363 (1510), 3715–3720. Available at: http://www.pubmedcentral.nih.gov/articlerender.fcgi?artid=2735506&tool=pmcentrez&rendertype=abstract (accessed May 18, 2016).

Lindquist, S., et al., 1995. The role of Hsp104 in stress tolerance and [PSI+] propagation in Saccharomyces cerevisiae. Cold Spring Harb. Symp. Quant. Biol. 60, 451–460. Available at: http://www.ncbi.nlm.nih.gov/pubmed/8824419 (accessed October 19, 2011).

Lugaresi, E., et al., 1986. Fatal familial insomnia and dysautonomia with selective degeneration of thalamic nuclei. N. Engl. J. Med. 315 (16), 997–1003. Available at: http://www.ncbi.nlm.nih.gov/pubmed/3762620 (accessed May 19, 2016).

Lund, P.M., Cox, B.S., 1981. Reversion analysis of [psi−] mutations in Saccharomyces cerevisiae. Genet. Res. 37 (02), 173–182. Available at: http://journals.cambridge.org/abstract_S0016672300020140 (accessed May 25, 2016).

Lyubchenko, Y.L., Krasnoslobodtsev, A.V., Luca, S., 2012. Fibrillogenesis of huntingtin and other glutamine containing proteins. Subcell. Biochem. 65, 225–251. Available at: http://www.pubmedcentral.nih.gov/articlerender.fcgi?artid=4226413&tool=pmcentrez&rendertype=abstract (accessed May 24, 2016).

MacLea, K.S., Ross, E.D., 2011. Strategies for identifying new prions in yeast. Prion 5 (4), 1–6. Available at: http://www.landesbioscience.com/journals/prion/article/17918/ (accessed October 19, 2011).

MacLea, K.S., et al., 2015. Distinct amino acid compositional requirements for formation and maintenance of the [PSI+] prion in yeast. Mol. Cell. Biol. 35 (5), 899–911. Available at: http://www.ncbi.nlm.nih.gov/pubmed/25547291 (accessed January 6, 2015).

Maji, S.K., et al., 2009. Structure-activity relationship of amyloid fibrils. FEBS Lett. 583 (16), 2610–2617. Available at: http://www.ncbi.nlm.nih.gov/pubmed/19596006 (accessed May 24, 2016).

Málaga-Trillo, E., et al., 2011. Fish models in prion biology: underwater issues. Biochim. Biophys. Acta 1812 (3), 402–414. Available at: http://www.sciencedirect.com/science/article/pii/S0925443910002139 (accessed April 10, 2016).

Manuelidis, L., 2007. A 25 nm virion is the likely cause of transmissible spongiform encephalopathies. J. Cell. Biochem. 100 (4), 897–915. Available at: http://www.ncbi.nlm.nih.gov/pubmed/17044041 (accessed May 19, 2016).

Manuelidis, L., 2013. Infectious particles, stress, and induced prion amyloids: a unifying perspective. Virulence 4 (5), 373–383. Available at: http://www.pubmedcentral.nih.gov/articlerender.fcgi?artid=3714129&tool=pmcentrez&rendertype=abstract (accessed July 3, 2013).

Manuelidis, L., Valley, S., Manuelidis, E.E., 1985. Specific proteins associated with Creutzfeldt-Jakob disease and scrapie share antigenic and carbohydrate determinants. Proc. Natl. Acad. Sci. U.S.A. 82 (12), 4263–4267. Available at: http://www.pubmedcentral.nih.gov/articlerender.fcgi?artid=397977&tool=pmcentrez&rendertype=abstract (accessed May 24, 2016).

Manuelidis, L., Liu, Y., Mullins, B., 2009. Strain-specific viral properties of variant Creutzfeldt-Jakob disease (vCJD) are encoded by the agent and not by host prion protein. J. Cell.

Biochem. 106 (2), 220–231. Available at: http://www.pubmedcentral.nih.gov/articlerender.fcgi?artid=2762821&tool=pmcentrez&rendertype=abstract (accessed May 19, 2016).
Marcelino-Cruz, A.M., et al., 2011. Site-specific structural analysis of a yeast prion strain with species-specific seeding activity. Prion 5 (3), 208–214. Available at: http://www.ncbi.nlm.nih.gov/pubmed/22048721 (accessed November 4, 2011).
Marsh, R.F., Hanson, R.P., 1969. Physical and chemical properties of the transmissible mink encephalopathy agent. J. Virol. 3 (2), 176–180. Available at: http://www.pubmedcentral.nih.gov/articlerender.fcgi?artid=375749&tool=pmcentrez&rendertype=abstract (accessed May 18, 2016).
Marzewski, D.J., et al., 1988. Creutzfeldt-Jakob disease following pituitary-derived human growth hormone therapy: a new American case. Neurology 38 (7), 1131–1133. Available at: http://www.ncbi.nlm.nih.gov/pubmed/3290703 (accessed May 19, 2016).
Masison, D.C., Wickner, R.B., 1995. Prion-inducing domain of yeast Ure2p and protease resistance of Ure2p in prion-containing cells. Science 270 (5233), 93–95. Available at: http://www.ncbi.nlm.nih.gov/pubmed/7569955.
Mathur, V., et al., 2012. Localization of HET-S to the cell periphery, and not to [Het-s] aggregates, is associated with [Het-s]-HET-S toxicity. Mol. Cell. Biol. 32 (1), 139–153. http://dx.doi.org/10.1128/MCB.06125-11. Available at: http://www.ncbi.nlm.nih.gov/pubmed/22037764 (accessed November 1, 2011).
Matthews, W.B., 1990. Bovine spongiform encephalopathy. BMJ 300 (6722), 412–413. Available at: http://www.pubmedcentral.nih.gov/articlerender.fcgi?artid=1662219&tool=pmcentrez&rendertype=abstract (accessed May 18, 2016).
Maurer-Stroh, S., et al., 2010. Exploring the sequence determinants of amyloid structure using position-specific scoring matrices. Nat. Methods 7 (3), 237–242. Available at: http://www.ncbi.nlm.nih.gov/pubmed/20154676 (accessed May 26, 2016).
McGlinchey, R.P., Kryndushkin, D., Wickner, R.B., 2011. Suicidal [PSI+] is a lethal yeast prion. Proc. Natl. Acad. Sci. U.S.A. 108 (13), 5337–5341. Available at: http://www.pubmedcentral.nih.gov/articlerender.fcgi?artid=3069153&tool=pmcentrez&rendertype=abstract (accessed October 13, 2014).
McKinley, M.P., Bolton, D.C., Prusiner, S.B., 1983. A protease-resistant protein is a structural component of the scrapie prion. Cell 35 (1), 57–62. Available at: http://www.ncbi.nlm.nih.gov/pubmed/6414721 (accessed May 24, 2016).
Merz, P.A., et al., 1983. Scrapie-associated fibrils in Creutzfeldt-Jakob disease. Nature 306 (5942), 474–476. Available at: http://www.ncbi.nlm.nih.gov/pubmed/6358899 (accessed May 14, 2016).
Mezey, E., et al., 1998. Alpha synuclein in neurodegenerative disorders: murderer or accomplice? Nat. Med. 4 (7), 755–757. Available at: http://www.ncbi.nlm.nih.gov/pubmed/9662355 (accessed May 23, 2016).
Michelitsch, M.D., Weissman, J.S., 2000. A census of glutamine/asparagine-rich regions: implications for their conserved function and the prediction of novel prions. Proc. Natl. Acad. Sci. U.S.A. 97 (22), 11910–11915. Available at: http://www.pubmedcentral.nih.gov/articlerender.fcgi?artid=17268&tool=pmcentrez&rendertype=abstract.
Mocsny, N., 1991. Precautions prevent spread of Creutzfeldt Jakob disease. J. Neurosci. Nurs. 23 (2), 116–119. Available at: http://www.ncbi.nlm.nih.gov/pubmed/1831471 (accessed May 19, 2016).
Montagna, P., et al., 2003. Familial and sporadic fatal insomnia. Lancet Neurol. 2 (3), 167–176. Available at: http://www.ncbi.nlm.nih.gov/pubmed/12849238 (accessed May 19, 2016).
Moody, K.M., et al., 2011. Sporadic fatal insomnia in a young woman: a diagnostic challenge: case report. BMC Neurol. 11 (1), 136. Available at: http://bmcneurol.biomedcentral.com/articles/10.1186/1471-2377-11-136 (accessed May 18, 2016).

Naeem, A., Fazili, N.A., 2011. Defective protein folding and aggregation as the basis of neurodegenerative diseases: the darker aspect of proteins. Cell Biochem. Biophys. 61 (2), 237–250. Available at: http://www.ncbi.nlm.nih.gov/pubmed/21573992 (accessed April 8, 2016).

Nakayashiki, T., et al., 2005. Yeast prions [URE3] and [PSI+] are diseases. Proc. Natl. Acad. Sci. U.S.A. 102 (30), 10575–10580. Available at: http://www.pnas.org/content/102/30/10575.full (accessed May 24, 2016).

Neumann, M., et al., 2006. Ubiquitinated TDP-43 in frontotemporal lobar degeneration and amyotrophic lateral sclerosis. Science (New York, N.Y.) 314 (5796), 130–133. Available at: http://science.sciencemag.org/content/314/5796/130.abstract (accessed July 10, 2014).

Oesch, B., et al., 1985. A cellular gene encodes scrapie PrP 27–30 protein. Cell 40 (4), 735–746. Available at: http://www.ncbi.nlm.nih.gov/pubmed/2859120 (accessed May 19, 2016).

Osherovich, L.Z., Weissman, J.S., 2001. Multiple Gln/Asn-rich prion domains confer susceptibility to induction of the yeast [PSI(+)] prion. Cell 106 (2), 183–194. Available at: http://www.ncbi.nlm.nih.gov/pubmed/11511346.

Osherovich, L.Z., et al., 2004. Dissection and design of yeast prions. PLoS Biol. 2 (4), E86. Available at: http://www.pubmedcentral.nih.gov/articlerender.fcgi?artid=374241&tool=pmcentrez&rendertype=abstract (accessed July 18, 2011).

Parham, S.N., Resende, C.G., Tuite, M.F., 2001. Oligopeptide repeats in the yeast protein Sup35p stabilize intermolecular prion interactions. EMBO J. 20 (9), 2111–2119. Available at: http://www.pubmedcentral.nih.gov/articlerender.fcgi?artid=125439&tool=pmcentrez&rendertype=abstract (accessed October 19, 2011).

Patel, B.K., Gavin-smyth, J., Liebman, S.W., 2009. The yeast global transcriptional co-repressor protein Cyc8 can propagate as a prion. Nat. Cell Biol. 11 (3), 344–349.

Patino, M.M., et al., 1996. Support for the prion hypothesis for inheritance of a phenotypic trait in yeast. Science 273 (5275), 622–626. Available at: http://science.sciencemag.org/content/273/5275/622.abstract (accessed May 23, 2016).

Pattison, I.H., Jones, K.M., 1967. The possible nature of the transmissible agent of scrapie. Vet. Rec. 80 (1), 2–9. Available at: http://www.ncbi.nlm.nih.gov/pubmed/4961994 (accessed May 14, 2016).

Paushkin, S.V., et al., 1996. Propagation of the yeast prion-like [psi+] determinant is mediated by oligomerization of the SUP35-encoded polypeptide chain release factor. EMBO J. 15 (12), 3127–3134. Available at: http://www.pubmedcentral.nih.gov/articlerender.fcgi?artid=450255&tool=pmcentrez&rendertype=abstract (accessed October 20, 2011).

Pearson, G.R., et al., 1991. Feline spongiform encephalopathy. Vet. Rec. 128 (22), 532. Available at: http://www.ncbi.nlm.nih.gov/pubmed/1866888 (accessed May 16, 2016).

Pearson, G.R., et al., 1992. Feline spongiform encephalopathy: fibril and PrP studies. Vet. Rec. 131 (14), 307–310. Available at: http://www.ncbi.nlm.nih.gov/pubmed/1279883 (accessed May 16, 2016).

Peretz, D., et al., 2001. Strain-specified relative conformational stability of the scrapie prion protein. Protein Sci. 10 (4), 854–863. Available at: http://www.ncbi.nlm.nih.gov/pubmed/11274476 (accessed July 21, 2016).

Prilusky, J., et al., 2005. FoldIndex: a simple tool to predict whether a given protein sequence is intrinsically unfolded. Bioinformatics 21 (16), 3435–3438. Available at: http://www.ncbi.nlm.nih.gov/pubmed/15955783 (accessed May 26, 2016).

Prusiner, S.B., 1982. Novel proteinaceous infectious particles cause scrapie. Science 216 (4542), 136–144. Available at: http://www.ncbi.nlm.nih.gov/pubmed/6801762 (accessed September 13, 2011).

Prusiner, S.B., 1989. Scrapie prions. Annu. Rev. Microbiol. 43, 345–374. Available at: http://www.ncbi.nlm.nih.gov/pubmed/2572197 (accessed May 18, 2016).

Prusiner, S.B., Hadlow, W.J., Garfin, D.E., et al., 1978a. Partial purification and evidence for multiple molecular forms of the scrapie agent. Biochemistry 17 (23), 4993–4999. Available at: http://www.ncbi.nlm.nih.gov/pubmed/102338 (accessed May 14, 2016).

Prusiner, S.B., Hadlow, W.J., Eklund, C.M., et al., 1978b. Sedimentation characteristics of the scrapie agent from murine spleen and brain. Biochemistry 17 (23), 4987–4992. Available at: http://www.ncbi.nlm.nih.gov/pubmed/214106 (accessed May 14, 2016).

Prusiner, S.B., Groth, D.F., Cochran, S.P., McKinley, M.P., et al., 1980a. Gel electrophoresis and glass permeation chromatography of the hamster scrapie agent after enzymatic digestion and detergent extraction. Biochemistry 19 (21), 4892–4898. Available at: http://www.ncbi.nlm.nih.gov/pubmed/6775698 (accessed May 14, 2016).

Prusiner, S.B., Groth, D.F., Cochran, S.P., Masiarz, F.R., et al., 1980b. Molecular properties, partial purification, and assay by incubation period measurements of the hamster scrapie agent. Biochemistry 19 (21), 4883–4891. Available at: http://www.ncbi.nlm.nih.gov/pubmed/6775697 (accessed May 14, 2016).

Prusiner, S.B., et al., 1981. Scrapie agent contains a hydrophobic protein. Proc. Natl. Acad. Sci. U.S.A. 78 (11), 6675–6679. Available at: http://www.pubmedcentral.nih.gov/articlerender.fcgi?artid=349112&tool=pmcentrez&rendertype=abstract (accessed May 14, 2016).

Prusiner, S.B., Gabizon, R., McKinley, M.P., 1987. On the biology of prions. Acta Neuropathol. 72 (4), 299–314. Available at: http://www.ncbi.nlm.nih.gov/pubmed/3554880 (accessed May 24, 2016).

Prusiner, S.B., et al., 1998. Prion protein biology. Cell 93 (3), 337–348. Available at: http://www.ncbi.nlm.nih.gov/pubmed/9590169 (accessed September 19, 2011).

Prusiner, S.B., et al., 2015. Evidence for α-synuclein prions causing multiple system atrophy in humans with parkinsonism. Proc. Natl. Acad. Sci. U.S.A. 112 (38), E5308–E5317. Available at: http://www.pnas.org/content/112/38/E5308.full (accessed May 16, 2016).

Rappaport, E.B., 1987. Iatrogenic Creutzfeldt-Jakob disease. Neurology 37 (9), 1520–1522. Available at: http://www.ncbi.nlm.nih.gov/pubmed/3306455 (accessed May 19, 2016).

Riek, R., et al., 1996. NMR structure of the mouse prion protein domain PrP(121–231). Nature 382 (6587), 180–182. Available at: http://www.ncbi.nlm.nih.gov/pubmed/8700211 (accessed July 21, 2016).

Riek, R., et al., 1998. Prion protein NMR structure and familial human spongiform encephalopathies. Proc. Natl. Acad. Sci. U.S.A. 95 (20), 11667–11672. Available at: http://www.ncbi.nlm.nih.gov/pubmed/9751723 (accessed July 21, 2016).

Roberts, B.T., Wickner, R.B., 2003. Heritable activity: a prion that propagates by covalent autoactivation. Genes Dev. 17 (17), 2083–2087. Available at: http://www.pubmedcentral.nih.gov/articlerender.fcgi?artid=196450&tool=pmcentrez&rendertype=abstract (accessed May 24, 2016).

Rogoza, T., et al., 2010. Non-Mendelian determinant [ISP+] in yeast is a nuclear-residing prion form of the global transcriptional regulator Sfp1. Proc. Natl. Acad. Sci. U.S.A. 107 (23), 10573–10577. Available at: http://www.pubmedcentral.nih.gov/articlerender.fcgi?artid=2890785&tool=pmcentrez&rendertype=abstract (accessed September 28, 2011).

Ross, E.D., Toombs, J.A., 2010. The effects of amino acid composition on yeast prion formation and prion domain interactions. Prion 4 (2), 60–65. Available at: http://www.pubmedcentral.nih.gov/articlerender.fcgi?artid=2933052&tool=pmcentrez&rendertype=abstract.

Ross, E.D., Baxa, U., Wickner, R.B., 2004. Scrambled prion domains form prions and amyloid. Mol. Cell. Biol. 24 (16), 7206–7213. Available at: http://www.pubmedcentral.nih.gov/articlerender.fcgi?artid=479727&tool=pmcentrez&rendertype=abstract.

Ross, E.D., Edskes, H.K., et al., 2005a. Primary sequence independence for prion formation. Proc. Natl. Acad. Sci. U.S.A. 102 (36), 12825–12830. Available at: http://www.pubmedcentral.nih.gov/articlerender.fcgi?artid=1200301&tool=pmcentrez&rendertype=abstract (accessed June 27, 2013).

Ross, E.D., Minton, A., Wickner, R.B., 2005b. Prion domains: sequences, structures and interactions. Nat. Cell Biol. 7 (11), 1039–1044. Available at: http://dx.doi.org/10.1038/ncb1105-1039.

Ross, E.D., et al., 2013. A bioinformatics method for identifying Q/N-rich prion-like domains in proteins. Methods Mol. Biol. 1017, 219–228. Available at: http://www.ncbi.nlm.nih.gov/pubmed/23719919 (accessed June 27, 2013).

Santoso, A., et al., 2000. Molecular basis of a yeast prion species barrier. Cell 100 (2), 277–288. Available at: http://www.ncbi.nlm.nih.gov/pubmed/10660050.

Schätzl, H.M., 2007. The phylogeny of mammalian and non-mammalian prion proteins. In: Hörnlimann, B., Riesner, D., Kretzschmar, H.A. (Eds.), Prions in Humans and Animals. Walter de, Gruyter, Berlin, pp. 119–133.

Schneider, K., et al., 2008. The early history of the transmissible spongiform encephalopathies exemplified by scrapie. Brain Res. Bull. 77 (6), 343–355. Available at: http://www.ncbi.nlm.nih.gov/pubmed/18951958 (accessed April 8, 2016).

Shewmaker, F., Wickner, R.B., Tycko, R., 2006. Amyloid of the prion domain of Sup35p has an in-register parallel beta-sheet structure. Proc. Natl. Acad. Sci. U.S.A. 103 (52), 19754–19759. Available at: http://www.pubmedcentral.nih.gov/articlerender.fcgi?artid=1750918&tool=pmcentrez&rendertype=abstract (accessed May 30, 2014).

Shewmaker, F., et al., 2009. Two prion variants of Sup35p have in-register parallel beta-sheet structures, independent of hydration. Biochemistry 48 (23), 5074–5082. Available at: http://www.pubmedcentral.nih.gov/articlerender.fcgi?artid=2744896&tool=pmcentrez&rendertype=abstract (accessed May 24, 2016).

Shkundina, I.S., et al., 2006. The role of the N-terminal oligopeptide repeats of the yeast Sup35 prion protein in propagation and transmission of prion variants. Genetics 172 (2), 827–835. Available at: http://www.pubmedcentral.nih.gov/articlerender.fcgi?artid=1456247&tool=pmcentrez&rendertype=abstract (accessed September 28, 2011).

Si, K., Kandel, E.R., 2016. The role of functional prion-like proteins in the persistence of memory. Cold Spring Harb. Perspect. Biol. 8 (4). http://dx.doi.org/10.1101/cshperspect.a021774. Available at: http://www.ncbi.nlm.nih.gov/pubmed/27037416 (accessed April 12, 2016).

Si, K., Giustetto, M., et al., 2003a. A neuronal isoform of CPEB regulates local protein synthesis and stabilizes synapse-specific long-term facilitation in aplysia. Cell 115 (7), 893–904. Available at: http://www.cell.com/article/S0092867403010213/fulltext (accessed May 23, 2016).

Si, K., Lindquist, S., Kandel, E.R., 2003b. A neuronal isoform of the aplysia CPEB has prion-like properties. Cell 115 (7), 879–891. Available at: http://www.cell.com/article/S0092867403010201/fulltext (accessed May 23, 2016).

Si, K., et al., 2010. Aplysia CPEB can form prion-like multimers in sensory neurons that contribute to long-term facilitation. Cell 140 (3), 421–435. Available at: http://www.cell.com/article/S0092867410000097/fulltext (accessed April 8, 2016).

Sigurðsson, B., 1954. Rida, a chronic encephalitis of sheep: with general remarks on infections which develop slowly and some of their special characteristics. Br. Vet. J. 110, 341–354.

Simoneau, S., et al., 2007. In vitro and in vivo neurotoxicity of prion protein oligomers. PLoS Pathog. 3 (8), e125. Available at: http://journals.plos.org/plospathogens/article?id=10.1371/journal.ppat.0030125 (accessed May 23, 2016).

Sipe, J.D., Cohen, A.S., 2000. Review: history of the amyloid fibril. J. Struct. Biol. 130 (2–3), 88–98. Available at: http://www.ncbi.nlm.nih.gov/pubmed/10940217 (accessed April 20, 2016).

Somerville, R.A., Gentles, N., 2011. Characterization of the effect of heat on agent strains of the transmissible spongiform encephalopathies. J. Gen. Virol. 92 (Pt. 7), 1738–1748. Available at: http://www.ncbi.nlm.nih.gov/pubmed/21471321 (accessed May 19, 2016).

Sondheimer, N., Lindquist, S., 2000. Rnq1: an epigenetic modifier of protein function in yeast. Mol. Cell 5 (1), 163–172. Available at: http://www.ncbi.nlm.nih.gov/pubmed/10678178 (accessed October 18, 2011).

Spillantini, M.G., et al., 1997. Alpha-synuclein in Lewy bodies. Nature 388 (6645), 839–840. Available at: http://www.ncbi.nlm.nih.gov/pubmed/9278044 (accessed January 22, 2015).

Stanley, W.M., 1935. Isolation of a crystalline protein possessing the properties of tobacco-mosaic virus. Science 81 (2113), 644–645. Available at: http://www.ncbi.nlm.nih.gov/pubmed/17743301 (accessed May 12, 2016).

Stephan, J.S., et al., 2015. The CPEB3 protein is a functional prion that interacts with the actin cytoskeleton. Cell Rep. 11 (11), 1772–1785. Available at: http://www.cell.com/article/S2211124715004921/fulltext (accessed June 16, 2015).

Sun, Z., et al., 2011. Molecular determinants and genetic modifiers of aggregation and toxicity for the ALS disease protein FUS/TLS. PLoS Biol. 9 (4), e1000614. Available at: http://journals.plos.org/plosbiology/article?id=10.1371/journal.pbio.1000614 (accessed May 26, 2016).

Sunde, M., et al., 1997. Common core structure of amyloid fibrils by synchrotron X-ray diffraction. J. Mol. Biol. 273 (3), 729–739. Available at: http://www.ncbi.nlm.nih.gov/pubmed/9356260 (accessed October 19, 2015).

Supattapone, S., 2015. Expanding the prion disease repertoire. Proc. Natl. Acad. Sci. U.S.A. 112 (38), 11748–11749. Available at: http://www.pubmedcentral.nih.gov/articlerender.fcgi?artid=4586830&tool=pmcentrez&rendertype=abstract (accessed May 22, 2016).

Suzuki, G., Tanaka, M., 2013. Active conversion to the prion state as a molecular switch for cellular adaptation to environmental stress. Bioessays 35 (1), 12–16. Available at: http://www.ncbi.nlm.nih.gov/pubmed/23175284 (accessed May 24, 2016).

Sweeny, E.A., Shorter, J., 2016. Mechanistic and structural insights into the prion-disaggregase activity of Hsp104. J. Mol. Biol. 428 (9 Pt. B), 1870–1885. Available at: http://www.ncbi.nlm.nih.gov/pubmed/26608812 (accessed May 24, 2016).

Tanaka, M., et al., 2004. Conformational variations in an infectious protein determine prion strain differences. Nature 428 (6980), 323–328. Available at: http://www.ncbi.nlm.nih.gov/pubmed/15029196 (accessed July 21, 2016).

Tanaka, M., et al., 2006. The physical basis of how prion conformations determine strain phenotypes. Nature 442 (7102), 585–589. Available at: http://www.ncbi.nlm.nih.gov/pubmed/16810177 (accessed May 24, 2013).

Tank, E.M.H., et al., 2007. Prion protein repeat expansion results in increased aggregation and reveals phenotypic variability. Mol. Cell. Biol. 27 (15), 5445–5455. Available at: http://www.pubmedcentral.nih.gov/articlerender.fcgi?artid=1952097&tool=pmcentrez&rendertype=abstract (accessed September 28, 2011).

Tartaglia, G.G., et al., 2008. Prediction of aggregation-prone regions in structured proteins. J. Mol. Biol. 380 (2), 425–436. Available at: http://www.ncbi.nlm.nih.gov/pubmed/18514226. accessed May 26, 2016.

Taylor, D.M., 1989. Bovine spongiform encephalopathy and human health. Vet. Rec. 125 (16), 413–415. Available at: http://www.ncbi.nlm.nih.gov/pubmed/2686150 (accessed May 18, 2016).

Taylor, K.L., et al., 1999. Prion domain initiation of amyloid formation in vitro from native Ure2p. Science 283 (5406), 1339–1343. Available at: http://www.ncbi.nlm.nih.gov/pubmed/10037606 (accessed May 24, 2016).

Telling, G.C., et al., 1994. Transmission of Creutzfeldt-Jakob disease from humans to transgenic mice expressing chimeric human-mouse prion protein. Proc. Natl. Acad. Sci. U.S.A. 91 (21), 9936–9940. Available at: http://www.pubmedcentral.nih.gov/articlerender.fcgi?artid=44932&tool=pmcentrez&rendertype=abstract (accessed May 19, 2016).

Ter-Avanesyan, M.D., et al., 1993. Deletion analysis of the SUP35 gene of the yeast Saccharomyces cerevisiae reveals two non-overlapping functional regions in the encoded protein. Mol. Microbiol. 7 (5), 683–692. Available at: http://www.ncbi.nlm.nih.gov/pubmed/8469113 (accessed October 19, 2011).

Ter-Avanesyan, M.D., et al., 1994. The SUP35 omnipotent suppressor gene is involved in the maintenance of the non-Mendelian determinant [psi+] in the yeast Saccharomyces cerevisiae. Genetics 137 (3), 671–676. Available at: http://www.pubmedcentral.nih.gov/articlerender.fcgi?artid=1206026&tool=pmcentrez&rendertype=abstract (accessed October 19, 2011).

Toombs, J.A., McCarty, B.R., Ross, E.D., 2010. Compositional determinants of prion formation in yeast. Mol. Cell. Biol. 30 (1), 319–332. Available at: http://www.pubmedcentral.nih.gov/articlerender.fcgi?artid=2798286&tool=pmcentrez&rendertype=abstract.

Toombs, J.A., et al., 2011. [PSI+] maintenance is dependent on the composition, not primary sequence, of the oligopeptide repeat domain. PLoS One 6 (7), e21953. Available at: http://www.pubmedcentral.nih.gov/articlerender.fcgi?artid=3132755&tool=pmcentrez&rendertype=abstract (accessed September 28, 2011).

True, H.L., Lindquist, S.L., 2000. A yeast prion provides a mechanism for genetic variation and phenotypic diversity. Nature 407 (6803), 477–483. Available at: http://dx.doi.org/10.1038/35035005 (accessed April 4, 2016).

Tuite, M.F., Mundy, C.R., Cox, B.S., 1981. Agents that cause a high frequency of genetic change from [psi+] to [psi-] in Saccharomyces cerevisiae. Genetics 98 (4), 691–711. Available at: http://www.pubmedcentral.nih.gov/articlerender.fcgi?artid=1214469&tool=pmcentrez&rendertype=abstract (accessed July 8, 2014).

Tuite, M.F., Staniforth, G.L., Cox, B.S., 2015. [PSI+] turns 50. Prion 9 (5), 318–322. Available at: http://www.tandfonline.com/doi/pdf/10.1080/19336896.2015.1111508 (accessed April 21, 2016).

Tycko, R., Wickner, R.B., 2013. Molecular structures of amyloid and prion fibrils: consensus versus controversy. Acc. Chem. Res. 46 (7), 1487–1496. Available at: http://www.pubmedcentral.nih.gov/articlerender.fcgi?artid=3632659&tool=pmcentrez&rendertype=abstract (accessed May 24, 2016).

Vance, C., et al., 2009. Mutations in FUS, an RNA processing protein, cause familial amyotrophic lateral sclerosis type 6. Science 323 (5918), 1208–1211. Available at: http://science.sciencemag.org/content/323/5918/1208.abstract (accessed March 11, 2015).

Verma, M., Vats, A., Taneja, V., 2015. Toxic species in amyloid disorders: oligomers or mature fibrils. Ann. Ind. Acad. Neurol. 18 (2), 138–145. Available at: http://www.pubmedcentral.nih.gov/articlerender.fcgi?artid=4445186&tool=pmcentrez&rendertype=abstract (accessed February 20, 2016).

Vitrenko, Y.A., et al., 2007. Propagation of the [PIN+] prion by fragments of Rnq1 fused to GFP. Curr. Genet. 51 (5), 309–319. Available at: http://www.pubmedcentral.nih.gov/

articlerender.fcgi?artid=2597802&tool=pmcentrez&rendertype=abstract (accessed October 28, 2011).
Wadsworth, J.D.F., et al., 2003. Molecular and clinical classification of human prion disease. Br. Med. Bull. 66, 241–254. Available at: http://www.ncbi.nlm.nih.gov/pubmed/14522862 (accessed October 28, 2011).
Walker, L.C., Jucker, M., 2015. Neurodegenerative diseases: expanding the prion concept. Annu. Rev. Neurosci. 38, 87–103. Available at: http://www.ncbi.nlm.nih.gov/pubmed/25840008 (accessed July 20, 2016).
Wan, W., Stubbs, G., 2014. Fungal prion HET-s as a model for structural complexity and self-propagation in prions. Proc. Natl. Acad. Sci. U.S.A. 111 (14), 5201–5206. Available at: http://www.pubmedcentral.nih.gov/articlerender.fcgi?artid=3986130&tool=pmcentrez&rendertype=abstract (accessed May 23, 2016).
Watts, J.C., et al., 2013. Transmission of multiple system atrophy prions to transgenic mice. Proc. Natl. Acad. Sci. U.S.A. 110 (48), 19555–19560. Available at: http://www.pnas.org/content/110/48/19555.long (accessed May 22, 2016).
Wells, G.A., et al., 1987. A novel progressive spongiform encephalopathy in cattle. Vet. Rec. 121 (18), 419–420. Available at: http://www.ncbi.nlm.nih.gov/pubmed/3424605 (accessed March 9, 2016).
Wickner, R.B., 1994. [URE3] as an altered URE2 protein: evidence for a prion analog in Saccharomyces cerevisiae. Science 264 (5158), 566–569. Available at: http://www.ncbi.nlm.nih.gov/pubmed/7909170 (accessed August 30, 2011).
Wickner, R.B., 2012. Discovering protein-based inheritance through yeast genetics. J. Biol. Chem. 287 (18), 14432–14442. Available at: http://www.jbc.org/content/287/18/14432 (accessed May 23, 2016).
Wickner, R.B., et al., 2008. Protein inheritance (prions) based on parallel in-register beta-sheet amyloid structures. Bioessays 30 (10), 955–964. Available at: http://www.pubmedcentral.nih.gov/articlerender.fcgi?artid=3086512&tool=pmcentrez&rendertype=abstract (accessed May 24, 2016).
Wickner, R.B., et al., 2011. The yeast prions [PSI+] and [URE3] are molecular degenerative diseases. Prion 5 (4), 258–262. Available at: http://www.pubmedcentral.nih.gov/articlerender.fcgi?artid=4012404&tool=pmcentrez&rendertype=abstract (accessed May 24, 2016).
Wickner, R.B., et al., 2016. Yeast and fungal prions: amyloid-handling systems, amyloid structure, and prion biology. Adv. Genet. 93, 191–236. Available at: http://www.ncbi.nlm.nih.gov/pubmed/26915272 (accessed March 10, 2016).
Will, R.G., et al., 1996. A new variant of Creutzfeldt-Jakob disease in the UK. Lancet 347 (9006), 921–925. Available at: http://www.ncbi.nlm.nih.gov/pubmed/8598754 (accessed May 18, 2016).
Wille, H., Prusiner, S.B., Cohen, F.E., 2000. Scrapie infectivity is independent of amyloid staining properties of the N-terminally truncated prion protein. J. Struct. Biol. 130 (2–3), 323–338. Available at: http://linkinghub.elsevier.com/retrieve/pii/S1047847700942424 (accessed July 21, 2016).
Williams, E.S., Young, S., 1980. Chronic wasting disease of captive mule deer: a spongiform encephalopathy. J. Wildl. Dis. 16 (1), 89–98. Available at: http://www.ncbi.nlm.nih.gov/pubmed/7373730 (accessed May 18, 2016).
Woerman, A.L., et al., 2015. Propagation of prions causing synucleinopathies in cultured cells. Proc. Natl. Acad. Sci. U.S.A. 112 (35), E4949–E4958. Available at: http://www.pnas.org/content/112/35/E4949.abstract (accessed May 15, 2016).
Wood, J., Lund, L., Done, S., 1992a. The natural occurrence of scrapie in moufflon. Vet Rec 130 (2), 25–27. Available at: http://europepmc.org/abstract/MED/1542978 (accessed May 16, 2016).

Wood, J.N., et al., 1992b. Natural scrapie in goats: case histories and clinical signs. Vet. Rec. 131 (4), 66–68. Available at: http://www.ncbi.nlm.nih.gov/pubmed/1529502 (accessed May 16, 2016).

Wyatt, J.M., et al., 1991. Naturally occurring scrapie-like spongiform encephalopathy in five domestic cats. Vet. Rec. 129 (11), 233–236. Available at: http://www.ncbi.nlm.nih.gov/pubmed/1957458 (accessed May 16, 2016).

Yokota, O., et al., 2002. NACP/alpha-synuclein, NAC, and beta-amyloid pathology of familial Alzheimer's disease with the E184D presenilin-1 mutation: a clinicopathological study of two autopsy cases. Acta Neuropathol. 104 (6), 637–648. Available at: http://www.ncbi.nlm.nih.gov/pubmed/12410385 (accessed May 23, 2016).

Zabel, M.D., Reid, C., 2015. A brief history of prions. Pathog. Dis. 73 (9), ftv087. Available at: http://www.ncbi.nlm.nih.gov/pubmed/26449713 (accessed January 29, 2016).

Zahn, R., et al., 2000. NMR solution structure of the human prion protein. Proc. Natl. Acad. Sci. U.S.A. 97 (1), 145–150. Available at: http://www.ncbi.nlm.nih.gov/pubmed/10618385 (accessed July 21, 2016).

Zhang, Z., et al., 2013. De novo generation of infectious prions with bacterially expressed recombinant prion protein. FASEB J 27 (12), 4768–4775. Available at: http://www.pubmedcentral.nih.gov/articlerender.fcgi?artid=3834773&tool=pmcentrez&rendertype=abstract (accessed May 19, 2016).

CHAPTER SEVEN

The Structure of Mammalian Prions and Their Aggregates

E. Vázquez-Fernández*,†, H.S. Young*, J.R. Requena‡, H. Wille*,†,1

*University of Alberta, Edmonton, AB, Canada
†Centre for Prions and Protein Folding Diseases, University of Alberta, Edmonton, AB, Canada
‡CIMUS Biomedical Research Institute, University of Santiago de Compostela-IDIS, Santiago de Compostela, Spain
[1]Corresponding author: e-mail address: wille@ualberta.ca

Contents

Abstract

Prion diseases, such as Creutzfeldt–Jakob disease in humans, bovine spongiform encephalopathy in cattle, chronic wasting disease in cervids (i.e., deer, elk, moose, and reindeer), and sheep scrapie, are caused by the misfolding of the cellular prion protein (PrP^{C}) into a disease-causing conformer (PrP^{Sc}). PrP^{C} is a normal, GPI-anchored protein that is expressed on the surface of neurons and other cell types. The structure of PrP^{C} is well understood, based on studies of recombinant PrP, which closely mimics the structure of native PrP^{C}. In contrast, PrP^{Sc} is prone to aggregate into a variety of quaternary structures, such as oligomers, amorphous aggregates, and amyloid fibrils. The propensity of PrP^{Sc} to assemble into these diverse forms of aggregates is also responsible for our limited knowledge about its structure. Then again, the repeating nature

International Review of Cell and Molecular Biology, Volume 329
ISSN 1937-6448
http://dx.doi.org/10.1016/bs.ircmb.2016.08.013

of certain regular PrP^{Sc} aggregates has allowed (lower resolution) insights into the structure of the infectious conformer, establishing a four-rung β-solenoid structure as a key element of its architecture.

1. INTRODUCTION

1.1 Infectious Mammalian Prions

The term "prion" was coined by Stanley Prusiner in 1982 and defined as a "proteinaceous infectious particle" (Prusiner, 1982), which, in retrospect, was an understatement. Initially met with widespread skepticism, the prion concept has gained virtually universal acceptance, as its validity has been confirmed in experiment after experiment. In fact, a prion can now be defined sensu strictiori as an infectious protein (Aguzzi and Lakkaraju, 2016; Prusiner, 1998, 2012). Therefore, the term "infectious prion" is redundant, given that being infectious is an intrinsic property of a prion. In any case, the adjective "infectious" should be taken as a descriptor, in the same way as the adjectives in the expressions "white snow" or "burning fire."

PrP^{Sc} was the first prion identified (Bolton et al., 1982). It was isolated as the major component of infectious proteinaceous material present in the brains of Syrian hamsters infected with scrapie, a lethal, transmissible neurodegenerative disease (Bolton et al., 1982). Ever since, several other prions have been identified in yeasts, filamentous fungi, and other organisms, but these prions are not disease-related, nor do they have any evolutionary relationship with the prion protein (Wickner et al., 2000). Besides, a number of mammalian brain proteins that cause neurodegenerative diseases and can be experimentally propagated among animals under certain circumstances, such as misfolded Aβ, tau, or α-synuclein, are strong candidates to being admitted to the list of mammalian prions (Jaunmuktane et al., 2015; Prusiner, 2012), although controversy exists whether they should carry the term "prion" (Aguzzi and Lakkaraju, 2016; Castilla and Requena, 2015). In this review, we will focus solely on PrP^{Sc}.

PrP^{Sc} is the causal agent of the transmissible spongiform encephalopathies a group of fatal neurodegenerative diseases that, among others, include Creutzfeldt–Jakob disease (CJD) and Kuru, both affecting humans (Prusiner, 1998), bovine spongiform encephalopathy (BSE), scrapie, which affects sheep and goats, and chronic wasting disease, which affects cervids (Aguzzi and Calella, 2009; Prusiner, 1998). In all cases, after devastating the brain in which it is hosted, by mechanisms that are still poorly understood, PrP^{Sc} has the potential to transmit to other individuals of the same species, or even other species. This may occur by an oral route, for example, as a consequence of ritual or industrial cannibalism, as happened in the Kuru epidemic and BSE epizootic, respectively (Aguzzi and Calella, 2009; Prusiner, 1998). Or by iatrogenic means, as in iatrogenic CJD caused by reutilization of neurosurgical instruments, dura mater grafts, or pituitary-derived growth hormone contaminated with PrP^{Sc} (Aguzzi and Calella, 2009).

How can a protein be infectious? The key characteristic of classic infectious agents is that they reproduce, and therefore, multiply, within their hosts because they carry with them a genetic blueprint that can be copied. Remarkably, the fact that such material is a nucleic acid and not a protein was not evident until the decisive experiments by Hershey and Chase (1952). In contrast to nucleic acids, whose propagation involves reproduction of their primary structure, the propagation of prions involves reproduction of their secondary, tertiary, and quaternary structures, i.e., their conformation (Prusiner, 1998, 2012). Thus, prions coerce a precursor of themselves, exhibiting a normal protein fold, into the prion conformation. In particular, PrP^{Sc} coerces PrP^{C}, a GPI-anchored membrane protein with a function that is not completely understood into the PrP^{Sc} conformation. What is the molecular mechanism of this transformation? To answer this question it is obviously essential to first know what the structure of PrP^{Sc} is, something we have not yet accomplished in sufficient detail. This review presents a comprehensive vision of what we currently know about the structure of PrP^{Sc}.

1.2 Protease Resistance, Insolubility, and Aggregation

Protease resistance, insolubility, and aggregation are three molecular characteristics of PrP^{Sc} (Prusiner, 1998) that distinguish it from native PrP^{C} (Fig. 1). Therefore, such properties quickly became operational surrogates to differentiate PrP^{Sc} from PrP^{C}, and more generally (vide supra) an infectious prion from other nonprionic misfolded forms of PrP. Of course, the gold standard to identify a prion is through its infectivity, but given the long incubation times of bioassays, even in very susceptible hosts such as Syrian hamsters (Prusiner, 1998) or bank voles (Di Bari et al., 2013), partial resistance to proteolysis and insolubility have often been used as a first surrogate for the presence of prions in a given sample. In particular, partial resistance to proteinase K (PK) was the basis of the tests developed and used to identify

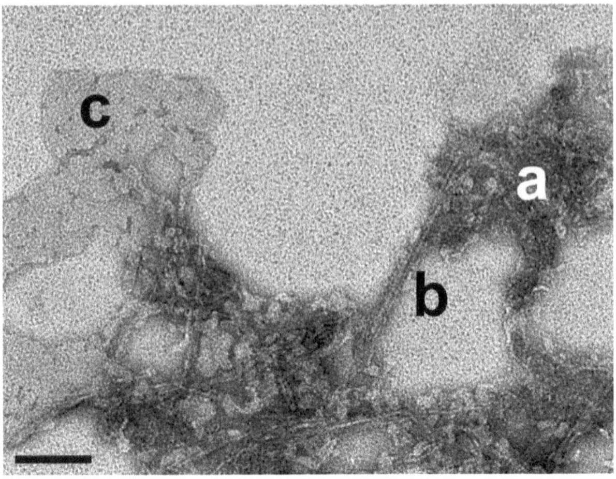

Fig. 1 Brain-derived prions form pleomorphic aggregation products. Upon purification from infected brain tissues, prions are prone to polymerize into a variety of different aggregation products (a–c). Here, RML prions were purified from the brains of sick FVB mice using limited proteolysis with proteinase K (PK) followed by precipitation with a Keggin-type polyoxometalate (Wille et al., 2009b). Negative-stain electron microscopy showed that the proteolytically truncated PrP^{Sc}, termed PrP 27–30, forms amorphous aggregates (a), amyloid fibrils (b), and two-dimensional (2D) crystals (c). Bar = 100 nm.

PrP^{Sc} in brain homogenates from cattle subjected to surveillance programs during the European BSE or "mad cow disease" crisis in the eighties and nineties. Detection of PK-resistant material by Western blot or ELISA allowed elimination of infected animals from the food supply and was a key tool in controlling the crisis and leading to the virtual eradication of BSE (Maheshwari et al., 2015). Typically, when PrP^{Sc} is subjected to a standard PK concentration of 20–100 μg/mL for 1 h at 37°C, a PK-resistant fragment comprising PrP residues ~80/90 to the C-terminus, and carrying the C-terminal GPI-anchor, remains intact (Fig. 2D). Considering the existence of di-, mono-, and nonglycosylated forms of PrP^{Sc}, three broad bands are typically seen in SDS PAGE or Western blots, collectively termed as PrP 27–30, in agreement with their apparent molecular mass (Prusiner, 1998).

However, it was soon discovered that this is an oversimplification. Early studies found that in some cases samples carrying infectivity did not contain any discernible PK-resistant material. This was erroneously interpreted as evidence against the presence of PrP^{Sc} in these prion disease cases (Lasmézas et al., 1997). Elegant studies by Safar et al. led to the conclusion that a sizeable fraction of PrP^{Sc} is not PK resistant (Safar et al., 1998). These researchers developed a conformation-dependent immunoassay (CDI) in

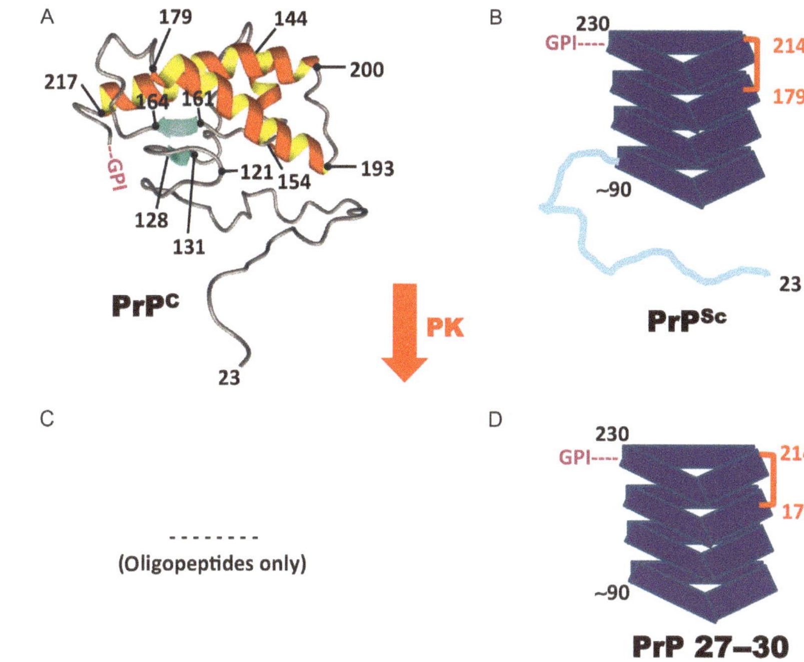

Fig. 2 Schematic representations of the domain structures of PrP^C and PrP^{Sc}. (A) PrP^C contains a folded domain spanning from position ~121 to the C-terminus and an unfolded N-terminal tail, which starts at residue 23, since the N-terminal signal-peptide spanning residues 1–22 is posttranslationally removed. The folded domain is dominated by three α-helices, residues ~144–154, ~179–193, and 200–217, and also contains two short, paired β-strands, residues 128–131 and 161–164 (Riek et al., 1996). A single disulfide bond, residues 179–214, connects the two most C-terminal α-helices (not shown for simplicity). A GPI-anchor is linked to its C-terminus, and one or two N-linked sugars can be attached to positions 181 and 197 (not shown for simplicity). (B) PrP^{Sc} also contains two domains, a slightly shorter and unfolded N-terminal tail, spanning up to position ~90, and a compact, C-terminal four-rung β-solenoid spanning from position ~90 up to the C-terminus. The three α-helices, short β-strands, loops, and part of the N-terminal tail that are present in PrP^C, have been completely refolded. The single disulfide bond is maintained. In this scheme, the shape of the β-solenoid rungs and the lengths of the β-strands are completely arbitrary. Any short, connecting loops and turns that must exist have been omitted. The N-linked sugars would necessarily point outward (not shown). (C) Treatment of PrP^C with PK results in its complete destruction with only short oligopeptides remaining. (D) In contrast, treatment of PrP^{Sc} with PK typically results in destruction of the N-terminal, unfolded domain, while the C-terminal compact β-solenoid is preserved (see text for additional considerations on PK resistance of PrP^{Sc}). The resulting PK-resistant fragment is termed PrP 27–30. PK also introduces additional nicks, likely at connecting loops, resulting in additional, shorter PK-resistant PrP^{Sc} fragments (Vázquez-Fernández et al., 2012), but these are much less abundant. Numbering is based on the murine PrP sequence. *Panel (A) Adapted from Chattopadhyay, M., Walter, E.D., Newell, D.J., Jackson, P.J., Aronoff-Spencer, E., Peisach, J., Gerfen, G.J., Bennett, B., Antholine, W.E., Millhauser, G.L., 2005. The octarepeat domain of the prion protein binds Cu(II) with three distinct coordination modes at pH 7.4. J. Am. Chem. Soc. 127, 12647–12656.*

which PrP^{Sc} could be detected and measured in the presence of PrP^{C} without relying on limited proteolysis by PK. Instead, CDI takes advantage of the increased reactivity of antibodies recognizing PrP epitopes that are cryptic in PrP^{Sc} but become exposed after denaturation with guanidine HCl. In contrast, these epitopes are fully exposed in PrP^{C} (Safar et al., 1998). In addition, Safar and coworkers found that treatment of many prion samples with PK prior to application of the CDI assay resulted in a substantial decrease of the signal unveiled by the guanidine HCl, which, in other words, demonstrated the existence of a PK-sensitive fraction of PrP^{Sc} in these samples (Safar et al., 1998). Subsequent studies by Tzaban et al. (2002) suggested that the quaternary structure affects the PK sensitivity of PrP^{Sc}, with the smaller aggregates within a given PrP^{Sc} ensemble being less resistant. Following this lead, Pastrana et al. isolated a PK-sensitive fraction of PrP^{Sc} (Pastrana et al., 2006), which proved to be infectious (Sajnani et al., 2012). While this fraction is indeed made up of smaller oligomers, it is not completely clear whether size is the only determinant of its properties, given that extensive manipulations, such as high-force centrifugation, extraction with detergents, and sonication were used to prepare it. Interestingly, it was possible to create novel prion strains in vitro that were entirely protease sensitive, indicating that protease resistance is not a necessary requirement for prion infectivity (Colby et al., 2010; Legname et al., 2004).

Besides the ~90–230 PK-resistant core fragment, treatment of PrP^{Sc} with PK produces other smaller, much less abundant fragments (Silva et al., 2015; Vázquez-Fernández et al., 2012). Such outcome can be readily explained if a substantial portion of the natively unfolded N-terminal "tail" of PrP^{C} would conserve its unfolded status in PrP^{Sc}, namely, the stretch spanning from the N-terminus to position ~90 (Donne et al., 1997). Such flexible tail would be easily cleaved off by PK, yielding the predominant ~90–230 residue PK-resistant fragment (Fig. 2). The other smaller and less abundant fragments would be the result of a less efficient cleavage by PK of short flexible stretches interspersed between more rigid β-strands (Silva et al., 2015). It should be noted, however, that thermolysin and pronase E, two proteases that are milder than PK, cleave the flexible N-terminus of PrP^{C} but not that of PrP^{Sc} (Cronier et al., 2008; D'Castro et al., 2010), which suggests subtle structural differences between the conformers of the N-terminus (Yam et al., 2010), even if both are flexible.

Insolubility in detergents is another classic property of PrP^{Sc}. In contrast, PrP^{C} is fully soluble under the same conditions. In consequence, this specific property of PrP^{Sc} has been extensively used to isolate it (Baron et al., 2011;

Bolton et al., 1987). Insolubility might be a consequence of the aggregated states that are typical of PrP^{Sc} (vide infra), but it might also be an intrinsic property of the PrP^{Sc} fold. In this respect, FTIR studies have determined that PrP^{Sc} contains a very substantial amount of β-sheet secondary structure (Smirnovas et al., 2011). This contrasts with PrP^{C}, whose structure has been solved by means of NMR spectroscopy and X-ray crystallography of recombinantly produced protein samples believed to be folded similarly as brain PrP^{C}. PrP^{C} is characterized by an α-helix-rich conformation (Donne et al., 1997; Riek et al., 1996), with just a very small, two-stranded β-sheet (Fig. 2A). Finally, PrP^{Sc} forms oligomers or higher-order aggregates, in contrast to PrP^{C}, which is believed to be monomeric. The aggregation status of PrP^{Sc} will be discussed in the following section.

2. AMORPHOUS AGGREGATES AND OLIGOMERS

2.1 Infectious Oligomers

As mentioned in the previous section, PrP^{Sc} is multimeric (Fig. 1). PrP^{Sc} monomers have never been isolated or detected nor was it possible to isolate infectious monomers through solubilization of aggregated PrP^{Sc} polymers (Wille and Prusiner, 1999). In a classic study, Bellinger-Kawahara et al. analyzed the characteristics and kinetics of prion infectivity deactivation by ionizing radiation and concluded that the target size of PrP^{Sc} corresponds to that of a dimer or trimer (Bellinger-Kawahara et al., 1988), suggesting that dimeric and/or trimeric assemblies of PrP^{Sc} might exist. However, PrP^{Sc} dimers or trimers have never been isolated either. Instead, PrP^{Sc} is typically isolated as an aggregate (Prusiner, 1998) and PrP^{Sc} molecules present in infected brain homogenates show the biophysical properties of oligomers, for example, in sucrose gradient sedimentation experiments (Tixador et al., 2010; Tzaban et al., 2002).

The first PrP^{Sc} aggregates to be isolated and characterized were fibrillar, rod-like structures as seen by negative-stain electron microscopy (Prusiner et al., 1983). It is not entirely clear to what extent these fibers were natural or artificially formed during the isolation of PrP^{Sc}. Initial experiments suggested that detergents and PK treatment were necessary to generate them (McKinley et al., 1991). However, a recent study has shown that while such treatment helps to visualize these "prion rods" by means of electron microscopy, the fibers might preexist such treatment (Terry et al., 2016). In the opposite direction, Gabizon et al. showed that PrP^{Sc} rods could be dispersed into liposomes formed by a combination of phospholipids and detergent (Gabizon et al., 1987). This treatment resulted in a nearly complete disappearance of the rods, with no discernible fragments being detectable by negative-stain electron microscopy (Gabizon et al., 1987), whereas the infectivity of the samples not only did not decrease but actually increased about 10-fold. Rods could be reassembled by elimination of the detergent and lipid, which means that there is likely an equilibrium between the fibrillar structures and smaller oligomers. With this in mind, Silveira et al. carried out studies to characterize the infectivity of PrP^{Sc} assemblies of different sizes (Silveira et al., 2005). These authors first isolated PrP 27–30 rods and then treated them with detergent and sonication or detergent alone. They separated fractions of different sizes by means of flow field-flow fractionation and assessed their infectivity in animal bioassays. These experiments showed that beyond a certain size threshold of ~5 units, aggregates are infectious, with oligomers of ~14–28 units being the most infectious ones (Silveira et al., 2005). These oligomers appeared as small globular particles in negative-stain electron micrographs. Once again, the fact that the samples were extracted with detergents, subjected to ultracentrifugation, PK treatment, and further treatments with detergent and sonication precludes any conclusion on whether the PrP^{Sc} species thus obtained represent aggregates present in infected brains. While in some prion diseases (i.e., in particular combinations of host species and prion strain), amyloid fibrils are present in prion-infected brain tissue in vivo (DeArmond et al., 1985), in other cases (e.g., RML prions in FVB wild-type mice) no amyloid fibrils were detected despite intense searches using electron microscopy (Godsave et al., 2013). In the latter case, fibrillization could be induced through addition of detergent or multivalent binding partners (Levine et al., 2015), indicating the fibrillization competence of PrP^{Sc} even in nonamyloidogenic disease forms. Thus, it is conceivable that in these prion disease cases a range of PrP^{Sc} oligomers of different sizes are present in the brain and are first compacted into larger fibers during PK and detergent assisted PrP^{Sc} isolation, and that later treatments with detergent and sonication partially fragment and disperse the fibers into smaller subunits that could be identical or more likely similar to those existing at the beginning of the process (Levine et al., 2015; Silveira et al., 2005).

2.2 Amorphous Aggregates

It is very easy to convert PrP^C into an amorphous aggregate. PrP^C is very susceptible to chemical and mechanical denaturing, including, but not limited to, changes of pH (Prigent and Rezaei, 2011), reduction of its single disulfide bond (Zhang et al., 1997), oxidation (Requena et al., 2001), or high isostatic pressure (Torrent et al., 2004). In particular, a β-sheet-rich oligomeric transition intermediate has been described under combinations of low pH and increasing amounts of denaturants (Baskakov et al., 2001; Swietnicki et al., 2000). Some of these aggregates can evolve into classic amyloid fibrils, while others are off-pathway with respect to fibrillization (Singh et al., 2012). Some aggregates generated by these methods exhibit insolubility and/or partial resistance to PK, the two operational characteristics of prions (vide supra) that have been proposed as "surrogates" for PrP^{Sc}, under the assumption that structural properties observed in these aggregates might provide clues on the structure of bona fide PrP^{Sc} (Jackson et al., 1999; Prigent and Rezaei, 2011; Tattum et al., 2006; Torrent et al., 2004). However, the usefulness of these surrogates toward elucidating the structure of PrP^{Sc} is questionable. Therefore, experimental efforts to study the structure of the infectious conformer should concentrate on material of proven infectivity.

3. TWO-DIMENSIONAL CRYSTALS

3.1 Inducing Order in Two Dimensions

It is not uncommon for membrane proteins, which reside in the more or less two-dimensional (2D) environment of the membrane, to form 2D crystals (Stahlberg et al., 2015). PrP 27–30 was first described to form amyloid fibrils, which at the time were termed "prion rods" (Prusiner et al., 1983). Upon closer inspection, one of the first published electron micrographs showing the prion rods also contains a small 2D crystal (Prusiner et al., 1983). It took a long time for these 2D crystals to be recognized and appreciated as alternative polymerization products of PrP 27–30 (Wille et al., 2002).

Immunogold labeling with monoclonal antibodies directed against PrP clearly established the presence of PrP in these 2D crystals (Wille et al., 2007). In order to ascertain that the 2D crystals contain the infectious conformer, bioassays with preparations consisting only of 2D crystals would be necessary. Despite numerous attempts, it proved to be impossible to separate the 2D crystals from the prion rods either during the purification of PrP 27–30 (Figs. 3 and 4) or afterward (Wille et al., 2009b), which emphasizes the similarity of these quaternary structures with respect to their biophysical and biochemical parameters. Alternatively, immunogold labeling with monoclonal antibodies specific for PrP^{Sc}/PrP 27–30 could prove the presence of the infectious conformer in these 2D crystals. Unfortunately, to date,

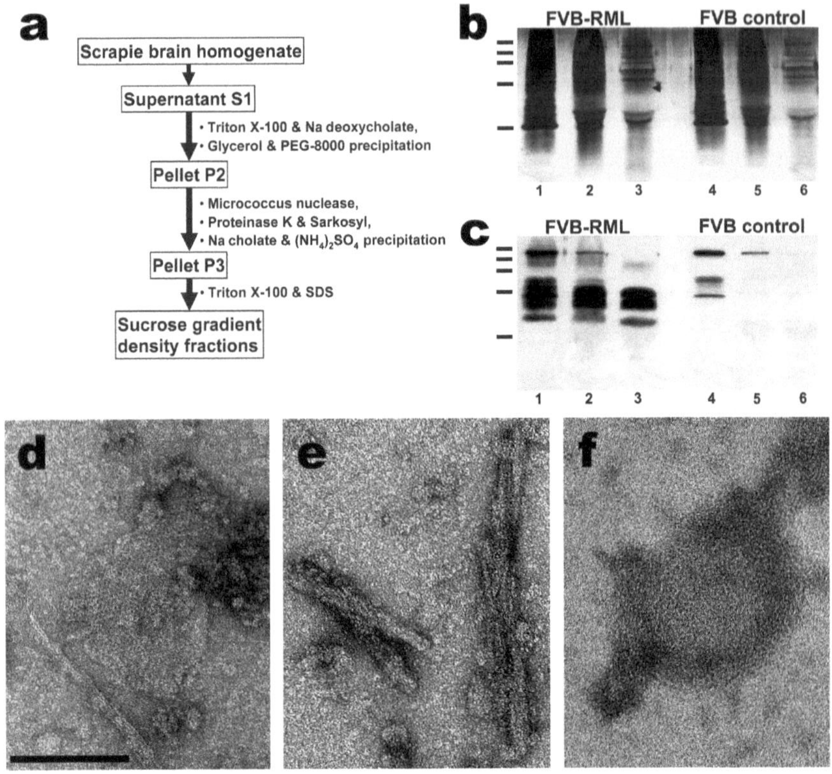

Fig. 3 Purification of brain-derived prions following a traditional protocol. Initially, the large-scale preparation of infectious prions followed a labor-intensive, multistep purification protocol (A) (Prusiner et al., 1982). SDS PAGE of gel samples from the purification procedure followed by silver staining (B) or Western blotting (C) highlights the step-wise enrichment of PrP 27–30. Clarified brain homogenate (A, supernatant S1) contained a large number of proteins (B, lane 1), including full-length PrP^{Sc} (C, lane 1). Upon limited proteolysis with PK many other proteins remain (B, lane 2), while PrP^{Sc} is N-terminally truncated into PrP 27–30 (C, lane 2). After repeated precipitations (A, pellet P3) PrP 27–30 is substantially enriched (C, lane 3), but numerous protein contaminants remain (B, lane 3). Negative-stain electron microscopy revealed that at this intermediate purification stage PrP 27–30 has already formed the main aggregation products: 2D crystals (D), amyloid fibrils (E), and amorphous aggregates (F). Bar in (D) = 100 nm, and applies to all micrograph panels. *Adapted from Wille, H., Shanmugam, M., Murugesu, M., Ollesch, J., Stubbs, G., Long, J.R., Safar, J.G., Prusiner, S.B., 2009b. Surface charge of polyoxometalates modulates polymerization of the scrapie prion protein. Proc. Natl. Acad. Sci. U.S.A. 106, 3740–3745.*

most if not all antibodies that were described to recognize only PrP^{Sc} or PrP 27–30 were found to also recognize other misfolded forms of PrP (Biasini et al., 2008). Therefore, it was not possible to use immunogold labeling as a direct route to prove that the 2D crystals contain the infectious

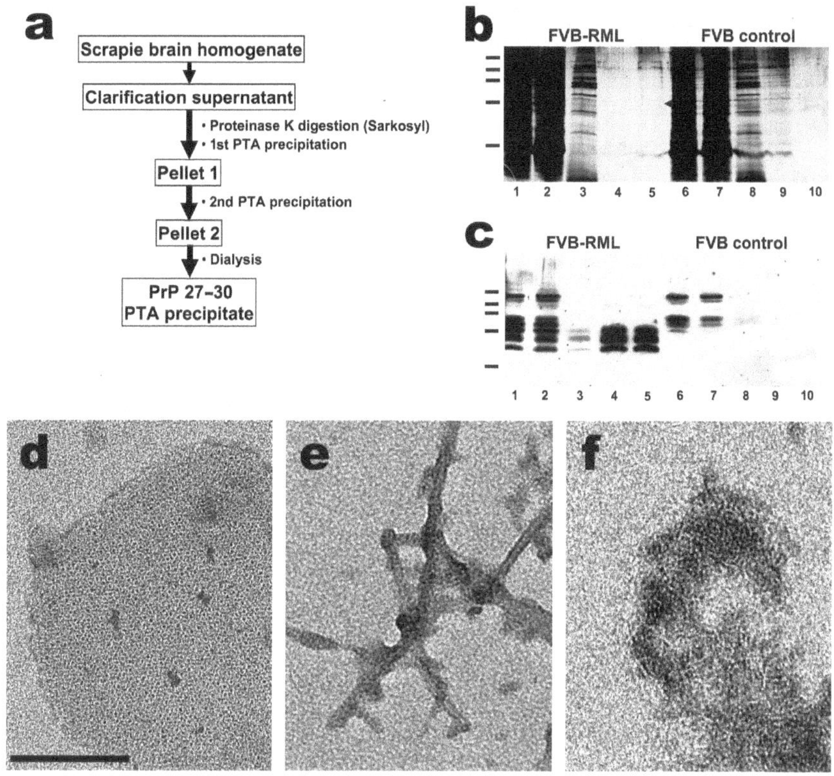

Fig. 4 Purification of brain-derived prions using an efficient protocol. The specific precipitation of PrP^{Sc} and PrP 27–30 by phosphotungstate anions (PTA) in the presence of Sarkosyl (A) forms the basis for this efficient purification protocol (Safar et al., 1998). SDS PAGE of gel samples from the purification procedure followed by silver staining (B) or Western blotting (C) highlights the step-wise enrichment of PrP 27–30. Brain homogenate (B and C, lane 1) and clarified brain homogenate (B and C, lane 2) contain many proteins, most of which are digested by a stringent proteolysis using PK (B, lane 3). The N-terminally truncated PrP 27–30 (C, lane 3) is concentrated through repeated precipitation with PTA (C, lanes 4 and 5). Negative-stain electron microscopy indicated that during this purification procedure PrP 27–30 also forms 2D crystals (D), amyloid fibrils (E), and amorphous aggregates (F). It is notable that the quality of the two-dimensional crystals and amyloid fibrils is superior to those originating from the traditional purification procedure (Fig. 3). Bar in (D) = 100 nm, and applies to all micrograph panels. *The figure was adapted from Wille, H., Shanmugam, M., Murugesu, M., Ollesch, J., Stubbs, G., Long, J.R., Safar, J.G., Prusiner, S.B., 2009b. Surface charge of polyoxometalates modulates polymerization of the scrapie prion protein. Proc. Natl. Acad. Sci. U.S.A. 106, 3740–3745.*

conformer (only). However, the fact that three monoclonal antibodies that *do not* recognize infectious PrP conformers failed to label the 2D crystals provided at least indirect support for the notion that the 2D crystals contain indeed the infectious conformer (Wille et al., 2007, 2009b). Furthermore, in situ denaturation of the 2D crystals on the electron microscopy grids with urea made these hidden epitopes accessible, again confirming the presence of PrP in the 2D crystals, and arguing that it is the infectious conformer that forms these 2D crystals (Wille et al., 2009b).

The comparatively low quality of the 2D crystals made it necessary to use image processing, in order to extract useful information about the underlying protein subunits (Wille et al., 2002). By combining a single-particle approach with subsequent crystallographic averaging PrP 27–30 trimers were determined as the likely subunits of the 2D crystals (Govaerts et al., 2004; Wille et al., 2007).

The fortuitous observation that miniprions ($PrP^{Sc}106$), which represent the shortest isoform of PrP that support the formation of infectious prions (Muramoto et al., 1996; Supattapone et al., 1999), form isomorphous 2D crystals to those from PrP 27–30 provided novel insights into the underlying molecular structures of PrP 27–30 and $PrP^{Sc}106$, respectively (Wille et al., 2002). By calculating difference maps between processed images of 2D crystals from PrP 27–30 and $PrP^{Sc}106$, it was possible to locate the internal deletion that was used to create PrP106 (deletion of residues 141–176, murine PrP numbering) on the 2D crystal lattice (Govaerts et al., 2004; Wille et al., 2002). FTIR spectroscopy had suggested that the residues of the internal deletion of $PrP^{Sc}106$ might constitute part of the β-structure in the infectious conformer (Supattapone et al., 1999), and therefore, the difference map would allow localization of β-structure within the 2D crystals of PrP 27–30 and $PrP^{Sc}106$ (Govaerts et al., 2004; Wille et al., 2002). These difference maps also revealed the location of the N-linked sugars, since $PrP^{Sc}106$ is always diglycosylated, while PrP 27–30 is a mixture of di-, mono-, and unglycosylated peptide chains (Supattapone et al., 1999). In addition, monoamino-nanogold was used to covalently label the N-linked sugars on the 2D crystals, which provided independent confirmation for their location on the crystal lattice (Wille et al., 2002). The difference mapping results, together with other data, formed a set of experimental constraints that were then used to model the structure of the infectious conformer (Govaerts et al., 2004; Wille et al., 2002).

3.2 β-Helical Models

The first model for the structure of PrP 27–30 was based on a limited number of experimental restraints (Huang et al., 1996) and could not accommodate the constraints that were derived from the 2D crystals and the difference maps, which provided the locations of the β-structure and the N-linked sugars. Therefore, a renewed effort was undertaken to devise a plausible model for the structure of PrP 27–30 that could accommodate the spatial limitations of the tightly packed 2D crystal lattice. A parallel β-helix was found to be the best fit for the new constraints (Wille et al., 2002), with a predicted molecular height corresponding to four rungs of β-helix structure (=1.9 nm) (Govaerts et al., 2004).

In the intervening years, a substantial number of alternative models for the structure of PrP 27–30 were proposed (for review, see Requena and Wille, 2014), some of which also included a β-helical architecture as part of the model. Parallel to the modeling efforts, a variety of experimental approaches contributed an ever increasing number of experimental constraints that restrict the fold space for the putative structure of the infectious conformer. At this moment in time, none of the molecular models for the structure of PrP 27–30 can accommodate all experimental constraints (Requena and Wille, 2014), which provides an opening for renewed efforts to predict the structure of the infectious conformer.

4. AMYLOID FIBRILS

4.1 The Helical Periodicity of Amyloid and Its Utility

Most amyloid fibrils consist of two or more protofilaments and display a clear helical twist, which originates from the winding of the individual protofilaments around the axis of the fibril. PrP 27–30 is prone to form amyloid fibrils (Fig. 5) with a helical twist that is readily noticeable in most cases (Sim and Caughey, 2009; Terry et al., 2016; Vázquez-Fernández et al., 2016). The protofilaments in PrP 27–30 were found to wind with both left- and right-handed orientations around the axis of the fibril, and in all cases only two protofilaments were observed (Sim and Caughey, 2009; Terry et al., 2016; Vázquez-Fernández et al., 2016).

Due to the chirality of the constituting amino acids, β-sheets often adopt a right-handed twist, with flat and left-handed varieties being somewhat disfavored but not uncommon (Chothia, 1973). At present, it is unclear how the molecular twist of individual β-sheets translates into the higher-order twist of the larger, multiprotofilament amyloid fibrils. Moreover, the presence of left- and right-handed twists in otherwise homogeneous amyloid preparations is puzzling, but not uncommon as demonstrated by

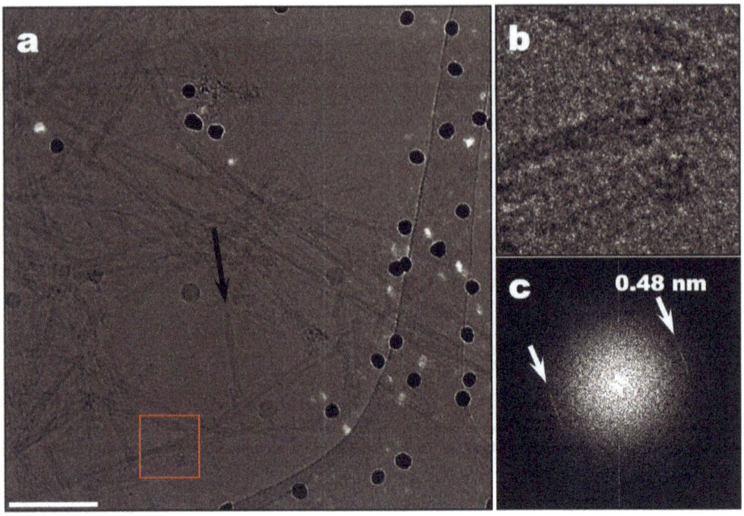

Fig. 5 High-resolution image of GPI-anchorless PrP 27–30. (A) An unprocessed cryo-EM micrograph showing GPI-anchorless PrP 27–30 amyloid fibrils. The *arrow* points to a fibril comprised of two clear protofilaments that twist around each other. The *black dots* are fiducial gold, which was added for tomography purposes. Scale bar = 100 nm. (B) Enlargement of the *red square* in (A), showing two short segments of PrP 27–30 amyloid fibrils. (C) Fast Fourier-transform (FFT) from the *red square*. The *arrows* point to the 0.48 nm repeat signal, which indicates the cross-β signal of the β-strands running perpendicular to the fiber axis.

the PrP 27–30 fibrils (Sim and Caughey, 2009; Terry et al., 2016; Vázquez-Fernández et al., 2016).

The repeating and regular nature of amyloid fibrils, with regard to both the helical periodicity and the regular stacking of individual subunits/proteins along the fibril axis, are useful features that help in elucidating their underlying molecular structure.

4.1.1 X-Ray Fiber Diffraction

X-ray fiber diffraction is a well-established technique, which provided the experimental data that led to the discovery of the structure of DNA (Franklin and Gosling, 1953; Watson and Crick, 1953) as well as the structure of complex assemblies such as intact tobacco mosaic virus (TMV) (Namba and Stubbs, 1986). The analysis of amyloid fibrils using X-ray fiber diffraction employs many of the same technical approaches that were used to analyze the structures of DNA or TMV, and the meridional cross-β diffraction signal at 0.48 nm has become a defining criterion in recognizing an amyloid (Astbury et al., 1935; Eanes and Glenner, 1968; Eisenberg and Jucker, 2012; Sunde et al., 1997).

Low-intensity X-ray sources and sample disorientation limited the results of the first X-ray fiber diffraction study on PrP 27–30 amyloid fibrils (Nguyen et al., 1995), making it difficult to discern the 0.48-nm diffraction on the meridian and thus no structural information was obtained. By using synchrotron-based X-ray sources, which provide much higher intensity X-ray beams, diffraction patterns from PrP 27–30 revealed an axial repeating unit of 1.92 nm per molecule, represented by a series of meridional diffraction intensities at 0.48, 0.64, and 0.96 nm, respectively (Fig. 6A, *solid arrows*). These results indicate that each molecule of PrP 27–30 in the amyloid fibrils contributes a height of 1.92 nm to the length of the fibril, which is equivalent to the height of four β-strands (Wille et al., 2009a). All diffraction intensities on the equator originated essentially from lipid and detergent molecules, which copurified with PrP 27–30. The absence of a strong ~10 Å equatorial diffraction, which is commonly seen with stacked β-sheet amyloids (compare Fig. 6B), implied the presence of a β-helical or β-solenoidal structure (Wan et al., 2012, 2015; Wille et al., 2009a).

In contrast, X-ray fiber diffraction from in vitro generated amyloids that were derived from recombinantly expressed forms of the prion protein (e.g., recPrP(23–230) and recPrP(89–230)) resulted in a diffraction pattern with distinct differences to the ones obtained from PrP 27–30 (Fig. 6B). Here, the diffraction patterns were dominated by a very strong meridional diffraction at 0.48 nm and a strong equatorial diffraction at ~1.0-nm, which are the characteristics of a stacked β-sheet structure (Sunde et al., 1997; Wan et al., 2015; Wille et al., 2009a). Since in vitro generated PrP amyloids in most cases lack infectivity, it is not surprising that their structure is different from that of the infectious conformer (Wille et al., 2009a).

4.1.2 Electron Cryomicroscopy and Image Processing

Electron cryomicroscopy (cryo-EM) allows to study the structure of proteins and protein aggregates in the frozen-hydrated state (Jiménez et al., 1999; Mizuno et al., 2011; Schmidt et al., 2015), which preserves their native structure and, under ideal circumstances, permits the collection of (near) atomic-resolution images (Cheng, 2015). Recent technical developments, i.e., the introduction of direct electron detectors to record electron micrographs, have drastically increased the amount of structural information that can be acquired from a limited number of (high-resolution) images when taken under low-dose conditions (Cheng et al., 2015).

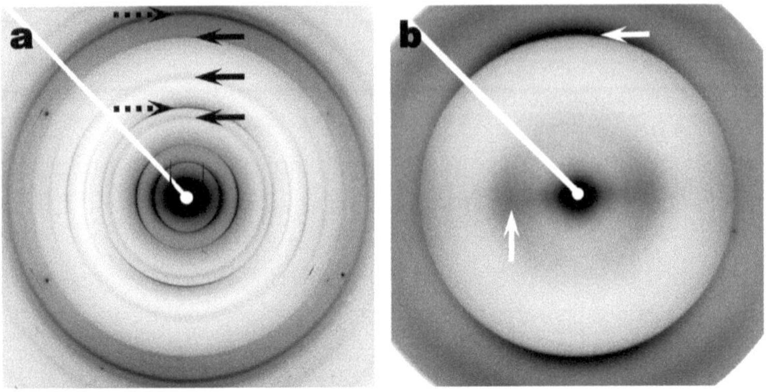

Fig. 6 X-ray fiber diffraction patterns from brain-derived PrP 27–30 amyloid and recombinant PrP amyloid. (A) Brain-derived PrP 27–30 amyloid produced X-ray fiber diffraction patterns containing a characteristic series of meridional diffraction at 0.48, 0.64, and 0.96 nm (*solid arrows*), which corresponds to the fourth-, third-, and second-order diffraction of a 1.92-nm repeating unit (Wille et al., 2009a). Together with the general cross-β signal at 0.48 nm, this diffraction pattern indicates the presence of a four-rung β-solenoid structure as a key feature of the infectious PrP conformer. In addition, the pattern contains strong meridional diffraction signals from the GPI-anchor and -associated lipids at 0.42 and 0.84 nm (*dotted line arrows*), as well as equatorial diffraction from the detergent that was used in the purification (not labeled). (B) In contrast, the diffraction pattern from recombinant PrP amyloid that was fibrillized in vitro is dominated by meridional diffraction at 0.48 nm (*horizontal arrow*) and equatorial diffraction at ~1.0 nm (*vertical arrow*). This simple pattern is indicative of a generic stacked β-sheet structure as the basis for recombinant PrP amyloid. *The figure was adapted from Wille, H., 2015. X-ray fiber diffraction from prions: of mice and beta-solenoids. In: Legname, G., Giachin, G. (Eds.), The Prion Phenomena in Neurodegenerative Diseases. New Frontiers in Neuroscience. Nova Biomedical, New York, pp. 15–27.*

In the past negative-stain electron microscopy was used to analyze the structure of recombinant PrP amyloid and to generate 3D reconstructions of individual amyloid fibrils (Tattum et al., 2006). These fibrils were observed to consist of two intertwined protofilaments with ~6 nm repeating units, but the imaging conditions limited the resolution that could be achieved to ~2.8 nm. In any case, the samples were not infectious, but rather represented a generic amyloid conformation (compare Fig. 6B).

Recently, the same approach was used to study the structure of PrP 27–30 that was prepared from the brains of prion-infected rodents (Terry et al., 2016). Again, the imaging conditions and, furthermore, the heterogeneity of the samples limited the resolution that could be achieved. Particularly, the morphological heterogeneity of the isolated prion fibrils made it

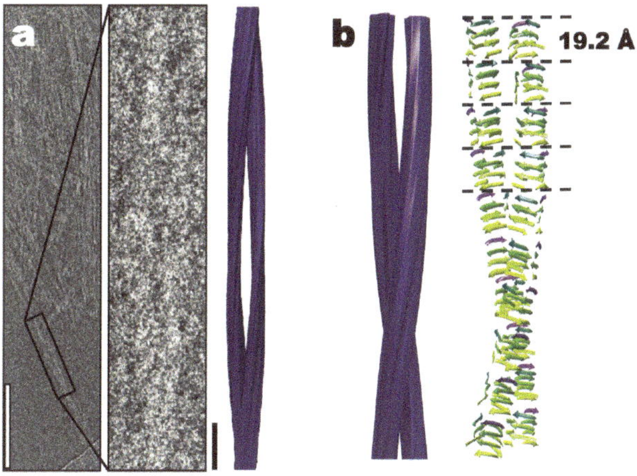

Fig. 7 3D reconstruction of a PrP 27–30 amyloid fibril and a cartoon to approximate the underlying molecular architecture. (A) Section of an electron micrograph showing GPI-anchorless PrP 27–30 amyloid fibrils. A single isolated and twisted fibril used for a 3D reconstruction is enclosed by a *black box* (*left*). Higher magnification and close-up view of the single prion fibril (*middle*). 3D reconstruction of the GPI-anchorless PrP 27–30 amyloid fibril (*right*). (B) 3D reconstruction of an individual GPI-anchorless PrP 27–30 amyloid fibril with two protofilaments (*left*). Cartoon depicting the potential configuration of the polypeptide chains in the protein monomers (*right*). Please note that this is *not* an atomistic model for the structure of the infectious prion, but rather a conceptual rendering to highlight its molecular architecture. *The figure was adapted from Vázquez-Fernández, E., Vos, M.R., Afanasyev, P., Cebey, L., Sevillano, A.M., Vidal, E., Rosa, I., Renault, L., Ramos, A., Peters, P.J., Fernández, J.J., van Heel, M., Young, H.S., Requena, J.R., Wille, H., 2016. The structural architecture of an infectious mammalian prion using electron cryomicroscopy. PLoS Pathog. 12, e1005835.*

impossible to produce clear 3D reconstructions that could resolve more than the presence of two separate protofilaments.

More recently, cryo-EM was used to investigate the structure of PrP 27–30 at higher resolution (Vázquez-Fernández et al., 2016). Fourier-transform analyses of raw, unprocessed images from individual PrP 27–30 amyloid fibrils routinely detected intensity peaks at 0.48 nm (Fig. 5), which correspond to the cross-β signals seen with X-ray fiber diffraction. 3D reconstructions of isolated PrP 27–30 amyloid fibrils were able to resolve the protofilaments in sufficient detail (Fig. 7A) to calculate the molecular volume and subsequently the height of individual PrP 27–30 molecules, which indicated a height of ~1.8 nm for each molecule of PrP 27–30 (Vázquez-Fernández et al., 2016). This height compared well with the repeating unit size that was determined by X-ray fiber diffraction of PrP 27–30 amyloid samples to be 1.92 nm (Wille et al., 2009a).

Moreover, an alternative image processing approach, which is averaging and classifying hundreds of images of short fibril sections (also known as single-particle averaging), has shown repetitive features along the length of the protofilament with an average height of ~2 nm (Vázquez-Fernández et al., 2016). Fourier-transform analyses of the averaged image classes indicated the presence of the canonical 0.48 nm signal associated with highly organized β-strands that are oriented perpendicular to the fibril axis. Furthermore, Fourier-transform analyses also detected additional, faint repeating signals (Vázquez-Fernández et al., 2016), indicating additional repetitive structures along the fibril axis: (1) a ~2-nm signal, corroborating the features that are visible in the 2D classes corresponding to the height of the folded protein monomer inside the protofilament. (2) Furthermore, a ~4-nm signal, consistent with a potential dimeric arrangement of the protein, which was also observed in other amyloid fibrils (Mizuno et al., 2011; White et al., 2009).

4.2 The Structure of the Infectious Conformer

The described structural constraints obtained, among other techniques, through electron crystallography on PrP 27–30 2D crystals (Govaerts et al., 2004; Wille et al., 2002, 2007), as well as X-ray fiber diffraction (Wille et al., 2009a) and cryo-EM (Vázquez-Fernández et al., 2016) on PrP 27–30 amyloid fibrils revealed that the infectious prion conformer contains a four-rung β-solenoid structure as the key feature of its architecture (Requena and Wille, 2014).

While important structural elements are still undefined, such as identification of specific residues participating in the β-strands that form each rung, what we know now about the structure of PrP^{Sc} allows us to formulate a sound hypothesis about how mammalian prions propagate in vivo. Templating based on a four-rung β-solenoid architecture must involve the upper- and lowermost β-solenoid rungs. These edge strands are inherently aggregation prone, as they are predestined to propagate their hydrogen-bonding pattern into any amyloidogenic peptide they encounter (Richardson and Richardson, 2002). Thus, these upper and lower rungs can template an incoming unfolded PrP molecule to create an additional β-solenoid rung. Once an additional β-rung has formed, a fresh "sticky" edge is available to continue templating until the incoming unfolded PrP molecule has been fully converted into a fresh PrP^{Sc} solenoid. The way in which templating occurs might feature a head-to-tail or a head-to-head/tail-to-tail orientation. In the former case, templating of β-sheets involves direct contact between different parts of the molecule, while in the latter case, the same protein stretches come into contact principally. The presence of a distinct ~4.0 nm dimer signal in Fourier-transform analyses of fibril segments supports a head-to-head/tail-to-tail arrangement (Vázquez-Fernández et al., 2016), but ultimately, higher resolution data will be needed to distinguish between these options.

5. OUTLOOK ON THE FUTURE STRUCTURAL BIOLOGY OF PRIONS

The structure of the infectious prion (in any of its forms: PrP^{Sc}, PrP 27–30, $PrP^{Sc}106$, or GPI-anchorless PrP 27–30) is slowly being unraveled by a variety of experimental approaches, but its general insolubility and propensity to aggregate have limited the quality of the available results and severely curtailed the applicability of many techniques. Since the structural conversion from α-helix-rich PrP^{C} to β-sheet-rich PrP^{Sc} forms the basis for infection, transmission, and pathogenesis, solving the structure of infectious conformer remains a key challenge in prion research.

Given the difficulties of deciphering the structures of amyloids and amyloidogenic proteins at high resolution, often shorter peptides are studied as surrogates for the full-length proteins, e.g., PrP(106–126) (Walsh et al., 2009) and α-synuclein(68–78) (Rodriguez et al., 2015). This approach works well for those cases where the amyloidogenic peptide is short by itself (e.g., Aβ(1–42); Xiao et al., 2015) but runs into substantial problems when the structures of longer amyloidogenic proteins lie at the heart of the problem (Tuttle et al., 2016; Wan et al., 2015).

Advances in a variety of research technologies will undoubtedly play a crucial role in ultimately deciphering the structure of the infectious prion. The ability to produce infectious prions with high efficiency in vitro from recombinantly produced PrP (Wang et al., 2010) will allow the generation of isotopically labeled protein for, among others, solid-state NMR spectroscopy. By elucidating the cofactors that aid this in vitro conversion (Miller et al., 2013), additional gains in conversion efficiency can be expected. With the introduction of direct electron detectors, cryo-EM is undergoing a dramatic shift in its ability to resolve the structures of proteins and protein aggregates (Fromm et al., 2015). Despite these imaging advances, the structures of heterogeneous samples such as amyloid fibrils will remain a challenge, as is demonstrated by the first study to investigate bona fide infectious prions using this technique (Vázquez-Fernández et al., 2016).

6. CONCLUDING REMARKS

Elucidation of the structure of PrP^{Sc} is a central challenge in prion research that remains to be completed. The intractable nature of this insoluble, polymeric, aggregation-prone conformer has made such endeavor a time-consuming and difficult quest. However, a number of breakthroughs, including recent cryo-EM studies taking advantage of images of unprecedented quality, and of the opportunities afforded by the repetitive nature of PrP^{Sc} fibers, have allowed deciphering the basic elements of the architecture of mammalian prions. In turn, this allows, for the first time, to begin to understand how mammalian prions propagate. It is noteworthy that the molecular forces responsible for the templating that underlies such process—hydrogen bonding, charge interactions, aromatic stacking, and steric constraints—are fundamentally similar to those operating during DNA replication, albeit lacking its exquisite precision and flexibility. We are close to the "eureka moment" in which an understanding of the structure allowed an understanding of DNA propagation (Watson and Crick, 1953). However, the higher complexity of the structure of PrP^{Sc} as compared to that of nucleic acids will require continued efforts to achieve a complete understanding of key phenomena associated to prion propagation such as transmission barriers between different species or the existence of strains.

ACKNOWLEDGMENTS

The authors would like to acknowledge support from European Commission Grant FP7 222887 "Priority," Spanish Ministry of Education Grant BFU2006-04588/BMC, and Spanish Ministry of Economy Grant BFU2013-48436-C2-1-P (all to J.R.R.) and grants from the Alberta Prion Research Institute (APRI 201100010, APRI 201100011, ABIBS 201300024, and ABIBS 201300012) and the Alberta Livestock & Meat Agency (ALMA 2012A001R) (all to H.W.).

REFERENCES

Aguzzi, A., Calella, A.M., 2009. Prions: protein aggregation and infectious diseases. Physiol. Rev. 89, 1105–1152.

Aguzzi, A., Lakkaraju, A.K.K., 2016. Cell biology of prions and prionoids: a status report. Trends Cell Biol. 26, 40–51.

Astbury, W.T., Dickinson, S., Bailey, K., 1935. The X-ray interpretation of denaturation and the structure of the seed globulins. Biochem. J. 29, 2351–2360.

Baron, G.S., Hughson, A.G., Raymond, G.J., Offerdahl, D.K., Barton, K.A., Raymond, L.D., Dorward, D.W., Caughey, B., 2011. Effect of glycans and the glycophosphatidylinositol anchor on strain dependent conformations of scrapie prion protein: improved purifications and infrared spectra. Biochemistry 50, 4479–4490.

Baskakov, I.V., Legname, G., Prusiner, S.B., Cohen, F.E., 2001. Folding of prion protein to its native α-helical conformation is under kinetic control. J. Biol. Chem. 276, 19687–19690.

Bellinger-Kawahara, C.G., Kempner, E., Groth, D., Gabizon, R., Prusiner, S.B., 1988. Scrapie prion liposomes and rods exhibit target sizes of 55,000 Da. Virology 164, 537–541.

Biasini, E., Seegulam, M.E., Patti, B.N., Solforosi, L., Medrano, A.Z., Christensen, H.M., Senatore, A., Chiesa, R., Williamson, R.A., Harris, D.A., 2008. Non-infectious aggregates of the prion protein react with several PrP^{Sc}-directed antibodies. J. Neurochem. 105, 2190–2204.

Bolton, D.C., McKinley, M.P., Prusiner, S.B., 1982. Identification of a protein that purifies with the scrapie prion. Science 218, 1309–1311.

Bolton, D.C., Bendheim, P.E., Marmorstein, A.D., Potempska, A., 1987. Isolation and structural studies of the intact scrapie agent protein. Arch. Biochem. Biophys. 258, 579–590.

Castilla, J., Requena, J.R., 2015. Prion-like diseases: looking for their niche in the realm of infectious diseases. Virus Res. 207, 1–4.

Cheng, Y., 2015. Single-particle cryo-EM at crystallographic resolution. Cell 161, 450–457.

Cheng, Y., Grigorieff, N., Penczek, P.A., Walz, T., 2015. A primer to single-particle cryo-electron microscopy. Cell 161, 438–449.

Chothia, C., 1973. Conformation of twisted beta-pleated sheets in proteins. J. Mol. Biol. 75, 295–302.

Colby, D.W., Wain, R., Baskakov, I.V., Legname, G., Palmer, C.G., Nguyen, H.O., Lemus, A., Cohen, F.E., DeArmond, S.J., Prusiner, S.B., 2010. Protease-sensitive synthetic prions. PLoS Pathog. 6, e1000736.

Cronier, S., Gros, N., Tattum, M.H., Jackson, G.S., Clarke, A.R., Collinge, J., Wadsworth, J.D.F., 2008. Detection and characterization of proteinase K-sensitive disease-related prion protein with thermolysin. Biochem. J. 416, 297–305.

D'Castro, L., Wenborn, A., Gros, N., Joiner, S., Cronier, S., Collinge, J., Wadsworth, J.D., 2010. Isolation of proteinase K-sensitive prions using pronase E and phosphotungstic acid. PLoS One 5, e15679.

DeArmond, S.J., McKinley, M.P., Barry, R.A., Braunfeld, M.B., McColloch, J.R., Prusiner, S.B., 1985. Identification of prion amyloid filaments in scrapie-infected brain. Cell 41, 221–235.

Di Bari, M.A., Nonno, R., Castilla, J., D'Agostino, C., Pirisinu, L., Riccardi, G., Conte, M., Richt, J., Kunkle, R., Langeveld, J., Vaccari, G., Agrimi, U., 2013. Chronic wasting disease in bank voles: characterisation of the shortest incubation time model for prion diseases. PLoS Pathog. 9, e1003219.

Donne, D.G., Viles, J.H., Groth, D., Mehlhorn, I., James, T.L., Cohen, F.E., Prusiner, S.B., Wright, P.E., Dyson, H.J., 1997. Structure of the recombinant full-length hamster prion protein PrP(29-231): the N terminus is highly flexible. Proc. Natl. Acad. Sci. U.S.A. 94, 13452–13457.

Eanes, E.D., Glenner, G.G., 1968. X-ray diffraction studies on amyloid filaments. J. Histochem. Cytochem. 16, 673–677.

Eisenberg, D., Jucker, M., 2012. The amyloid state of proteins in human diseases. Cell 148, 1188–1203.

Franklin, R.E., Gosling, R.G., 1953. Molecular configuration in sodium thymonucleate. Nature 171, 740–741.

Fromm, S.A., Bharat, T.A., Jakobi, A.J., Hagen, W.J., Sachse, C., 2015. Seeing tobacco mosaic virus through direct electron detectors. J. Struct. Biol. 189, 87–97.

Gabizon, R., McKinley, M.P., Prusiner, S.B., 1987. Purified prion proteins and scrapie infectivity copartition into liposomes. Proc. Natl. Acad. Sci. U.S.A. 84, 4017–4021.

Godsave, S.F., Wille, H., Pierson, J., Prusiner, S.B., Peters, P.J., 2013. Plasma membrane invaginations containing clusters of full-length PrP^{Sc} are an early form of prion-associated neuropathology *in vivo*. Neurobiol. Aging 34, 1621–1631.
Govaerts, C., Wille, H., Prusiner, S.B., Cohen, F.E., 2004. Evidence for assembly of prions with left-handed beta-helices into trimers. Proc. Natl. Acad. Sci. U.S.A. 101, 8342–8347.
Hershey, A.D., Chase, M., 1952. Independent functions of viral protein and nucleic acid in growth of bacteriophage. J. Gen. Physiol. 36, 39–56.
Huang, Z., Prusiner, S.B., Cohen, F.E., 1996. Scrapie prions: a three-dimensional model of an infectious fragment. Fold. Des. 1, 13–19.
Jackson, G.S., Hosszu, L.L.P., Power, A., Hill, A.F., Kenney, J., Saibil, H., Craven, C.J., Waltho, J.P., Clarke, A.R., Collinge, J., 1999. Reversible conversion of monomeric human prion protein between native and fibrilogenic conformations. Science 283, 1935–1937.
Jaunmuktane, Z., Mead, S., Ellis, M., Wadsworth, J.D., Nicoll, A.J., Kenny, J., Launchbury, F., Linehan, J., Richard-Loendt, A., Walker, A.S., Rudge, P., Collinge, J., Brandner, S., 2015. Evidence for human transmission of amyloid-β pathology and cerebral amyloid angiopathy. Nature 525, 247–250.
Jiménez, J.L., Guijarro, J.I., Orlova, E., Zurdo, J., Dobson, C.M., Sunde, M., Saibil, H.R., 1999. Cryo-electron microscopy structure of an SH3 amyloid fibril and model of the molecular packing. EMBO J. 18, 815–821.
Lasmézas, C.I., Deslys, J.P., Robain, O., Jaegly, A., Beringue, V., Peyrin, J.M., Fournier, J.G., Hauw, J.J., Rossier, J., Dormont, D., 1997. Transmission of the BSE agent to mice in the absence of detectable abnormal prion protein. Science 275, 402–405.
Legname, G., Baskakov, I.V., Nguyen, H.O., Riesner, D., Cohen, F.E., DeArmond, S.J., Prusiner, S.B., 2004. Synthetic mammalian prions. Science 305, 673–676.
Levine, D.J., Stöhr, J., Falese, L.E., Ollesch, J., Wille, H., Prusiner, S.B., Long, J.R., 2015. Mechanism of scrapie prion precipitation with phosphotungstate anions. ACS Chem. Biol. 10, 1269–1277.
Maheshwari, A., Fischer, M., Gambetti, P., Parker, A., Ram, A., Soto, C., Concha-Marambio, L., Cohen, Y., Belay, E.D., Maddox, R.A., Mead, S., Goodman, C., Kass, J.S., Schonberger, L.B., Hussein, H.M., 2015. Recent US case of variant Creutzfeldt-Jakob disease—global implications. Emerg. Infect. Dis. 21, 750–759.
McKinley, M.P., Meyer, R.K., Kenaga, L., Rahbar, F., Cotter, R., Serban, A., Prusiner, S.B., 1991. Scrapie prion rod formation in vitro requires both detergent extraction and limited proteolysis. J. Virol. 65, 1340–1351.
Miller, M.B., Wang, D.W., Wang, F., Noble, G.P., Ma, J., Woods Jr., V.L., Li, S., Supattapone, S., 2013. Cofactor molecules induce structural transformation during infectious prion formation. Structure 21, 2061–2068.
Mizuno, N., Baxa, U., Steven, A.C., 2011. Structural dependence of HET-s amyloid fibril infectivity assessed by cryoelectron microscopy. Proc. Natl. Acad. Sci. U.S.A. 108, 3252–3257.
Muramoto, T., Scott, M., Cohen, F.E., Prusiner, S.B., 1996. Recombinant scrapie-like prion protein of 106 amino acids is soluble. Proc. Natl. Acad. Sci. U.S.A. 93, 15457–15462.
Namba, K., Stubbs, G., 1986. Structure of tobacco mosaic virus at 3.6 Å resolution: implications for assembly. Science 231, 1401–1406.
Nguyen, J.T., Inouye, H., Baldwin, M.A., Fletterick, R.J., Cohen, F.E., Prusiner, S.B., Kirschner, D.A., 1995. X-ray diffraction of scrapie prion rods and PrP peptides. J. Mol. Biol. 252, 412–422.

Pastrana, M.A., Sajnani, G., Onisko, B., Castilla, J., Morales, R., Soto, C., Requena, J.R., 2006. Isolation and characterization of a proteinase K-sensitive PrP^{Sc} fraction. Biochemistry 45, 15710–15717.
Prigent, S., Rezaei, H., 2011. PrP assemblies: spotting the responsible regions in prion propagation. Prion 5, 69–75.
Prusiner, S.B., 1982. Novel proteinaceous infectious particles cause scrapie. Science 216, 136–144.
Prusiner, S.B., 1998. Prions. Proc. Natl. Acad. Sci. U.S.A. 95, 13363–13383.
Prusiner, S.B., 2012. A unifying role for prions in neurodegenerative diseases. Science 336, 1511–1513.
Prusiner, S.B., Bolton, D.C., Groth, D.F., Bowman, K.A., Cochran, S.P., McKinley, M.P., 1982. Further purification and characterization of scrapie prions. Biochemistry 21, 6942–6950.
Prusiner, S.B., McKinley, M.P., Bowman, K.A., Bolton, D.C., Bendheim, P.E., Groth, D.F., Glenner, G.G., 1983. Scrapie prions aggregate to form amyloid-like birefringent rods. Cell 35, 349–358.
Requena, J.R., Wille, H., 2014. The structure of the infectious prion protein: experimental data and molecular models. Prion 8, 60–66.
Requena, J.R., Groth, D., Legname, G., Stadtman, E.R., Prusiner, S.B., Levine, R.L., 2001. Copper-catalyzed oxidation of the recombinant SHa(29-231) prion protein. Proc. Natl. Acad. Sci. U.S.A. 98, 7170–7175.
Richardson, J.S., Richardson, D.C., 2002. Natural beta-sheet proteins use negative design to avoid edge-to-edge aggregation. Proc. Natl. Acad. Sci. U.S.A. 99, 2754–2759.
Riek, R., Hornemann, S., Wider, G., Billeter, M., Glockshuber, R., Wüthrich, K., 1996. NMR structure of the mouse prion protein domain PrP(121-231). Nature 382, 180–182.
Rodriguez, J.A., Ivanova, M.I., Sawaya, M.R., Cascio, D., Reyes, F.E., Shi, D., Sangwan, S., Guenther, E.L., Johnson, L.M., Zhang, M., Jiang, L., Arbing, M.A., Nannenga, B.L., Hattne, J., Whitelegge, J., Brewster, A.S., Messerschmidt, M., Boutet, S., Sauter, N.K., Gonen, T., Eisenberg, D.S., 2015. Structure of the toxic core of α-synuclein from invisible crystals. Nature 525, 486–490.
Safar, J., Wille, H., Itri, V., Groth, D., Serban, H., Torchia, M., Cohen, F.E., Prusiner, S.B., 1998. Eight prion strains have PrP(Sc) molecules with different conformations. Nat. Med. 4, 1157–1165.
Sajnani, G., Silva, C.J., Ramos, A., Pastrana, M.A., Onisko, B.C., Erickson, M.L., Antaki, E.M., Dynin, I., Vázquez-Fernández, E., Sigurdson, C.J., Carter, J.M., Requena, J.R., 2012. PK-sensitive PrP is infectious and shares basic structural features with PK-resistant PrP. PLoS Pathog. 8, e1002547.
Schmidt, M., Rohou, A., Lasker, K., Yadav, J.K., Schiene-Fischer, C., Fändrich, M., Grigorieff, N., 2015. Peptide dimer structure in an Aβ(1-42) fibril visualized with cryo-EM. Proc. Natl. Acad. Sci. U.S.A. 112, 11858–11863.
Silva, C.J., Vázquez-Fernández, E., Onisko, B., Requena, J.R., 2015. Proteinase K and the structure of PrP^{Sc}: the good, the bad and the ugly. Virus Res. 207, 120–126.
Silveira, J.R., Raymond, G.J., Hughson, A.G., Race, R.E., Sim, V.L., Hayes, S.F., Caughey, B., 2005. The most infectious prion protein particles. Nature 437, 257–261.
Sim, V.L., Caughey, B., 2009. Ultrastructures and strain comparison of under-glycosylated scrapie prion fibrils. Neurobiol. Aging 30, 2031–2042.
Singh, J., Sabareesan, A.T., Mathew, M.K., Udgaonkar, J.B., 2012. Development of the structural core and of conformational heterogeneity during the conversion of oligomers of the mouse prion protein to worm-like amyloid fibrils. J. Mol. Biol. 423, 217–231.

Smirnovas, V., Baron, G.S., Offerdahl, D.K., Raymond, G.J., Caughey, B., Surewicz, W.K., 2011. Structural organization of brain-derived mammalian prions examined by hydrogen-deuterium exchange. Nat. Struct. Mol. Biol. 18, 504–506.

Stahlberg, H., Biyani, N., Engel, A., 2015. 3D reconstruction of two-dimensional crystals. Arch. Biochem. Biophys. 581, 68–77.

Sunde, M., Serpell, L.C., Bartlam, M., Fraser, P.E., Pepys, M.B., Blake, C.C., 1997. Common core structure of amyloid fibrils by synchrotron X-ray diffraction. J. Mol. Biol. 273, 729–739.

Supattapone, S., Bosque, P., Muramoto, T., Wille, H., Aagaard, C., Peretz, D., Nguyen, H.O., Heinrich, C., Torchia, M., Safar, J., Cohen, F.E., DeArmond, S.J., Prusiner, S.B., Scott, M., 1999. Prion protein of 106 residues creates an artificial transmission barrier for prion replication in transgenic mice. Cell 96, 869–878.

Swietnicki, S., Morillas, M., Chen, S.G., Gambetti, P., Surewicz, W.K., 2000. Aggregation and fibrillization of the recombinant human prion protein huPrP90-231. Biochemistry 39, 424–431.

Tattum, M.H., Cohen-Krausz, S., Thumanu, K., Wharton, C.W., Khalili-Shirazi, A., Jackson, G.S., Orlova, E.V., Collinge, J., Clarke, A.R., Saibil, H.R., 2006. Elongated oligomers assemble into mammalian PrP amyloid fibrils. J. Mol. Biol. 357, 975–985.

Terry, C., Wenborn, A., Gros, N., Sells, J., Joiner, S., Hosszu, L.L.P., Tattum, H., Panico, S., Clare, D.K., Collinge, J., Saibil, H.R., Wadsworth, J.D.F., 2016. *Ex vivo* mammalian prions are formed of paired double helical prion protein fibrils. Open Biol. 6, 160035.

Tixador, P., Herzog, L., Reine, F., Jaumain, E., Chapuis, J., Le Dur, A., Laude, H., Béringue, V., 2010. The physical relationship between infectivity and prion protein aggregates is strain-dependent. PLoS Pathog. 6, e1000859.

Torrent, J., Alvarez-Martinez, M.T., Harricane, M.C., Heitz, F., Liautard, J.P., Balny, C., Lange, R., 2004. High pressure induces scrapie-like prion protein misfolding and amyloid fibril formation. Biochemistry 43, 7162–7170.

Tuttle, M.D., Comellas, G., Nieuwkoop, A.J., Covell, D.J., Berthold, D.A., Kloepper, K.D., Courtney, J.M., Kim, J.K., Barclay, A.M., Kendall, A., Wan, W., Stubbs, G., Schwieters, C.D., Lee, V.M., George, J.M., Rienstra, C.M., 2016. Solid-state NMR structure of a pathogenic fibril of full-length human α-synuclein. Nat. Struct. Mol. Biol. 23, 409–415.

Tzaban, T., Friedlander, G., Schonberger, O., Horonchik, L., Yedidia, Y., Shaked, G., Gabizon, R., Taraboulos, A., 2002. Protease-sensitive scrapie prion protein in aggregates of heterogeneous sizes. Biochemistry 41, 12868–12875.

Vázquez-Fernández, E., Alonso, J., Pastrana, M.A., Ramos, A., Stitz, L., Vidal, E., Dynin, I., Petsch, B., Silva, C.J., Requena, J.R., 2012. Structural organization of mammalian prions as probed by limited proteolysis. PLoS One 7, e50111.

Vázquez-Fernández, E., Vos, M.R., Afanasyev, P., Cebey, L., Sevillano, A.M., Vidal, E., Rosa, I., Renault, L., Ramos, A., Peters, P.J., Fernández, J.J., van Heel, M., Young, H.S., Requena, J.R., Wille, H., 2016. The structural architecture of an infectious mammalian prion using electron cryomicroscopy. PLoS Pathog. 12, e1005835.

Walsh, P., Simonetti, K., Sharpe, S., 2009. Core structure of amyloid fibrils formed by residues 106–126 of the human prion protein. Structure 17, 417–426.

Wan, W., Wille, H., Stöhr, J., Baxa, U., Prusiner, S.B., Stubbs, G., 2012. Degradation of fungal prion HET-s(218-289) induces formation of a generic amyloid fold. Biophys. J. 102, 2339–2344.

Wan, W., Wille, H., Stöhr, J., Kendall, A., Bian, W., McDonald, M., Tiggelaar, S., Watts, J.C., Prusiner, S.B., Stubbs, G., 2015. Structural studies of truncated forms of the prion protein PrP. Biophys. J. 108, 1548–1554.

Wang, F., Wang, X., Yuan, C.G., Ma, J., 2010. Generating a prion with bacterially expressed recombinant prion protein. Science 327, 1132–1135.

Watson, J.D., Crick, F.H., 1953. Molecular structure of nucleic acids; a structure for deoxyribose nucleic acid. Nature 171, 737–738.
White, H.E., Hodgkinson, J.L., Jahn, T.R., Cohen-Krausz, S., Gosal, W.S., Müller, S., Orlova, E.V., Radford, S.E., Saibil, H.R., 2009. Globular tetramers of beta(2)-microglobulin assemble into elaborate amyloid fibrils. J. Mol. Biol. 389, 48–57.
Wickner, R.B., Taylor, K.L., Edskes, H.K., Maddelein, M.L., Moriyama, H., Roberts, B.T., 2000. Prions of yeast as heritable amyloidoses. J. Struct. Biol. 130, 310–322.
Wille, H., Prusiner, S.B., 1999. Ultrastructural studies on scrapie prion protein crystals obtained from reverse micellar solutions. Biophys. J. 76, 1048–1062.
Wille, H., Michelitsch, M.D., Guenebaut, V., Supattapone, S., Serban, A., Cohen, F.E., Agard, D.A., Prusiner, S.B., 2002. Structural studies of the scrapie prion protein by electron crystallography. Proc. Natl. Acad. Sci. U.S.A. 99, 3563–3568.
Wille, H., Govaerts, C., Borovinskiy, A., Latawiec, D., Downing, K.H., Cohen, F.E., Prusiner, S.B., 2007. Electron crystallography of the scrapie prion protein complexed with heavy metals. Arch. Biochem. Biophys. 467, 239–248.
Wille, H., Bian, W., McDonald, M., Kendall, A., Colby, D.W., Bloch, L., Ollesch, J., Borovinskiy, A.L., Cohen, F.E., Prusiner, S.B., Stubbs, G., 2009a. Natural and synthetic prion structure from X-ray fiber diffraction. Proc. Natl. Acad. Sci. U.S.A. 106, 16990–16995.
Wille, H., Shanmugam, M., Murugesu, M., Ollesch, J., Stubbs, G., Long, J.R., Safar, J.G., Prusiner, S.B., 2009b. Surface charge of polyoxometalates modulates polymerization of the scrapie prion protein. Proc. Natl. Acad. Sci. U.S.A. 106, 3740–3745.
Xiao, Y., Ma, B., McElheny, D., Parthasarathy, S., Long, F., Hoshi, M., Nussinov, R., Ishii, Y., 2015. Aβ(1-42) fibril structure illuminates self-recognition and replication of amyloid in Alzheimer's disease. Nat. Struct. Mol. Biol. 22, 499–505.
Yam, A.Y., Gao, C.M., Wang, X., Wu, P., Peretz, D., 2010. The octarepeat region of the prion protein is conformationally altered in PrP(Sc). PLoS One 5, e9316.
Zhang, H., Stöckel, J., Mehlhorn, I., Groth, D., Baldwin, M.A., Prusiner, S.B., James, T.L., Cohen, F.E., 1997. Physical studies of conformational plasticity in a recombinant prion protein. Biochemistry 36, 3543–3553.

INDEX

Note: Page numbers followed by "*f*" indicate figures, and "*t*" indicate tables.

D

E

F

G

FSC
www.fsc.org
MIX
Papier | Fördert
gute Waldnutzung
FSC® C083411

Kolibri 360 GmbH
Ferdinand Jühlke Straße 7
99095 Erfurt, Deutschland
produktsicherheit@kolibri360.de